MW01469200

Just The

facts101
Textbook Key Facts

Textbook Outlines, Highlights, and Practice Quizzes

Microbiology: An Evolving Science

by Joan L. Slonczewski, 2nd Edition

All "Just the Facts101" Material Written or Prepared by Cram101 Publishing

Title Page

"Just the Facts101" is a Cram101 publication and tool designed to give you all the facts from your textbooks. Visit Cram101.com for the full practice test for each of your chapters for virtually any of your textbooks.

Cram101 has built custom study tools specific to your textbook. We provide all of the factual testable information and unlike traditional study guides, we will never send you back to your textbook for more information.

YOU WILL NEVER HAVE TO HIGHLIGHT A BOOK AGAIN!

Cram101 StudyGuides

All of the information in this StudyGuide is written specifically for your textbook. We include the key terms, places, people, and concepts... the information you can expect on your next exam!

Want to take a practice test?

Throughout each chapter of this StudyGuide you will find links to cram101.com where you can select specific chapters to take a complete test on, or you can subscribe and get practice tests for up to 12 of your textbooks, along with other exclusive cram101.com tools like problem solving labs and reference libraries.

Cram101.com

Only cram101.com gives you the outlines, highlights, and PRACTICE TESTS specific to your textbook. Cram101.com is an online application where you'll discover study tools designed to make the most of your limited study time.

By purchasing this book, you get 50% off the normal subscription free!. Just enter the promotional code **'DK73DW20451'** on the Cram101.com registration screen.

www.Cram101.com

Learning System

Microbiology: An Evolving Science
Joan L. Slonczewski, 2nd

CONTENTS

1. Part 1: The Microbial Cell 5
2. Part 2: Genes and Genomes 107
3. Part 3: Metabolism and Biochemistry 189
4. Part 4: Microbial Diversity and Ecology 253
5. Part 5: Medicine and Immunology 354

Calvin cycle

Interleukin

Reporter gene

Anaerobic infection

Index case

Nystatin

Pathogenicity island

Borrelia burgdorferi

Copepod

Haemophilus influenzae

Herpes simplex

NADH dehydrogenase

Toll-like receptor

Vibrio cholerae

Confocal microscopy

Herpes simplex virus

Sigma factor

Operon

Cell biology

Chapter 1. Part 1: The Microbial Cell

CHAPTER OUTLINE: KEY TERMS, PEOPLE, PLACES, CONCEPTS

______ Essential nutrient

______ DNA polymerase

______ Genetic engineering

______ Microbial ecology

______ Nitrogen fixation

______ Amino acid

______ Biosynthesis

______ Chain reaction

______ Food chain

______ Polymerase chain reaction

______ Archaea

______ Bacteria

______ Bifidobacterium

______ Caulerpa taxifolia

______ Epstein-Barr virus

______ Eukaryote

______ Microbial genetics

______ Nitrogen cycle

______ Prokaryote

RNA virus

Thiomargarita namibiensis

Life cycle

Paul Berg

DNA sequencing

Genome

Genomic island

Sequencing

Extreme environment

Bubonic plague

Mycobacterium tuberculosis

Palmaria palmata

Yersinia pestis

Tuberculosis

Robert Boyle

Medical statistics

Isaac Newton

Smallpox

Robert Hooke

Micrographia

Louis Pasteur

Putrefaction

Ribosomal RNA

Lazzaro Spallanzani

Spontaneous generation

Dairy product

Molecular clock

Autoclave

Endospore

Germ theory of disease

Medical microbiology

Extraterrestrial life

Early Earth

Sidney Altman

Bacillus anthracis

Agar

Petri dish

Red algae

Cowpox

Encephalitozoon intestinalis

Helicobacter pylori

Lyme disease

Simian immunodeficiency virus

Vaccination

Cholera toxin

Inoculation

Polio vaccine

Smallpox vaccine

Antiseptic

Chlorine

Rabies

Antibiotic resistance

Drug discovery

Staphylococcus

Prion

Tobacco mosaic virus

Viroid

Chapter 1. Part 1: The Microbial Cell

Drug development

Drug resistance

Mosaic virus

Transfer RNA

Enrichment culture

Food web

Lithotroph

Photosynthesis

Winogradsky column

Bacteroides

Biofilm

Endosymbiont

Haber process

Purple sulfur bacteria

Sulfate-reducing bacteria

Vibrio fischeri

Rhizobia

Immune system

Monera

Chapter 1. Part 1: The Microbial Cell

CHAPTER OUTLINE: KEY TERMS, PEOPLE, PLACES, CONCEPTS

Nucleus

Organelle

Blood cell

Iron-sulfur cluster

Cell envelope

Electron microscope

Chlorosome

Cytoplasm

Recombinant DNA

James D. Watson

X-ray crystallography

Double helix

Agrobacterium tumefaciens

DNA-binding protein

Taq polymerase

Thermus aquaticus

Cloning

Bioremediation

Oil spill

_____ | Airy disk

_____ | Human eye

_____ | Bacilli

_____ | Slime mold

_____ | Spirillum

_____ | Trypanosoma brucei

_____ | Atomic force microscopy

_____ | Transmission electron microscopy

_____ | Frequency

_____ | Wavelength

_____ | Speed of light

_____ | Fluorescence

_____ | Reflection

_____ | Refractive index

_____ | Scattering

_____ | Electromagnetic spectrum

_____ | Acquired immunodeficiency syndrome

_____ | Numerical aperture

_____ | Rhizopus oligosporus

Bacillus thuringiensis

Crystal violet

Gram-positive bacteria

Hans Christian

Iodine

Staining

Stains

Insecticide

Methylene blue

Starvation response

Cell wall

Clostridium botulinum

Counterstain

Mycolic acid

Proteobacteria

Peptidoglycan

Safranin

Negative stain

Substrate-level phosphorylation

Chapter 1. Part 1: The Microbial Cell

CHAPTER OUTLINE: KEY TERMS, PEOPLE, PLACES, CONCEPTS

_______________ Syphilis

_______________ Organic acid

_______________ Entamoeba histolytica

_______________ Bacillus subtilis

_______________ Interference microscopy

_______________ Fluorophore

_______________ Origin of replication

_______________ Confocal laser scanning microscopy

_______________ Complementary DNA

_______________ Cryo-electron tomography

_______________ Pseudomonas aeruginosa

_______________ Magnetosome

_______________ Magnetotactic bacteria

_______________ John Desmond Bernal

_______________ ATP synthase

_______________ Borrelia afzelii

_______________ Cell physiology

_______________ Genetic analysis

_______________ Inner membrane

Chapter 1. Part 1: The Microbial Cell

CHAPTER OUTLINE: KEY TERMS, PEOPLE, PLACES, CONCEPTS

- Lipopolysaccharide
- Shine-Dalgarno sequence
- Staphylococcus aureus
- Cell membrane
- Disinfectant
- Messenger RNA
- Isoelectric focusing
- Isoelectric point
- Northern blot
- Polyamine
- Proteome
- Gel electrophoresis
- Polyacrylamide gel
- Protein
- Proteomics
- Nucleic acid
- Mixotricha paradoxa
- Subatomic particle
- Membrane protein

_____ Electrochemical potential _____

_____ Polyribosome _____

_____ Ribosomal protein _____

_____ Ribosome _____

_____ Fusion protein _____

_____ Lac operon _____

_____ Protein A _____

_____ Cytoskeleton _____

_____ Phosphatidylethanolamine _____

_____ Aquaporin _____

_____ Aspirin _____

_____ Brugia malayi _____

_____ Osmotic pressure _____

_____ Passive transport _____

_____ Active transport _____

_____ Virulence factor _____

_____ Weak acid _____

_____ Weak base _____

_____ ATP hydrolysis _____

Chapter 1. Part 1: The Microbial Cell

CHAPTER OUTLINE: KEY TERMS, PEOPLE, PLACES, CONCEPTS

______ Energy carrier

______ Chemical reaction

______ Cardiolipin

______ Membrane lipids

______ Oleic acid

______ Phosphatidylglycerol

______ Streptococcus pneumoniae

______ Trichophyton rubrum

______ Fatty acid

______ Side chain

______ Cholesterol

______ Hyperthermophile

______ Terpenoid

______ Hopanoids

______ Glycan

______ N-Acetylglucosamine

______ N-Acetylmuramic acid

______ Clostridium difficile

______ Toxic shock syndrome

Chapter 1. Part 1: The Microbial Cell

CHAPTER OUTLINE: KEY TERMS, PEOPLE, PLACES, CONCEPTS

______ S-layer

______ Teichoic acid

______ Ethambutol

______ Galactan

______ Mycoplasma

______ Gram-negative bacteria

______ Lipoprotein

______ Density gradient

______ Glucosamine

______ Porin

______ Beta barrel

______ Endotoxin

______ Chitin

______ Osmotic shock

______ Amp resistance

______ Coccidioides

______ DNA gyrase

______ DNA replication

______ Okazaki fragment

_____ | Quinolone

_____ | RNA polymerase

_____ | Southern blot

_____ | Biocrystallization

_____ | Protein synthesis inhibitor

_____ | Cell division

_____ | Endoplasmic reticulum

_____ | Replication fork

_____ | Replisome

_____ | Septum

_____ | Caulobacter crescentus

_____ | Protein Z

_____ | Permissive temperature

_____ | Verrucomicrobia

_____ | Thylakoid

_____ | Carboxysome

_____ | Holdfast

_____ | Magnetotaxis

_____ | Motility

Periodontal disease

Pilin

Porphyromonas gingivalis

Sulfur-reducing bacteria

Biomineralization

Flagellin

Chemotaxis

Myxobacteria

Lactic acid

Marine habitats

Algal bloom

Nutrition

Generation time

Legionella pneumophila

Cobalt

Growth factor

MHC restriction

Micronutrient

Molybdenum

Chapter 1. Part 1: The Microbial Cell

CHAPTER OUTLINE: KEY TERMS, PEOPLE, PLACES, CONCEPTS

Autotroph

Biogeochemical cycle

Carbon cycle

Q fever

Rickettsia prowazekii

Heterotroph

Parasitism

Membrane potential

Microbial metabolism

Bradyrhizobium

Denitrification

Denitrifying bacteria

Nitrification

Rhizobium

Sinorhizobium meliloti

Lignin

Facilitated diffusion

Permease

Lactose permease

Chapter 1. Part 1: The Microbial Cell
CHAPTER OUTLINE: KEY TERMS, PEOPLE, PLACES, CONCEPTS

Lactococcus

Neisseria gonorrhoeae

Siderophore

Ames test

Salmonella enterica

Hemocytometer

Propidium iodide

Cell counting

Cell growth

Binary fission

Budding

Deinococcus radiodurans

Mitosis

DNA Research

Plasmodium falciparum

Reproduction

Quorum sensing

Turbidostat

Cystic fibrosis

Visit Cram101.com for full Practice Exams

Chapter 1. Part 1: The Microbial Cell

CHAPTER OUTLINE: KEY TERMS, PEOPLE, PLACES, CONCEPTS

- Exopolysaccharide
- Marine snow
- Aphotic zone
- Dictyostelium discoideum
- Anabaena
- Clostridium tetani
- Dipicolinic acid
- Germination
- Heterocyst
- Sporangium
- Suspended animation
- Gliding motility
- Myxococcus xanthus
- Starvation
- Lactobacillus plantarum
- Extremophile
- Photosystem I
- Photosystem II
- On Plants

Sequence analysis

Arrhenius equation

Listeria monocytogenes

Mesophile

Psychrophile

Food processing

Erwinia

Pseudomonas syringae

Enzyme

Thermophile

Piezophile

Hypertonic

Osmolarity

Water activity

Halophile

Hypotonic

Yogurt

Alkaliphile

Cyclodextrin

Chapter 1. Part 1: The Microbial Cell

CHAPTER OUTLINE: KEY TERMS, PEOPLE, PLACES, CONCEPTS

______ Halobacterium salinarum

______ Ammonia production

______ Electron transport chain

______ Electrochemical gradient

______ Electron acceptor

______ Anaerobic respiration

______ Hydrogen peroxide

______ Hydroxyl radical

______ Campylobacter jejuni

______ Oligotroph

______ Eutrophication

______ Bacteriostatic agent

______ Fuel cell

______ Sanitation

______ Botulinum toxin

______ Ethanol

______ Toxin

______ Laminar flow

______ Pasteurization

_____ | Botulism

_____ | Yersinia enterocolitica

_____ | Irradiation

_____ | Phenol coefficient

_____ | Aldehyde

_____ | Ethylene oxide

_____ | Molecular structure

_____ | Streptomycin

_____ | Phage therapy

_____ | Phytophthora cinnamomi

_____ | Probiotic

_____ | Rous sarcoma virus

_____ | Pandemic

_____ | Bacteriophage

_____ | Capsid

_____ | Viral envelope

_____ | West Nile virus

_____ | Zidovudine

_____ | Cloning vector

Chapter 1. Part 1: The Microbial Cell

CHAPTER OUTLINE: KEY TERMS, PEOPLE, PLACES, CONCEPTS

Protease inhibitor

Reading frame

Rice dwarf virus

Inhibitor protein

Scrapie

Serum albumin

Filamentous phage

Propionibacterium freudenreichii

Tegument

International Committee on Taxonomy of Viruses

Rabies virus

Reverse transcriptase

Virus classification

Hepatitis B

Inverted repeat

RNA-dependent RNA polymerase

Cauliflower mosaic virus

Evolution

Molecular evolution

Visit Cram101.com for full Practice Exams

Plant cell

Corynebacterium diphtheriae

Hin recombinase

Lysogeny

Prophage

Shiga toxin

Shigella

Site-specific recombination

Lysis

Lytic cycle

Human genome

Ebola virus

Endocytosis

Hepatitis C

Tropism

Sialic acid

Tissue tropism

Wart

Keratinocyte

_____ | Oncogene _____

_____ | Plant virus _____

_____ | Latent period _____

_____ | Virulence _____

_____ | Tissue culture _____

_____ | Plaque forming unit _____

_____ | Reservoir _____

_____ | Population density _____

Calvin cycle	The Calvin cycle, is a series of biochemical redox reactions that take place in the stroma of chloroplasts in photosynthetic organisms. It is also known as the dark reactions. The cycle was discovered by Melvin Calvin, James Bassham, and Andrew Benson at the University of California, Berkeley by using the radioactive isotope carbon-14. It is one of the light-independent (dark) reactions used for carbon fixation.
Interleukin	Interleukins are a group of cytokines (secreted proteinssignaling molecules) that were first seen to be expressed by white blood cells (leukocytes). The term interleukin derives from (inter-) 'as a means of communication', and (-leukin) 'deriving from the fact that many of these proteins are produced by leukocytes and act on leukocytes'. The name is something of a relic, though (the term was coined by Dr. Vern Paetkau, University of Victoria); it has since been found that interleukins are produced by a wide variety of body cells.
Reporter gene	In molecular biology, a reporter gene is a gene that researchers attach to a regulatory sequence of another gene of interest in bacteria, cell culture, animals or plants.

Chapter 1. Part 1: The Microbial Cell

	Certain genes are chosen as reporters because the characteristics they confer on oanisms expressing them are easily identified and measured, or because they are selectable markers. Reporter genes are often used as an indication of whether a certain gene has been taken up by or expressed in the cell or oanism population.
Anaerobic infection	Anaerobic infections are caused by anaerobic bacteria. Anaerobic bacteria do not grow on solid media in room air (10% carbon dioxide and 18% oxygen); facultative anaerobic bacteria can grow in the presence as well as in the absence of air. Microaerophilic bacteria do not grow at all aerobically or grow poorly, but grow better under 10% carbon dioxide or anaerobically.
Index case	The index case is the initial patient in the population of an epidemiological investigation. The index case may indicate the source of the disease, the possible spread, and which reservoir holds the disease in between outbreaks. The index case is the first patient that indicates the existence of an outbreak.
Nystatin	Nystatin is a polyene antifungal medication to which many molds and yeast infections are sensitive, including Candida. Due to its toxicity profile, there are currently no injectable formulations of this drug on the US market. However, nystatin may be safely given orally as well as applied topically due to its minimal absorption through mucocutaneous membranes such as the gut and the skin.
Pathogenicity island	Pathogenicity islands (PAIs) are a distinct class of genomic islands acquired by microorganisms through horizontal gene transfer. They are incorporated in the genome of pathogenic organisms, but are usually absent from those nonpathogenic organisms of the same or closely related species. These mobile genetic elements may range from 10-200 kb and encode genes which contribute to the virulence of the respective pathogen.
Borrelia burgdorferi	Borrelia burgdorferi is a species of Gram negative bacteria of the spirochete class of the genus Borrelia. B. burgdorferi is predominant in North America, but also exists in Europe, and is the agent of Lyme disease. It is a zoonotic, vector-borne disease transmitted by ticks and is named after the researcher Willy Burgdorfer who first isolated the bacterium in 1982. B. burgdorferi is one of the few pathogenic bacteria that can survive without iron, having replaced all of its iron-sulfur cluster enzymes with enzymes that use manganese, thus avoiding the problem many pathogenic bacteria face in acquiring iron.
Copepod	Copepods (; meaning 'oar-feet') are a group of small crustaceans found in the sea and nearly every freshwater habitat.

Some species are planktonic (drifting in sea waters), some are benthic (living on the ocean floor), and some continental species may live in limno-terrestrial habitats and other wet terrestrial places, such as swamps, under leaf fall in wet forests, bogs, springs, ephemeral ponds and puddles, damp moss, or water-filled recesses (phytotelmata) of plants such as bromeliads and pitcher plants. Many live underground in marine and freshwater caves, sinkholes, or stream beds.

Haemophilus influenzae	Haemophilus influenzae, formerly called Pfeiffer's bacillus or Bacillus influenzae, Gram-negative, rod-shaped bacterium first described in 1892 by Richard Pfeiffer during an influenza pandemic. A member of the Pasteurellaceae family, it is generally aerobic, but can grow as a facultative anaerobe. H. influenzae was mistakenly considered to be the cause of influenza until 1933, when the viral etiology of the flu became apparent; the bacterium is colloquially known as bacterial influenza.
Herpes simplex	Herpes simplex is a viral disease caused by both Herpes simplex virus type 1 (HSV-1) and type 2 (HSV-2). Infection with the herpes virus is categorized into one of several distinct disorders based on the site of infection. Oral herpes, the visible symptoms of which are colloquially called cold sores or fever blisters, infects the face and mouth.
NADH dehydrogenase	NADH-Ubiquinone/plastoquinone (complex I), various chains

NADH dehydrogenase (also referred to as NADH:ubiquinone reductase or Complex I) is an enzyme located in the inner mitochondrial membrane that catalyzes the transfer of electrons from NADH to coenzyme Q (CoQ). It is one of the 'entry enzymes' of oxidative phosphorylation in the mitochondria. Function

NADH Dehydrogenase is the first enzyme (Complex I) of the mitochondrial electron transport chain. |
| Toll-like receptor | Toll-like receptors are a class of proteins that play a key role in the innate immune system. They are single, membrane-spanning, non-catalytic receptors that recognize structurally conserved molecules derived from microbes. Once these microbes have breached physical barriers such as the skin or intestinal tract mucosa, they are recognized by Toll like receptors, which activate immune cell responses. |
| Vibrio cholerae | Vibrio cholerae is a gram negative comma-shaped bacterium with a polar flagellum that causes cholera in humans. V. cholerae and other species of the genus Vibrio belong to the gamma subdivision of the Proteobacteria. There are two major biotypes of V. cholerae identified by hemagglutination testing, classical and El Tor, and numerous serogroups. |

Chapter 1. Part 1: The Microbial Cell

Confocal microscopy	Confocal microscopy is an optical imaging technique used to increase optical resolution and contrast of a micrograph by using point illumination and a spatial pinhole to eliminate out-of-focus light in specimens that are thicker than the focal plane. It enables the reconstruction of three-dimensional structures from the obtained images. This technique has gained popularity in the scientific and industrial communities and typical applications are in life sciences, semiconductor inspection and materials science.
Herpes simplex virus	Herpes simplex virus 1 and 2 are two members of the herpes virus family, Herpesviridae, that infect humans. Both Herpes simplex virus-1 (which produces most cold sores) and Herpes simplex virus-2 (which produces most genital herpes) are ubiquitous and contagious. They can be spread when an infected person is producing and shedding the virus.
Sigma factor	A sigma factor is a protein needed only for initiation of RNA synthesis. It is a bacterial transcription initiation factor that enables specific binding of RNA polymerase to gene promoters. The specific sigma factor used to initiate transcription of a given gene will vary, depending on the gene and on the environmental signals needed to initiate transcription of that gene.
Operon	In genetics, an operon is a functioning unit of genomic DNA containing a cluster of genes under the control of a single regulatory signal or promoter. The genes are transcribed together into an mRNA strand and either translated together in the cytoplasm, or undergo trans-splicing to create monocistronic mRNAs that are translated separately, i.e. several strands of mRNA that each encode a single gene product. The result of this is that the genes contained in the operon are either expressed together or not at all.
Cell biology	Cell biology is a scientific discipline that studies cells - their physiological properties, their structure, the organelles they contain, interactions with their environment, their life cycle, division and death. This is done both on a microscopic and molecular level. Cell biology research encompasses both the great diversity of single-celled organisms like bacteria and protozoa, as well as the many specialized cells in multicellular organisms such as humans.
Essential nutrient	An essential nutrient is a nutrient required for normal body functioning that either cannot be synthesized by the body at all, or cannot be synthesized in amounts adequate for good health (e.g. niacin, choline), and thus must be obtained from a dietary source. Essential nutrients are also defined by the collective physiological evidence for their importance in the diet, as represented in e.g. US government approved tables for Dietary Reference Intake. Some categories of essential nutrients include vitamins, dietary minerals, essential fatty acids, and essential amino acids.

DNA polymerase	A DNA polymerase is an enzyme (the suffix -ase is used to identify enzymes) that helps catalyze the polymerization of deoxyribonucleotides into a DNA strand. DNA polymerases are best known for their feedback role in DNA replication, in which the polymerase 'reads' an intact DNA strand as a template and uses it to synthesize the new strand. This process copies a piece of DNA. The newly polymerized molecule is complementary to the template strand and identical to the template's original partner strand.
Genetic engineering	Genetic engineering, is the direct human manipulation of an organism's genome using modern DNA technology. It involves the introduction of foreign DNA or synthetic genes into the organism of interest. The introduction of new DNA does not require the use of classical genetic methods, however traditional breeding methods are typically used for the propagation of recombinant organisms.
Microbial ecology	Microbial ecology is the ecology of microorganisms: their relationship with one another and with their environment. It concerns the three major domains of life -- Eukaryota, Archaea, and Bacteria -- as well as viruses. Microorganisms, by their omnipresence, impact the entire biosphere.
Nitrogen fixation	Nitrogen fixation is a process by which nitrogen (N_2) in the atmosphere is converted into ammonium (NH_{4+}). Atmospheric nitrogen or elemental nitrogen (N_2) is relatively inert: it does not easily react with other chemicals to form new compounds. Fixation processes free up the nitrogen atoms from their diatomic form (N_2) to be used in other ways.
Amino acid	Amino acids are molecules containing an amine group, a carboxylic acid group, and a side-chain that is specific to each amino acid. The key elements of an amino acid are carbon, hydrogen, oxygen, and nitrogen. They are particularly important in biochemistry, where the term usually refers to alpha-amino acids.
Biosynthesis	Biosynthesis is an enzyme-catalyzed process in cells of living organisms by which substrates are converted to more complex products. The biosynthesis process often consists of several enzymatic steps in which the product of one step is used as substrate in the following step. Examples for such multi-step biosynthetic pathways are those for the production of amino acids, fatty acids, and natural products.
Chain reaction	A chain reaction is a sequence of reactions where a reactive product or by-product causes additional reactions to take place. In a chain reaction, positive feedback leads to a self-amplifying chain of events.

Chapter 1. Part 1: The Microbial Cell

Food chain	A food chain is somewhat a linear sequence of links in a food web starting from a trophic species that eats no other species in the web and ends at a trophic species that is eaten by no other species in the web. A food chain differs from a food web, because the complex polyphagous network of feeding relations are aggregated into trophic species and the chain only follows linear monophagous pathways. A common metric used to quantify food web trophic structure is food chain length.
Polymerase chain reaction	The polymerase chain reaction is a scientific technique in molecular biology to amplify a single or a few copies of a piece of DNA across several orders of magnitude, generating thousands to millions of copies of a particular DNA sequence. Developed in 1983 by Kary Mullis, PCR is now a common and often indispensable technique used in medical and biological research labs for a variety of applications. These include DNA cloning for sequencing, DNA-based phylogeny, or functional analysis of genes; the diagnosis of hereditary diseases; the identification of genetic fingerprints (used in forensic sciences and paternity testing); and the detection and diagnosis of infectious diseases.
Archaea	The Archaea are a group of single-celled microorganisms. A single individual or species from this domain is called an archaeon (sometimes spelled 'archeon'). They have no cell nucleus or any other membrane-bound organelles within their cells.
Bacteria	Bacteria are a large domain of prokaryotic microorganisms. Typically a few micrometres in length, bacteria have a wide range of shapes, ranging from spheres to rods and spirals. Bacteria are present in most habitats on Earth, growing in soil, acidic hot springs, radioactive waste, water, and deep in the Earth's crust, as well as in organic matter and the live bodies of plants and animals, providing outstanding examples of mutualism in the digestive tracts of humans, termites and cockroaches.
Bifidobacterium	Bifidobacterium is a genus of Gram-positive, non-motile, often branched anaerobic bacteria. They are ubiquitous, endosymbiotic inhabitants of the gastrointestinal tract, vagina and mouth (B. dentium) of mammals and other animals. Bifidobacteria are one of the major genera of bacteria that make up the colon flora in mammals.
Caulerpa taxifolia	Caulerpa taxifolia is a species of seaweed, an alga of the genus Caulerpa. Native to the Indian Ocean, it has been widely used ornamentally in aquariums. The alga has a stem which spreads horizontally just above the seafloor, and from this stem grow vertical fern-like pinnae, whose blades are flat like yew, hence the species name 'taxifolia' (the genus of yew is 'Taxus').
Epstein-Barr virus	The Epstein-Barr virus , also called human herpesvirus 4 (HHV-4), is a virus of the herpes family, which includes herpes simplex virus 1 and 2, and is one of the most common viruses in humans. It is best known as the cause of infectious mononucleosis.

CHAPTER HIGHLIGHTS & NOTES: KEY TERMS, PEOPLE, PLACES, CONCEPTS

It is also associated with particular forms of cancer, particularly Hodgkin's lymphoma, Burkitt's lymphoma, nasopharyngeal carcinoma, and central nervous system lymphomas associated with HIV. Finally, there is evidence that infection with the virus is associated with a higher risk of certain autoimmune diseases, especially dermatomyositis, systemic lupus erythematosus, rheumatoid arthritis, Sjögren's syndrome, and multiple sclerosis.

Eukaryote	A eukaryote is an organism whose cells contain complex structures enclosed within membranes. Eukaryotes may more formally be referred to as the taxon Eukarya or Eukaryota. The defining membrane-bound structure that sets eukaryotic cells apart from prokaryotic cells is the nucleus, or nuclear envelope, within which the genetic material is carried.
Microbial genetics	Microbial genetics is a subject area within microbiology and genetic engineering. It studies the genetics of very small (micro) organisms. This involves the study of the genotype of microbial species and also the expression system in the form of phenotypes.It also involves the study of genetic processes taking place in these micro organisms i.e., recombination etc.
Nitrogen cycle	The nitrogen cycle is the process by which nitrogen is converted between its various chemical forms. This transformation can be carried out to both biological and non-biological processes. Important processes in the nitrogen cycle include fixation, mineralization, nitrification, and denitrification.
Prokaryote	The prokaryotes are a group of organisms that lack a cell nucleus (karyon), or any other membrane-bound organelles. The organisms that have a cell nucleus are called eukaryotes. Most prokaryotes are unicellular, but a few such as myxobacteria have multicellular stages in their life cycles. Prokaryotes do not have a nucleus, mitochondria, or any other membrane-bound organelles. In other words, neither their DNA nor any of their other sites of metabolic activity are collected together in a discrete membrane-enclosed area.
RNA virus	An RNA virus is a virus that has RNA (ribonucleic acid) as its genetic material. This nucleic acid is usually single-stranded RNA (ssRNA), but may be double-stranded RNA (dsRNA). Notable human diseases caused by RNA viruses include SARS, influenza, hepatitis C and polio.
Thiomargarita namibiensis	Thiomargarita namibiensis is a gram-negative coccoid Proteobacterium, found in the ocean sediments of the continental shelf of Namibia. It is the largest bacterium ever discovered, in general, 0.1-0.3 mm (100-300 μm) wide, but sometimes up to 0.75 mm (750 μm). Its size is large enough to be seen by the naked eye.
Life cycle	A life cycle is a period involving all different generations of a species succeeding each other through means of reproduction, whether through asexual reproduction or sexual reproduction (a period from one generation of organisms to the same identical).

Chapter 1. Part 1: The Microbial Cell

Paul Berg	Paul Berg is an American biochemist and professor emeritus at Stanford University. He was the recipient of the Nobel Prize in Chemistry in 1980, along with Walter Gilbert and Frederick Sanger. The award recognized their contributions to basic research involving nucleic acids.
DNA sequencing	DNA sequencing includes several methods and technologies that are used for determining the order of the nucleotide bases--adenine, guanine, cytosine, and thymine--in a molecule of DNA. Knowledge of DNA sequences has become indispensable for basic biological research, other research branches utilizing DNA sequencing, and in numerous applied fields such as diagnostic, biotechnology, forensic biology and biological systematics. The advent of DNA sequencing has significantly accelerated biological research and discovery. The rapid speed of sequencing attained with modern DNA sequencing technology has been instrumental in the sequencing of the human genome, in the Human Genome Project.
Genome	In modern molecular biology and genetics, the genome is the entirety of an organism's hereditary information. It is encoded either in DNA or, for many types of virus, in RNA. The genome includes both the genes and the non-coding sequences of the DNA/RNA. The term was adapted in 1920 by Hans Winkler, Professor of Botany at the University of Hamburg, Germany. The Oxford English Dictionary suggests the name to be a blend of the words gene and chromosome.
Genomic island	A Genomic island is part of a genome that has evidence of horizontal origins. The term is usually used in microbiology, especially with regard to bacteria. A GI can code for many functions, can be involved in symbiosis or pathogenesis, and may help an organism's adaptation.
Sequencing	In genetics and biochemistry, sequencing means to determine the primary structure (sometimes falsely called primary sequence) of an unbranched biopolymer. Sequencing results in a symbolic linear depiction known as a sequence which succinctly summarizes much of the atomic-level structure of the sequenced molecule. DNA sequencing is the process of determining the nucleotide order of a given DNA fragment.
Extreme environment	An extreme environment exhibits extreme conditions which are challenging to most life forms. These may be extremely high or low ranges of temperature, radiation, pressure, acidity, alkalinity, air, water, salt, sugar, carbon dioxide, sulphur, petroleum and many others. An extreme environment is one place where humans generally do not live or could die there.

Bubonic plague	Bubonic plague is a zoonotic disease, circulating mainly among small rodents and their fleas, and is one of three types of infections caused by Yersinia pestis (formerly known as Pasteurella pestis), which belongs to the family Enterobacteriaceae. Without treatment, the bubonic plague kills about two out of three infected humans within 4 days.
	The term bubonic plague is derived from the Greek word βουβ?v, meaning 'groin.' Swollen lymph nodes (buboes) especially occur in the armpit and groin in persons suffering from bubonic plague.
Mycobacterium tuberculosis	Mycobacterium tuberculosis is a pathogenic bacterial species in the genus Mycobacterium and the causative agent of most cases of tuberculosis. First discovered in 1882 by Robert Koch, M. tuberculosis has an unusual, waxy coating on the cell surface (primarily mycolic acid), which makes the cells impervious to Gram staining so acid-fast detection techniques are used instead. The physiology of M. tuberculosis is highly aerobic and requires high levels of oxygen.
Palmaria palmata	Palmaria palmata Kuntze, also called dulse, dillisk, dilsk, red dulse, sea lettuce flakes or creathnach, is a red alga (Rhodophyta) previously referred to as Rhodymenia palmata (Linnaeus) Greville. It grows on the northern coasts of the Atlantic and Pacific oceans. It is a well-known snack food, and in Iceland, where it is known as söl, it has been an important source of fiber throughout the centuries.
Yersinia pestis	Yersinia pestis is a Gram-negative rod-shaped bacterium. It is a facultative anaerobe that can infect humans and other animals.
	Human Y. pestis infection takes three main forms: pneumonic, septicemic, and the notorious bubonic plagues.
Tuberculosis	Tuberculosis is a common and often deadly infectious disease caused by various strains of mycobacteria, usually Mycobacterium tuberculosis in humans. Tuberculosis usually attacks the lungs but can also affect other parts of the body. It is spread through the air when people who have the disease cough, sneeze, or spit.
Robert Boyle	Robert Boyle, FRS, (25 January 1627 - 31 December 1691) was a 17th century natural philosopher, chemist, physicist, and inventor, also noted for his writings in theology. He has been variously described as Irish, English and Anglo-Irish, his father having come to Ireland from England during the time of the Plantations.
	Although his research clearly has its roots in the alchemical tradition, Boyle is largely regarded today as the first modern chemist, and therefore one of the founders of modern chemistry, and one of the pioneers of modern experimental scientific method.

Chapter 1. Part 1: The Microbial Cell

Medical statistics	Medical statistics deals with applications of statistics to medicine and the health sciences, including epidemiology, public health, forensic medicine, and clinical research. Medical statistics has been a recognized branch of statistics in the UK for more than 40 years but the term does not appear to have come into general use in North America, where the wider term 'biostatistics' is more commonly used. However, 'biostatistics' more commonly connotes all applications of statistics to biology.
Isaac Newton	Sir Isaac Newton PRS MP (25 December 1642 - 20 March 1727 [NS: 4 January 1643 - 31 March 1727]) was an English physicist, mathematician, astronomer, natural philosopher, alchemist, and theologian, who has been 'considered by many to be the greatest and most influential scientist who ever lived.' His monograph Philosophiæ Naturalis Principia Mathematica, published in 1687, lays the foundations for most of classical mechanics. In this work, Newton described universal gravitation and the three laws of motion, which dominated the scientific view of the physical universe for the next three centuries. Newton showed that the motions of objects on Earth and of celestial bodies are governed by the same set of natural laws, by demonstrating the consistency between Kepler's laws of planetary motion and his theory of gravitation, thus removing the last doubts about heliocentrism and advancing the Scientific Revolution.
Smallpox	Smallpox is an infectious disease unique to humans, caused by either of two virus variants, Variola major and Variola minor. The disease is also known by the Latin names Variola or Variola vera, which is a derivative of the Latin varius, meaning 'spotted', or varus, meaning 'pimple'. The term 'smallpox' was first used in Europe in the 15th century to distinguish variola from the 'great pox' (syphilis).
Robert Hooke	Robert Hooke FRS (18 July 1635 - 3 March 1703) was an English natural philosopher, architect and polymath who played an important role in the scientific revolution, through both experimental and theoretical work. His adult life comprised three distinct periods: as a brilliant scientific inquirer lacking money; achieving great wealth and standing through his reputation for hard work and scrupulous honesty following the great fire of 1666 (section: Hooke the architect), but eventually becoming ill and party to jealous intellectual disputes. These issues may have contributed to his relative historical obscurity (section: Personality and disputes).
Micrographia	Micrographia is a historic book by Robert Hooke, detailing the then thirty-year-old Hooke's observations through various lenses. Published in September 1665, the first major publication of the Royal Society, it was the first scientific best-seller, inspiring a wide public interest in the new science of microscopy. It is also notable for coining the biological term cell.
Louis Pasteur	Louis Pasteur was a French chemist and microbiologist born in Dole. He is remembered for his remarkable breakthroughs in the causes and preventions of diseases.

	His discoveries reduced mortality from puerperal fever, and he created the first vaccine for rabies and anthrax. His experiments supported the germ theory of disease. He was best known to the general public for inventing a method to stop milk and wine from causing sickness, a process that came to be called pasteurization. He is regarded as one of the three main founders of microbiology, together with Ferdinand Cohn and Robert Koch.
Putrefaction	Putrefaction is one of seven stages in the decomposition of the body of a dead animal. It can be viewed, in broad terms, as the decomposition of proteins, in a process that results in the eventual breakdown of cohesion between tissues and the liquefaction of most organs. In terms of thermodynamics, all organic tissue is a stored source of chemical energy and when not maintained by the constant biochemical efforts of the living organism it will break down into simpler products.
Ribosomal RNA	Ribosomal RNA is the RNA component of the ribosome, the enzyme that is the site of protein synthesis in all living cells. Ribosomal RNA provides a mechanism for decoding mRNA into amino acids and interacts with tRNAs during translation by providing peptidyl transferase activity. The tRNAs bring the necessary amino acids corresponding to the appropriate mRNA codon.
Lazzaro Spallanzani	Lazzaro Spallanzani was an Italian Catholic priest, biologist and physiologist who made important contributions to the experimental study of bodily functions, animal reproduction, and essentially discovered echolocation. His research of biogenesis paved the way for the investigations of Louis Pasteur. Career He was born in Scandiano in the modern province of Reggio Emilia and died in Pavia, Italy.
Spontaneous generation	Spontaneous generation is an obsolete principle regarding the origin of life from inanimate matter, which held that this process was a commonplace and everyday occurrence, as distinguished from univocal generation, or reproduction from parent(s). The hypothesis was synthesized by Aristotle, who compiled and expanded the work of prior natural philosophers and the various ancient explanations of the appearance of organisms; it held sway for two millennia. It is generally accepted to have been ultimately disproven in the 19th century by the experiments of Louis Pasteur, expanding upon the experiments of other scientists before him (such as Francesco Redi who had performed similar experiments in the 17th century).
Dairy product	Dairy products are generally defined as foods produced from cow's or domestic buffalo's milk. They are usually high-energy-yielding food products. A production plant for such processing is called a dairy or a dairy factory.

Chapter 1. Part 1: The Microbial Cell

Molecular clock	The molecular clock (based on the molecular clock hypothesis (MCH)) is a technique in molecular evolution that uses fossil constraints and rates of molecular change to deduce the time in geologic history when two species or other taxa diverged. It is used to estimate the time of occurrence of events called speciation or radiation. The molecular data used for such calculations is usually nucleotide sequences for DNA or amino acid sequences for proteins.
Autoclave	An autoclave is an instrument used to sterilize equipment and supplies by subjecting them to high pressure saturated steam at 121 °C for around 15-20 minutes depending on the size of the load and the contents. It was invented by Charles Chamberland in 1879, although a precursor known as the steam digester was created by Denis Papin in 1679. The name comes from Greek auto-, ultimately meaning self, and Latin clavis meaning key -- a self-locking device. Autoclaves are widely used in microbiology, medicine, tattooing, body piercing, veterinary science, mycology, dentistry, chiropody and prosthetics fabrication.
Endospore	An endospore is a dormant, tough, and non-reproductive structure produced by certain bacteria from the Firmicute phylum. The name 'endospore' is suggestive of a spore or seed-like form (endo means within), but it is not a true spore (i.e. not an offspring). It is a stripped-down, dormant form to which the bacterium can reduce itself.
Germ theory of disease	The germ theory of disease, is a theory that proposes that microorganisms are the cause of many diseases. Although highly controversial when first proposed, germ theory was validated in the late 19th century and is now a fundamental part of modern medicine and clinical microbiology, leading to such important innovations as antibiotics and hygienic practices. The ancient historical view was that disease was spontaneously generated instead of being created by microorganisms that grow by reproduction.
Medical microbiology	Medical microbiology is both a branch of medicine and microbiology which deals with the study of microorganisms including bacteria, viruses, fungi and parasites which are of medical importance and are capable of causing infectious diseases in human beings. It includes the study of microbial pathogenesis and epidemiology and is related to the study of disease pathology and immunology. This branch of microbiology is amongst the most widely studied and followed branches due to its great importance to medicine.
Extraterrestrial life	Extraterrestrial life is defined as life that does not originate from Earth. Referred to as alien life, or simply aliens these hypothetical forms of life range from simple bacteria-like organisms to beings far more complex than humans.

Early Earth	The 'Early Earth' is a term usually defined as Earth's first billion years, or gigayear. On the geologic time scale, the 'early Earth' comprises all of the Hadean eon (itself unofficially defined), as well as the Eoarchean and part of the Paleoarchean eras of the Archean eon. This period of Earth's history, being its earliest, involved the planet's condensation from a solar nebula and accretion from meteorites, as well as the formation of the earliest atmosphere and hydrosphere.
Sidney Altman	Sidney Altman is a Canadian American molecular biologist, who is currently the Sterling Professor of Molecular, Cellular, and Developmental Biology and Chemistry at Yale University. In 1989 he shared the Nobel Prize in Chemistry with Thomas R. Cech for their work on the catalytic properties of RNA. Altman was born on May 7, 1939 in Montreal, Quebec, Canada. His parents were immigrants to Canada, each coming from Eastern Europe as a young adult, in the 1920s.
Bacillus anthracis	Bacillus anthracis is a Gram-positive spore-forming, rod-shaped bacterium, with a width of 1-1.2μm and a length of 3-5μm. It can be grown in an ordinary nutrient medium under aerobic or anaerobic conditions. It is the only bacterium known to synthesize a protein capsule (D-glutamate), and the only pathogenic bacterium to carry its own adenylyl cyclase virulence factor (edema factor).
Agar	Agar is a gelatinous substance derived from a polysaccharide that accumulates in the cell walls of agarophyte red algae. Historically and in a modern context, it is chiefly used as an ingredient in desserts throughout Asia and also as a solid substrate to contain culture medium for microbiological work. The gelling agent is an unbranched polysaccharide obtained from the cell walls of some species of red algae, primarily from the genera Gelidium and Gracilaria, or seaweed (Sphaerococcus euchema).
Petri dish	A Petri dish is a shallow glass or plastic cylindrical lidded dish that biologists use to culture cells or small moss plants. It was named after German bacteriologist Julius Richard Petri, who invented it when working as an assistant to Robert Koch. Glass Petri dishes can be reused by sterilization (for example, in an autoclave or by dry heating in a hot air oven at 160 °C for one hour).
Red algae	The red algae, thus red plant), are one of the oldest groups of eukaryotic algae, and also one of the largest, with about 5,000-6,000 species of mostly multicellular, marine algae, including many notable seaweeds. Other references indicate as many as 10,000 species; more detailed counts indicate ~4,000 in ~600 genera (3,738 marine spp in 546 genera and 10 orders (plus the unclassifiable); 164 freshwater spp in 30 genera in 8 orders).

Chapter 1. Part 1: The Microbial Cell

Cowpox	Cowpox is a skin disease caused by a virus known as the Cowpox virus. The pox is related to the vaccinia virus and got its name from the distribution of the disease when dairymaids touched the udders of infected cows. The ailment manifests itself in the form of red blisters and is transmitted by touch from infected animals to humans.
Encephalitozoon intestinalis	Encephalitozoon intestinalis is a parasite. It can cause microsporidiosis. It is notable as having one of the smallest genome among known eukaryotic organisms, containing only 2.25 million base pairs.
Helicobacter pylori	Helicobacter pylori previously named Campylobacter pyloridis, is a Gram-negative, microaerophilic bacterium found in the stomach. It was identified in 1982 by Barry Marshall and Robin Warren, who found that it was present in patients with chronic gastritis and gastric ulcers, conditions that were not previously believed to have a microbial cause. It is also linked to the development of duodenal ulcers and stomach cancer.
Lyme disease	Lyme disease, is an emerging infectious disease caused by at least three species of bacteria belonging to the genus Borrelia. Borrelia burgdorferi sensu stricto is the main cause of Lyme disease in the United States, whereas Borrelia afzelii and Borrelia garinii cause most European cases. he town of Lyme, Connecticut, USA, where a number of cases were identified in 1975. Although Allen Steere realized that Lyme disease was a tick-borne disease in 1978, the cause of the disease remained a mystery until 1981, when B. burgdorferi was identified by Willy Burgdorfer.
Simian immunodeficiency virus	Simian immunodeficiency virus also known as African Green Monkey virus and also as Monkey AIDS is a retrovirus able to infect at least 33 species of African primates. Based on analysis of strains found in four species of monkeys from Bioko Island, which was isolated from the mainland by rising sea levels about 11,000 years ago, it has been concluded that Simian immunodeficiency virus has been present in monkeys and apes for at least 32,000 years, and probably much longer. Virus strains from two of these primate species, Simian immunodeficiency virussmm in sooty mangabeys and Simian immunodeficiency viruscpz in chimpanzees, are believed to have crossed the species barrier into humans, resulting in HIV-2 and HIV-1, respectively.
Vaccination	Vaccination is the administration of antigenic material (a vaccine) to stimulate the immune system of an individual to develop adaptive immunity to a disease. Vaccines can prevent or ameliorate the effects of infection by many pathogens. The efficacy of vaccination has been widely studied and verified; for example, the influenza vaccine, the HPV vaccine, and the chicken pox vaccine among others.

Cholera toxin	Cholera toxin is a protein complex secreted by the bacterium Vibrio cholerae. CTX is responsible for the massive, watery diarrhea characteristic of cholera infection. The cholera toxin is an oligomeric complex made up of six protein subunits: a single copy of the A subunit (part A, enzymatic), and five copies of the B subunit (part B, receptor binding).
Inoculation	Inoculation is the placement of something that will grow or reproduce, and is most commonly used in respect of the introduction of a serum, vaccine, or antigenic substance into the body of a human or animal, especially to produce or boost immunity to a specific disease. It can also be used to refer to the communication of a disease to a living organism by transferring its causative agent into the organism, the implanting of microorganisms or infectious material into a culture medium such as a brewers vat or a petri dish, or the placement of microorganisms or viruses at a site where infection is possible. The verb to inoculate is from Middle English inoculaten, which meant 'to graft a scion' (a plant part to be grafted onto another plant); which in turn is from Latin inoculare, past participle inoculat-.
Polio vaccine	Two polio vaccines are used throughout the world to combat poliomyelitis (or polio). The first was developed by Jonas Salk and first tested in 1952. Announced to the world by Salk on April 12, 1955, it consists of an injected dose of inactivated (dead) poliovirus. An oral vaccine was developed by Albert Sabin using attenuated poliovirus.
Smallpox vaccine	The smallpox vaccine was the first successful vaccine to be developed. The process of vaccination was first publicised by Edward Jenner in 1796, who acted upon his observation that milkmaids who caught the cowpox virus did not catch smallpox. Prior to widespread vaccination, mortality rates in individuals with smallpox were high--up to 35% in some cases.
Antiseptic	Antiseptics are antimicrobial substances that are applied to living tissueskin to reduce the possibility of infection, sepsis, or putrefaction. Antiseptics are generally distinguished from antibiotics by the latter's ability to be transported through the lymphatic system to destroy bacteria within the body, and from disinfectants, which destroy microorganisms found on non-living objects. Some antiseptics are true germicides, capable of destroying microbes (bacteriocidal), while others are bacteriostatic and only prevent or inhibit their growth.
Chlorine	Chlorine is the chemical element with atomic number 17 and symbol Cl. It is the second lightest halogen, with fluorine being the lightest. Chlorine is found in the periodic table in group 17. The element forms diatomic molecules under standard conditions, called dichlorine.

Chapter 1. Part 1: The Microbial Cell

Rabies	Rabies is a viral disease that causes acute encephalitis (inflammation of the brain) in warm-blooded animals. It is zoonotic (i.e., transmissible from animals to humans), most commonly by a bite from an infected animal. For a human, rabies is almost invariably fatal if post-exposure prophylaxis is not administered prior to the onset of severe symptoms.
Antibiotic resistance	Antibiotic resistance is a type of drug resistance where a microorganism is able to survive exposure to an antibiotic. While a spontaneous or induced genetic mutation in bacteria may confer resistance to antimicrobial drugs, genes that confer resistance can be transferred between bacteria in a horizontal fashion by conjugation, transduction, or transformation. Thus, a gene for antibiotic resistance that evolves via natural selection may be shared.
Drug discovery	In the fields of medicine, biotechnology and pharmacology, drug discovery is the process by which drugs are discovered or designed. In the past most drugs have been discovered either by identifying the active ingredient from traditional remedies or by serendipitous discovery. As our understanding of disease has increased to the extent that we know how disease and infection are controlled at the molecular and physiological level, scientists are now able to try to find compounds that specifically modulate those molecules, for instance via high throughput screening.
Staphylococcus	Staphylococcus is a genus of Gram-positive bacteria. Under the microscope, they appear round (cocci), and form in grape-like clusters. The Staphylococcus genus includes at least 40 species.
Prion	A prion is an infectious agent composed of protein in a misfolded form. This is in contrast to all other known infectious agents (virusbacteriafungusparasite) which must contain nucleic acids (either DNA, RNA, or both). The word prion, coined in 1982 by Stanley B. Prusiner, is derived from the words protein and infection.
Tobacco mosaic virus	Tobacco mosaic virus is a positive-sense single stranded RNA virus that infects plants, especially tobacco and other members of the family Solanaceae. The infection causes characteristic patterns (mottling and discoloration) on the leaves (hence the name). TMV was the first virus to be discovered.
Viroid	In 1971 T.O Diener discovered a new infectious agent smaller than virus and caused potato spindle tuber disease. It was found to be a free RNA lacked the protein coat that is found in viruses, hence named viroid.

Drug development	Drug development is a blanket term used to define the process of bringing a new drug to the market once a lead compound has been identified through the process of drug discovery. It includes pre-clinical research (microorganisms/animals) and clinical trials (on humans) and may include the step of obtaining regulatory approval to market the drug. New Chemical Entity (NCE) development Broadly the process can be divided into pre-clinical and clinical work.
Drug resistance	Drug resistance is the reduction in effectiveness of a drug such as an antimicrobial or an antineoplastic in curing a disease or condition. When the drug is not intended to kill or inhibit a pathogen, then the term is equivalent to dosage failure or drug tolerance. More commonly, the term is used in the context of resistance acquired by pathogens.
Mosaic virus	Mosaic viruses are plant viruses that cause the leaves to have a speckled appearance. Mosaic virus is not a taxon. Species include:•beet mosaic virus•plum pox virus (in the potyvirus genus)•tobacco mosaic virus•cassava mosaic virus•Cucumber mosaic virus•Alfalfa mosaic virus•Panicum mosaic satellite virus•Tulip breaking virus.
Transfer RNA	Transfer RNA is an adaptor molecule composed of RNA, typically 73 to 93 nucleotides in length, that is used in biology to bridge the four-letter genetic code (ACGU) in messenger RNA (mRNA) with the twenty-letter code of amino acids in proteins. The role of tRNA as an adaptor is best understood by considering its three-dimensional structure. One end of the tRNA carries the genetic code in a three-nucleotide sequence called the anticodon.
Enrichment culture	An enrichment culture is a medium with specific and known qualities that favors the growth of a particular microorganism. The enrichment culture's environment will support the growth of a selected microorganism, while inhibiting the growth of others. Lourens Bass Becking succinctly summarized enrichment cultures' abilities when he said 'everything is everywhere; the environment selects.' The botanist Martinus Beijerinck is credited with developing the first enrichment cultures.
Food web	A food web depicts feeding connections (what eats what) in an ecological community. Ecologists can broadly lump all life forms into one of two categories called trophic levels: 1) the autotrophs, and 2) the heterotrophs. To maintain their bodies, grow, develop, and to reproduce, autotrophs produce organic matter from inorganic substances, including both minerals and gases such as carbon dioxide.

Chapter 1. Part 1: The Microbial Cell

Lithotroph	A lithotroph is an organism that uses an inorganic substrate (usually of mineral origin) to obtain reducing equivalents for use in biosynthesis (e.g., carbon dioxide fixation) or energy conservation via aerobic or anaerobic respiration. Known chemolithotrophs are exclusively microbes; No known macrofauna possesses the ability to utilize inorganic compounds as energy sources. Macrofauna and lithotrophs can form symbiotic relationships, in which case the lithotrophs are called 'prokaryotic symbionts.' An example of this is chemolithotrophic bacteria in deep sea worms or plastids, which are organelles within plant cells that may have evolved from photolithotrophic cyanobacteria-like organisms.
Photosynthesis	Photosynthesis is a process used by plants and other organisms to capture the sun's energy to split off water's hydrogen from oxygen. Hydrogen is combined with carbon dioxide (absorbed from air or water) to form glucose and release oxygen. All living cells in turn use fuels derived from glucose and oxidize the hydrogen and carbon to release the sun's energy and reform water and carbon dioxide in the process (cellular respiration).
Winogradsky column	The Winogradsky column is a simple device for culturing a large diversity of microorganisms. Invented by Sergei Winogradsky, the device is a column of pond mud and water mixed with a carbon source such as newspaper (containing cellulose) blackened marshmallows or egg-shells (containing calcium carbonate) and a sulfur source such as gypsum (calcium sulfate) or egg-yolk. Incubating the column in sunlight for months results in an aerobicanaerobic gradient as well as a sulfide gradient.
Bacteroides	Bacteroides is a genus of Gram-negative, bacillus bacteria. Bacteroides species are non-endospore-forming, anaerobes, and may be either motile or non-motile, depending on the species. The DNA base composition is 40-48% GC. Unusual in bacterial organisms, Bacteroides membranes contain sphingolipids.
Biofilm	A biofilm is an aggregate of microorganisms in which cells adhere to each other on a surface. These adherent cells are frequently embedded within a self-produced matrix of extracellular polymeric substance (EPS). Biofilm EPS, which is also referred to as slime (although not everything described as slime is a biofilm), is a polymeric conglomeration generally composed of extracellular DNA, proteins, and polysaccharides.
Endosymbiont	An endosymbiont is any organism that lives within the body or cells of another organism, i.e. forming an endosymbiosis . Examples are nitrogen-fixing bacteria (called rhizobia) which live in root nodules on legume roots, single-celled algae inside reef-building corals, and bacterial endosymbionts that provide essential nutrients to about 10-15% of insects.

Haber process	The Haber process, is the nitrogen fixation reaction of nitrogen gas and hydrogen gas, over an enriched iron or ruthenium catalyst, which is used to industrially produce ammonia. Despite the fact that 78.1% of the air we breathe is nitrogen, the gas is relatively unavailable because it is so unreactive: nitrogen molecules are held together by strong triple bonds. It was not until the early 20th century that the Haber process was developed to harness the atmospheric abundance of nitrogen to create ammonia, which can then be oxidized to make the nitrates and nitrites essential for the production of nitrate fertilizer and explosives.
Purple sulfur bacteria	The purple sulfur bacteria are a group of Proteobacteria capable of photosynthesis, collectively referred to as purple bacteria. They are anaerobic or microaerophilic, and are often found in hot springs or stagnant water. Unlike plants, algae, and cyanobacteria, they do not use water as their reducing agent, and so do not produce oxygen.
Sulfate-reducing bacteria	Sulfate-reducing bacteria are those bacteria and archaea that can obtain energy by oxidizing organic compounds or molecular hydrogen (H_2) while reducing sulfate ($SO2-4$) to hydrogen sulfide (H_2S). In a sense, these organisms 'breathe' sulfate rather than oxygen, in a form of anaerobic respiration. Sulfate-reducing bacteria can be traced back to 3.5 billion years ago and are considered to be among the oldest forms of microorganisms, having contributed to the sulfur cycle soon after life emerged on Earth.
Vibrio fischeri	Vibrio fischeri is a Gram-negative, rod-shaped bacterium found globally in marine environments. V. fischeri has bioluminescent properties, and is found predominantly in symbiosis with various marine animals, such as the bobtail squid. It is heterotrophic and moves by means of flagella.
Rhizobia	Rhizobia are soil bacteria that fix nitrogen (diazotrophs) after becoming established inside root nodules of legumes (Fabaceae). Rhizobia require a plant host; they cannot independently fix nitrogen. In general, they are Gram-negative, motile, non-sporulating rods.
Immune system	An immune system is a system of biological structures and processes within an organism that protects against disease. To function properly, an immune system must detect a wide variety of agents, from viruses to parasitic worms, and distinguish them from the organism's own healthy tissue. Pathogens can rapidly evolve and adapt to avoid detection and neutralization by the immune system.
Monera	Monera is a kingdom that contains unicellular organisms without a nucleus (i.e., a prokaryotic cell organization), such as bacteria. The kingdom is considered superseded.

Chapter 1. Part 1: The Microbial Cell

The taxon Monera was first proposed as a phylum by Ernst Haeckel in 1866; subsequently, the taxon was raised to the rank of kingdom in 1925 by Édouard Chatton, gaining common acceptance, and the last commonly accepted mega-classification with the taxon Monera was the five-kingdom classification system established by Robert Whittaker in 1969. Under the three-domain system of taxonomy, which was established in 1990 and reflects the evolutionary history of life as currently understood, the organisms found in kingdom Monera have been divided into two domains, Archaea and Bacteria (with Eukarya as the third domain).

Nucleus

In neuroanatomy, a nucleus is a brain structure consisting of a relatively compact cluster of neurons. It is one of the two most common forms of nerve cell organization, the other being layered structures such as the cerebral cortex or cerebellar cortex. In anatomical sections, a nucleus shows up as a region of gray matter, often bordered by white matter.

Organelle

In cell biology, an organelle is a specialized subunit within a cell that has a specific function, and is usually separately enclosed within its own lipid bilayer.

The name organelle comes from the idea that these structures are to cells what an organ is to the body (hence the name organelle, the suffix -elle being a diminutive). Organelles are identified by microscopy, and can also be purified by cell fractionation.

Blood cell

A blood cell, is a cell of any type normally found in blood. In mammals, these fall into three general categories:•red blood cells -- Erythrocytes•white blood cells -- Leukocytes•platelets -- Thrombocytes

Together, these three kinds of blood cells sum up for a total 45% of blood tissue by volume (and the remaining 55% is plasma). This is called the hematocrit and can be determined by centrifuge or flow cytometry.

Iron-sulfur cluster

Iron-sulfur clusters are ensembles of iron and sulfide centres. Fe-S clusters are most often discussed in the context of the biological role for iron-sulfur proteins. Many Fe-S clusters are known in the area of organometallic chemistry and as precursors to synthetic analogues of the biological clusters .

Cell envelope

The cell envelope is the cell membrane and cell wall plus an outer membrane, if one is present.

Most bacterial cell envelopes fall into two major categories: Gram positive and Gram negative. These are differentiated by their Gram staining characteristics.

Electron microscope	An electron microscope is a type of microscope that uses a beam of electrons to illuminate the specimen and produce a magnified image. Electron microscopes (EM) have a greater resolving power than a light-powered optical microscope, because electrons have wavelengths about 100,000 times shorter than visible light (photons), and can achieve better than 50 pm resolution and magnifications of up to about 10,000,000x, whereas ordinary, non-confocal light microscopes are limited by diffraction to about 200 nm resolution and useful magnifications below 2000x. The electron microscope uses electrostatic and electromagnetic 'lenses' to control the electron beam and focus it to form an image.
Chlorosome	A Chlorosome is a photosynthetic antenna complex found in green sulfur bacteria (GSB) and some green filamentous anoxygenic phototrophs (FAP) (Chloroflexaceae, Oscillochloridaceae). They differ from other antenna complexes by their large size and lack of protein matrix supporting the photosynthetic pigments. Green sulfur bacteria are a group of organisms that generally live in extremely low-light environments, such as at depths of 100 meters in the Black Sea.
Cytoplasm	The cytoplasm is the gel-like substance residing between the cell membrane holding all the cell's internal sub-structures (called organelles), except for the nucleus. All the contents of the cells of prokaryote organisms (which lack a cell nucleus) are contained within the cytoplasm. Within the cells of eukaryote organisms the contents of the cell nucleus are separated from the cytoplasm, and are then called the nucleoplasm.
Recombinant DNA	Recombinant DNA molecules are DNA sequences that result from the use of laboratory methods (molecular cloning) to bring together genetic material from multiple sources, creating sequences that would not otherwise be found in biological organisms. Recombinant DNA is possible because DNA molecules from all organisms share the same chemical structure; they differ only in the sequence of nucleotides within that identical overall structure. Consequently, when DNA from a foreign source is linked to host sequences that can drive DNA replication and then introduced into a host organism, the foreign DNA is replicated along with the host DNA. Recombinant DNA molecules are sometimes called chimeric DNA, because they are usually made of material from two different species, like the mythical chimera.
James D. Watson	James D. Watson is an American molecular biologist, geneticist, and zoologist, best known as one of the discoverers of the structure of DNA in 1953 with Francis Crick and Rosalind Franklin. Watson, Crick, and Maurice Wilkins were awarded the 1962 Nobel Prize in Physiology or Medicine 'for their discoveries concerning the molecular structure of nucleic acids and its significance for information transfer in living material'.

Chapter 1. Part 1: The Microbial Cell

X-ray crystallography	X-ray crystallography is a method of determining the arrangement of atoms within a crystal, in which a beam of X-rays strikes a crystal and causes the beam of light to spread into many specific directions. From the angles and intensities of these diffracted beams, a crystallographer can produce a three-dimensional picture of the density of electrons within the crystal. From this electron density, the mean positions of the atoms in the crystal can be determined, as well as their chemical bonds, their disorder and various other information.
Double helix	In molecular biology, the term double helix refers to the structure formed by double-stranded molecules of nucleic acids such as DNA and RNA. The double helical structure of a nucleic acid complex arises as a consequence of its secondary structure, and is a fundamental component in determining its tertiary structure. The term entered popular culture with the publication in 1968 of The Double Helix: A Personal Account of the Discovery of the Structure of DNA, by James Watson. The DNA double helix is a spiral polymer of nucleic acids, held together by nucleotides which base pair together.
Agrobacterium tumefaciens	Agrobacterium tumefaciens is the causal agent of crown gall disease (the formation of tumours) in over 140 species of dicot. It is a rod shaped, Gram negative soil bacterium (Smith et al., 1907). Symptoms are caused by the insertion of a small segment of DNA (known as the T-DNA, for 'transfer DNA'), from a plasmid, into the plant cell, which is incorporated at a semi-random location into the plant genome.
DNA-binding protein	DNA-binding proteins are proteins that are composed of DNA-binding domains and thus have a specific or general affinity for either single or double stranded DNA. Sequence-specific DNA-binding proteins generally interact with the major groove of B-DNA, because it exposes more functional groups that identify a base pair. However there are some known minor groove DNA-binding ligands such as Netropsin, Distamycin, Hoechst 33258, Pentamidine and others. DNA-binding proteins include transcription factors which modulate the process of transcription, various polymerases, nucleases which cleave DNA molecules, and histones which are involved in chromosome packaging and transcription in the cell nucleus.
Taq polymerase	Taq polymerase, exonuclease Taq polymerase is a thermostable DNA polymerase named after the thermophilic bacterium Thermus aquaticus from which it was originally isolated by Thomas D. Brock in 1965. It is often abbreviated to 'Taq Pol' (or simply 'Taq'), and is frequently used in polymerase chain reaction (PCR), a method for greatly amplifying short segments of DNA. T.

aquaticus is a bacterium that lives in hot springs and hydrothermal vents, and Taq polymerase was identified as an enzyme able to withstand the protein-denaturing conditions (high temperature) required during PCR. Therefore it replaced the DNA polymerase from E. coli originally used in PCR. Taq's optimum temperature for activity is 75-80°C, with a half-life of greater than 2 hours at 92.5°C, 40 minutes at 95°C and 9 minutes at 97.5°C, and can replicate a 1000 base pair strand of DNA in less than 10 seconds at 72°C.

One of Taq's drawbacks is its relatively low replication fidelity. It lacks a 3' to 5' exonuclease proofreading activity, and has an error rate measured at about 1 in 9,000 nucleotides. The remaining two domains however may act in coordination, via coupled domain motion.

Thermus aquaticus	Thermus aquaticus is a species of bacterium that can tolerate high temperatures, one of several thermophilic bacteria that belong to the Deinococcus-Thermus group. It is the source of the heat-resistant enzyme Taq DNA polymerase, one of the most important enzymes in molecular biology because of its use in the polymerase chain reaction (PCR) DNA amplification technique.

When studies of biological organisms in hot springs began in the 1960s, scientists thought that the life of thermophilic bacteria could not be sustained in temperatures above about 55° Celsius (131° Fahrenheit). |
Cloning	Cloning in biology is the process of producing similar populations of genetically identical individuals that occurs in nature when organisms such as bacteria, insects or plants reproduce asexually. Cloning in biotechnology refers to processes used to create copies of DNA fragments (molecular cloning), cells (cell cloning), or organisms. The term also refers to the production of multiple copies of a product such as digital media or software.
Bioremediation	Bioremediation is the use of microorganismal metabolism to remove pollutants. Technologies can be generally classified as in situ or ex situ. In situ bioremediation involves treating the contaminated material at the site, while ex situ involves the removal of the contaminated material to be treated elsewhere.
Oil spill	An oil spill is a release of a liquid petroleum hydrocarbon into the environment due to human activity, and is a form of pollution. The term often refers to marine oil spills, where oil is released into the ocean or coastal waters. Oil spills include releases of crude oil from tankers, offshore platforms, drilling rigs and wells, as well as spills of refined petroleum products (such as gasoline, diesel) and their by-products, and heavier fuels used by large ships such as bunker fuel, or the spill of any oily white substance refuse or waste oil.
Airy disk	In optics, the Airy disk and Airy pattern are descriptions of the best focused spot of light that a perfect lens with a circular aperture can make, limited by the diffraction of light.

Chapter 1. Part 1: The Microbial Cell

	The diffraction pattern resulting from a uniformly-illuminated circular aperture has a bright region in the center, known as the Airy disk which together with the series of concentric bright rings around is called the Airy pattern. Both are named after George Biddell Airy.
Human eye	The human eye is an organ which reacts to light for several purposes. As a conscious sense organ, the mammalian eye allows vision. Rod and cone cells in the retina allow conscious light perception and vision including color differentiation and the perception of depth.
Bacilli	Bacilli refers to a taxonomic class of bacteria. It includes two orders, Bacillales and Lactobacillales, which contain several well-known pathogens like Bacillus anthracis (the cause of anthrax).
	Ambiguity
	There are several related concepts that make use of similar words, and the ambiguity can create considerable confusion.
Slime mold	Slime mold is a broad term describing protists that use spores to reproduce. Slime molds were formerly classified as fungi, but are no longer considered part of this kingdom.
	Their common name refers to part of some of these organisms' life cycles where they can appear as gelatinous 'slime'.
Spirillum	Spirillum in microbiology refers to a bacterium with a cell body that twists like a spiral. It is the third distinct bacterial cell shape type besides coccus and bacillus cells.
	Spermillum is the bactreria of a genus of Gram-negative bacteria (family Spirillaceae).
Trypanosoma brucei	Trypanosoma brucei is a parasitic protist species that causes African trypanosomiasis (or sleeping sickness) in humans and nagana in animals in Africa. There are 3 sub-species of T. brucei: T. b. brucei, T. b. gambiense and T. b.
Atomic force microscopy	Atomic force microscopy or scanning force microscopy (SFM) is a very high-resolution type of scanning probe microscopy, with demonstrated resolution on the order of fractions of a nanometer, more than 1000 times better than the optical diffraction limit. The precursor to the AFM, the scanning tunneling microscope, was developed by Gerd Binnig and Heinrich Rohrer in the early 1980s at IBM Research - Zurich, a development that earned them the Nobel Prize for Physics in 1986. Binnig, Quate and Gerber invented the first atomic force microscope in 1986. The first commercially available atomic force microscope was introduced in 1989. The AFM is one of the foremost tools for imaging, measuring, and manipulating matter at the nanoscale.

Transmission electron microscopy	Transmission electron microscopy is a microscopy technique whereby a beam of electrons is transmitted through an ultra thin specimen, interacting with the specimen as it passes through. An image is formed from the interaction of the electrons transmitted through the specimen; the image is magnified and focused onto an imaging device, such as a fluorescent screen, on a layer of photographic film, or to be detected by a sensor such as a CCD camera. TEMs are capable of imaging at a significantly higher resolution than light microscopes, owing to the small de Broglie wavelength of electrons.
Frequency	Frequency is the number of occurrences of a repeating event per unit time. It is also referred to as temporal frequency. The period is the duration of one cycle in a repeating event, so the period is the reciprocal of the frequency.
Wavelength	In physics, the wavelength of a sinusoidal wave is the spatial period of the wave--the distance over which the wave's shape repeats. It is usually determined by considering the distance between consecutive corresponding points of the same phase, such as crests, troughs, or zero crossings, and is a characteristic of both traveling waves and standing waves, as well as other spatial wave patterns. Wavelength is commonly designated by the Greek letter lambda (λ).
Speed of light	The speed of light in vacuum, usually denoted by c, is a universal physical constant important in many areas of physics. Its value is 299,792,458 metres per second, a figure that is exact since the length of the metre is defined from this constant and the international standard for time. In imperial units this speed is approximately 186,282 miles per second.
Fluorescence	Fluorescence is the emission of light by a substance that has absorbed light or other electromagnetic radiation. It is a form of luminescence. In most cases, emitted light has a longer wavelength, and therefore lower energy, than the absorbed radiation.
Reflection	Reflection is the change in direction of a wavefront at an interface between two different media so that the wavefront returns into the medium from which it originated. Common examples include the reflection of light, sound and water waves. The law of reflection says that for specular reflection the angle at which the wave is incident on the surface equals the angle at which it is reflected.
Refractive index	In optics the refractive index n of a substance (optical medium) is a number that describes how light, or any other radiation, propagates through that medium. Its most elementary occurrence (and historically the first one) is in Snell's law of refraction, $n_1 \sin\theta_1 = n_2 \sin\theta_2$, where θ_1 and θ_2 are the angles of incidence of a ray crossing the interface between two media with refractive indices n_1 and n_2.

Chapter 1. Part 1: The Microbial Cell

Scattering	Scattering is a general physical process where some forms of radiation, such as light, sound, or moving particles, are forced to deviate from a straight trajectory by one or more localized non-uniformities in the medium through which they pass. In conventional use, this also includes deviation of reflected radiation from the angle predicted by the law of reflection. Reflections that undergo scattering are often called diffuse reflections and unscattered reflections are called specular (mirror-like) reflections
	The types of non-uniformities which can cause scattering, sometimes known as scatterers or scattering centers, are too numerous to list, but a small sample includes particles, bubbles, droplets, density fluctuations in fluids, crystallites in polycrystalline solids, defects in monocrystalline solids, surface roughness, cells in organisms, and textile fibers in clothing.
Electromagnetic spectrum	The electromagnetic spectrum is the range of all possible frequencies of electromagnetic radiation. The 'electromagnetic spectrum' of an object is the characteristic distribution of electromagnetic radiation emitted or absorbed by that particular object.
	The electromagnetic spectrum extends from low frequencies used for modern radio communication to gamma radiation at the short-wavelength (high-frequency) end, thereby covering wavelengths from thousands of kilometers down to a fraction of the size of an atom.
Acquired immunodeficiency syndrome	Acquired immunodeficiency syndrome is a disease of the human immune system caused by the human immunodeficiency virus (HIV). This condition progressively reduces the effectiveness of the immune system and leaves individuals susceptible to opportunistic infections and tumors. HIV is transmitted through direct contact of a mucous membrane or the bloodstream with a bodily fluid containing HIV, such as blood, semen, vaginal fluid, preseminal fluid, and breast milk.
Numerical aperture	In optics, the numerical aperture of an optical system is a dimensionless number that characterizes the range of angles over which the system can accept or emit light. By incorporating index of refraction in its definition, NA has the property that it is constant for a beam as it goes from one material to another provided there is no optical power at the interface. The exact definition of the term varies slightly between different areas of optics.
Rhizopus oligosporus	Rhizopus oligosporus is a fungus of the family Mucoraceae that is a widely used starter culture for the home production of tempeh. The spores produce fluffy, white mycelia, binding the beans together to create an edible 'cake' of partly fermented soybeans.
	Rhizopus oligosporus produces an antibiotic that inhibits gram-positive bacteria, including the potentially harmful Staphylococcus aureus and the beneficial Bacillus subtilis (present in natto), even after the Rhizopus is consumed.

Bacillus thuringiensis	Bacillus thuringiensis is a Gram-positive, soil-dwelling bacterium, commonly used as a biological pesticide; alternatively, the Cry toxin may be extracted and used as a pesticide. B. thuringiensis also occurs naturally in the gut of caterpillars of various types of moths and butterflies, as well as on the dark surface of plants. During sporulation many Bt strains produce crystal proteins (proteinaceous inclusions), called δ-endotoxins, that have insecticidal action.
Crystal violet	Crystal violet, hexamethyl pararosaniline chloride, or pyoctanin(e)) is a triarylmethane dye. The dye is used as a histological stain and in Gram's method of classifying bacteria. Crystal violet has antibacterial, antifungal, and anthelmintic properties and was formerly important as a topical antiseptic.
Gram-positive bacteria	Gram-positive bacteria are those that are stained dark blue or violet by Gram staining. This is in contrast to Gram-negative bacteria, which cannot retain the crystal violet stain, instead taking up the counterstain (safranin or fuchsine) and appearing red or pink. Gram-positive organisms are able to retain the crystal violet stain because of the high amount of peptidoglycan in the cell wall.
Hans Christian	Hans Christian is a German-born musician and producer now based in the U.S. in Sturgeon Bay, Wisconsin. Christian is a multi-instrumentalist (often bass guitar, cello, nyckelharpa, and sarangi, but also balalaika, banjo, santoor, sitar, tambura, etc). usually associated with Indian world music, ethnic fusion, chamber jazz, and sometimes New Age music, but who also plays classical music.
Iodine	Iodine is a chemical element with the symbol I and atomic number 53. The name is from Greek ? οειδ?ς ioeides, meaning violet or purple, due to the color of elemental iodine vapor. Iodine and its compounds are primarily used in nutrition, and industrially in the production of acetic acid and certain polymers.
Staining	Staining is an auxiliary technique used in microscopy to enhance contrast in the microscopic image. Stains and dyes are frequently used in biology and medicine to highlight structures in biological tissues for viewing, often with the aid of different microscopes. Stains may be used to define and examine bulk tissues (highlighting, for example, muscle fibers or connective tissue), cell populations (classifying different blood cells, for instance), or organelles within individual cells.
Stains	

Chapter 1. Part 1: The Microbial Cell

	Stains is a commune in the northern suburbs of Paris, France. It is located 11.6 km (7.2 mi) from the center of Paris. Moreover, the city is notorious for its drug traffic and crime which plague the city.
Insecticide	An insecticide is a pesticide used against insects. They include ovicides and larvicides used against the eggs and larvae of insects respectively. Insecticides are used in agriculture, medicine, industry and the household.
Methylene blue	Methylene blue is a heterocyclic aromatic chemical compound with the molecular formula $C_{16}H_{18}N_3SCl$. It has many uses in a range of different fields, such as biology and chemistry. At room temperature it appears as a solid, odorless, dark green powder, that yields a blue solution when dissolved in water.
Starvation response	Starvation response in animals is a set of adaptive biochemical and physiological changes that reduce metabolism in response to a lack of food.
Cell wall	The cell wall is the tough, usually flexible but sometimes fairly rigid layer that surrounds some types of cells. It is located outside the cell membrane and provides these cells with structural support and protection, in addition to acting as a filtering mechanism. A major function of the cell wall is to act as a pressure vessel, preventing over-expansion when water enters the cell.
Clostridium botulinum	Clostridium botulinum is a Gram-positive, rod-shaped bacterium that produces several toxins. The best known are its neurotoxins, subdivided in types A-G, that cause the flaccid muscular paralysis seen in botulism. It is also the main paralytic agent in botox.
Counterstain	A counterstain is a stain with color contrasting to the principal stain, making the stained structure more easily visible.
	An example is the malachite green counterstain to the fuchsine stain in the Gimenez staining technique.
	Another example is eosin counterstain to haematoxylin in the H&E stain.
Mycolic acid	Mycolic acids are long fatty acids found in the cell walls of the mycolata taxon, a group of bacteria that includes Mycobacterium tuberculosis, the causative agent of the disease tuberculosis. They form the major component of the cell wall of mycolata species. Despite their name, mycolic acids have no biological link to fungi; the name arises from the filamentous appearance their presence gives mycolata under high magnification.
Proteobacteria	The Proteobacteria are a major group (phylum) of bacteria.

	They include a wide variety of pathogens, such as Escherichia, Salmonella, Vibrio, Helicobacter, and many other notable genera. Others are free-living, and include many of the bacteria responsible for nitrogen fixation.
Peptidoglycan	Peptidoglycan, is a polymer consisting of sugars and amino acids that forms a mesh-like layer outside the plasma membrane of bacteria (but not Archaea), forming the cell wall. The sugar component consists of alternating residues of β-(1,4) linked N-acetylglucosamine and N-acetylmuramic acid. Attached to the N-acetylmuramic acid is a peptide chain of three to five amino acids.
Safranin	Safranin is a biological stain used in histology and cytology. Safranin is used as a counterstain in some staining protocols, colouring all cell nuclei red. This is the classic counterstain in a Gram stain.
Negative stain	Negative staining is an established method, often used in diagnostic microscopy, for contrasting a thin specimen with an optically opaque fluid. For bright field microscopy, negative staining is typically performed using a black ink fluid such as nigrosin. The specimen, such as a wet bacterial culture spread on a glass slide, is mixed with the negative stain and allowed to dry.
Substrate-level phosphorylation	Substrate-level phosphorylation is a type of metabolism that results in the formation and creation of adenosine triphosphate (ATP) or guanosine triphosphate (GTP) by the direct transfer and donation of a phosphoryl (PO_3) group to adenosine diphosphate (ADP) or guanosine diphosphate (GDP) from a phosphorylated reactive intermediate. Note that the phosphate group does not have to directly come from the substrate. By convention, the phosphoryl group that is transferred is referred to as a phosphate group.
Syphilis	Syphilis is a sexually transmitted infection caused by the spirochete bacterium Treponema pallidum subspecies pallidum. The primary route of transmission is through sexual contact; however, it may also be transmitted from mother to fetus during pregnancy or at birth, resulting in congenital syphilis. Other human diseases caused by related Treponema pallidum include yaws (subspecies pertenue), pinta (subspecies carateum) and bejel (subspecies endemicum).
Organic acid	An organic acid is an organic compound with acidic properties. The most common organic acids are the carboxylic acids, whose acidity is associated with their carboxyl group -COOH. Sulfonic acids, containing the group $-SO_2OH$, are relatively stronger acids. Alcohols, with -OH, can act as acids but they are usually very weak.
Entamoeba histolytica	Entamoeba histolytica is an anaerobic parasitic protozoan, part of the genus Entamoeba. Predominantly infecting humans and other primates, E.

histolytica is estimated to infect about 50 million people worldwide. Previously, it was thought that 10% of the world population was infected, but these figures predate the recognition that at least 90% of these infections were due to a second species, E. dispar.

Bacillus subtilis	Bacillus subtilis, known also as the hay bacillus or grass bacillus, is a Gram-positive, catalase-positive bacterium commonly found in soil. A member of the genus Bacillus, B. subtilis is rod-shaped, and has the ability to form a tough, protective endospore, allowing the organism to tolerate extreme environmental conditions. Unlike several other well-known species, B. subtilis has historically been classified as an obligate aerobe, though recent research has demonstrated that this is not strictly correct.
Interference microscopy	Interference microscopy involving measurements of differences in the path between two beams of light that have been split. Types include:•Classical interference microscopy•Differential interference contrast microscopy•Fluorescence interference contrast microscopy.
Fluorophore	A fluorophore, in analogy to a chromophore, is a component of a molecule which causes a molecule to be fluorescent. It is a functional group in a molecule which will absorb energy of a specific wavelength and re-emit energy at a different (but equally specific) wavelength. The amount and wavelength of the emitted energy depend on both the fluorophore and the chemical environment of the fluorophore.
Origin of replication	The origin of replication is a particular sequence in a genome at which replication is initiated. This can either involve the replication of DNA in living organisms such as prokaryotes and eukaryotes, or that of DNA or RNA in viruses, such as double-stranded RNA viruses. DNA replication may proceed from this point bidirectionally or unidirectionally.
Confocal laser scanning microscopy	Confocal laser scanning microscopy is a technique for obtaining high-resolution optical images with depth selectivity. The key feature of confocal microscopy is its ability to acquire in-focus images from selected depths, a process known as optical sectioning. Images are acquired point-by-point and reconstructed with a computer, allowing three-dimensional reconstructions of topologically complex objects.
Complementary DNA	In genetics, complementary DNA is DNA synthesized from a messenger RNA (mRNA) template in a reaction catalyzed by the enzyme reverse transcriptase and the enzyme DNA polymerase. cDNA is often used to clone eukaryotic genes in prokaryotes. When scientists want to express a specific protein in a cell that does not normally express that protein (i.e., heterologous expression), they will transfer the cDNA that codes for the protein to the recipient cell.

Cryo-electron tomography	Cryo-electron tomography is a type of electron cryomicroscopy where tomography is used to obtain a 3D reconstruction of a sample from tilted 2D images at cryogenic temperatures. A cryoelectron tomography can be used to obtain structural details of complex cellular organizations at subnanometer resolutions.
Pseudomonas aeruginosa	Pseudomonas aeruginosa is a common bacterium which can cause disease in animals, including humans. It is found in soil, water, skin flora, and most man-made environments throughout the world. It thrives not only in normal atmospheres, but also in hypoxic atmospheres, and has thus colonized many natural and artificial environments.
Magnetosome	Magnetosome chains are membranous prokaryotic organelles present in magnetotactic bacteria. They contain 15 to 20 magnetite crystals that together act like a compass needle to orient magnetotactic bacteria in geomagnetic fields, thereby simplifying their search for their preferred microaerophilic environments. Each magnetite crystal within a magnetosome is surrounded by a lipid bilayer, and specific soluble and transmembrane proteins are sorted to the membrane.
Magnetotactic bacteria	Magnetotactic bacteria are a polyphyletic group of bacteria discovered by Richard P. Blakemore in 1975, that orient along the magnetic field lines of Earth's magnetic field. To perform this task, these bacteria have organelles called magnetosomes that contain magnetic crystals. The biological phenomenon of microorganisms tending to move in response to the environment's magnetic characteristics is known as magnetotaxis (although this term is misleading in that every other application of the term taxis involves a stimulus-response mechanism).
John Desmond Bernal	John Desmond Bernal FRS (b. 10 May 1901, Nenagh, Co. Tipperary, Ireland, d. London, 15 September 1971) was one of Britain's best known and most controversial scientists, called 'Sage' by his friends, and known for pioneering X-ray crystallography in molecular biology.
ATP synthase	ATP synthase is an important enzyme that provides energy for the cell to use through the synthesis of adenosine triphosphate (ATP). ATP is the most commonly used 'energy currency' of cells from most organisms. It is formed from adenosine diphosphate (ADP) and inorganic phosphate (P_i), and needs energy.
Borrelia afzelii	Borrelia afzelii is a species of Borrelia that can infect various species of vertebrate and invertebrates. Among thirty Borrelia known species, it is one of 4 which is likely to infect humans causing a variant of Lyme disease.

Chapter 1. Part 1: The Microbial Cell

Cell physiology	Cell physiology is the biological study of the cell's mechanism and interaction in its environment. The term 'physiology' refers to all the normal functions that take place in a living organism. Absorption of water by roots, production of food in the leaves, and growth of shoots towards light are examples of plant physiology.
Genetic analysis	Genetic analysis can be used generally to describe methods both used in and resulting from the sciences of genetics and molecular biology, or to applications resulting from this research. Genetic analysis may be done to identify genetic/inherited disorders and also to make a differential diagnosis in certain somatic diseases such as cancer. Genetic analyses of cancer include detection of mutations, fusion genes, and DNA copy number changes.
Inner membrane	The inner membrane is the biological membrane (phospholipid bilayer) of an organelle or Gram-negative bacteria that is within an outer membrane. In eukaryotic cells, this inner membrane is present within the nuclear envelope, mitochondria and plastids like the chloroplast. The lumen between the inner and outer membranes is referred to as intermembrane space.
Lipopolysaccharide	Lipopolysaccharides (LPS), also known as lipoglycans, are large molecules consisting of a lipid and a polysaccharide joined by a covalent bond; they are found in the outer membrane of Gram-negative bacteria, act as endotoxins and elicit strong immune responses in animals. LPS is the major component of the outer membrane of Gram-negative bacteria, contributing greatly to the structural integrity of the bacteria, and protecting the membrane from certain kinds of chemical attack. LPS also increases the negative charge of the cell membrane and helps stabilize the overall membrane structure.
Shine-Dalgarno sequence	The Shine-Dalgarno sequence proposed by Australian scientists John Shine (b.1946) and Lynn Dalgarno (b.1935), is a ribosomal binding site in the mRNA, generally located 8 basepairs upstream of the start codon AUG. The Shine-Dalgarno sequence exists both in bacteria and archaea, being also present in some chloroplastic and mitochondial transcripts. The six-base consensus sequence is AGGAGG; in E. coli, for example, the sequence is AGGAGGU. This sequence helps recruit the ribosome to the mRNA to initiate protein synthesis by aligning it with the start codon. The complementary sequence (CCUCCU), is called the anti-Shine-Dalgarno sequence and is located at the 3' end of the 16S rRNA in the ribosome.
Staphylococcus aureus	Staphylococcus aureus is a bacterial species named from Greek σταφυλ?κοκκος meaning the 'golden grape-cluster berry'. Also known as 'golden staph' and Oro staphira, it is a facultative anaerobic Gram-positive coccal bacterium.

Cell membrane	The cell membrane is a biological membrane that separates the interior of all cells from the outside environment. The cell membrane is selectively permeable to ions and organic molecules and controls the movement of substances in and out of cells. It basically protects the cell from outside forces.
Disinfectant	Disinfectants are substances that are applied to non-living objects to destroy microorganisms that are living on the objects. Disinfection does not necessarily kill all microorganisms, especially nonresistant bacterial spores; it is less effective than sterilisation, which is an extreme physical and/or chemical process that kills all types of life. Disinfectants are different from other antimicrobial agents such as antibiotics, which destroy microorganisms within the body, and antiseptics, which destroy microorganisms on living tissue.
Messenger RNA	Messenger RNA is a molecule of RNA that encodes a chemical 'blueprint' for a protein product. mRNA is transcribed from a DNA template, and carries coding information to the sites of protein synthesis, the ribosomes. In the ribosomes, the mRNA is translated into a polymer of amino acids: a protein.
Isoelectric focusing	Isoelectric focusing also known as electrofocusing, is a technique for separating different molecules by their electric charge differences. It is a type of zone electrophoresis, usually performed on proteins in a gel, that takes advantage of the fact that overall charge on the molecule of interest is a function of the pH of its surroundings. IEF involves adding an ampholyte solution into immobilized pH gradient (IPG) gels.
Isoelectric point	The isoelectric point sometimes abbreviated to IEP, is the pH at which a particular molecule or surface carries no net electrical charge. Amphoteric molecules called zwitterions contain both positive and negative charges depending on the functional groups present in the molecule. The net charge on the molecule is affected by pH of their surrounding environment and can become more positively or negatively charged due to the loss or gain of protons (H^+).
Northern blot	The northern blot is a technique used in molecular biology research to study gene expression by detection of RNA in a sample. With northern blotting it is possible to observe cellular control over structure and function by determining the particular gene expression levels during differentiation, morphogenesis, as well as abnormal or diseased conditions. Northern blotting involves the use of electrophoresis to separate RNA samples by size and detection with a hybridization probe complementary to part of or the entire target sequence.

Polyamine	A polyamine is an organic compound having two or more primary amino groups –NH_2. This class of compounds includes several synthetic substances that are important feedstocks for the chemical industry, such as ethylene diamine $H2N–CH_2–CH_2–NH_2$, 1,3-diaminopropane $H_2N–(CH_2)_3–NH_2$, and hexamethylenediamine $H_2N–(CH_2)_6–NH_2$. It also includes many substances that play important roles in both eukaryotic and prokaryotic cells, such as putrescine $H_2N–(CH_2)_4–NH_2$, cadaverine $H_2N–(CH_2)_5–NH_2$, spermidine $H_2N–((CH_2)_4–NH–)_2–H$, and spermine $H_2N–((CH_2)_4–NH–)_3–H$.
Proteome	The proteome is the entire set of proteins expressed by a genome, cell, tissue or organism. More specifically, it is the set of expressed proteins in a given type of cells or an organism at a given time under defined conditions. The term is a portmanteau of proteins and genome.
Gel electrophoresis	Gel electrophoresis is a method used in clinical chemistry to separate proteins by charge and or size (IEF agarose, essentially size independent) and in biochemistry and molecular biology to separate a mixed population of DNA and RNA fragments by length, to estimate the size of DNA and RNA fragments or to separate proteins by charge. Nucleic acid molecules are separated by applying an electric field to move the negatively charged molecules through an agarose matrix. Shorter molecules move faster and migrate farther than longer ones because shorter molecules migrate more easily through the pores of the gel.
Polyacrylamide gel	A Polyacrylamide Gel is a separation matrix used in electrophoresis of biomolecules, such as proteins or DNA fragments. Traditional DNA sequencing techniques such as Maxam-Gilbert or Sanger methods used polyacrylamide gels to separate DNA fragments differing by a single base-pair in length so the sequence could be read. Most modern DNA separation methods now use agarose gels, except for particularly small DNA fragments.
Protein	Proteins are biochemical compounds consisting of one or more polypeptides typically folded into a globular or fibrous form, facilitating a biological function. A polypeptide is a single linear polymer chain of amino acids bonded together by peptide bonds between the carboxyl and amino groups of adjacent amino acid residues. The sequence of amino acids in a protein is defined by the sequence of a gene, which is encoded in the genetic code.
Proteomics	Proteomics is the large-scale study of proteins, particularly their structures and functions. Proteins are vital parts of living organisms, as they are the main components of the physiological metabolic pathways of cells. The term 'proteomics' was first coined in 1997 to make an analogy with genomics, the study of the genes.

Nucleic acid	Nucleic acids are biological molecules essential for known forms of life on this planet; they include DNA (deoxyribonucleic acid) and RNA (ribonucleic acid). Together with proteins, nucleic acids are the most important biological macromolecules; each is found in abundance in all living things, where they function in encoding, transmitting and expressing genetic information. Nucleic acids were discovered by Friedrich Miescher in 1869. Experimental studies of nucleic acids constitute a major part of modern biological and medical research, and form a foundation for genome and forensic science, as well as the biotechnology and pharmaceutical industries.
Mixotricha paradoxa	Mixotricha paradoxa is a species of protozoan that lives inside the termite species Mastotermes darwiniensis and has multiple bacterial symbionts. The name, given by the Australian biologist J.L. Sutherland, who first described Mixotricha in 1933, means 'the paradoxical being with mixed-up hairs'. Mixotricha forms many symbiotic relationships.
Subatomic particle	In physics or chemistry, subatomic particles are the smaller particles composing nucleons and atoms. There are two types of subatomic particles: elementary particles, which are not made of other particles, and composite particles. Particle physics and nuclear physics study these particles and how they interact.
Membrane protein	A membrane protein is a protein molecule that is attached to, or associated with the membrane of a cell or an organelle. More than half of all proteins interact with membranes. Biological membranes consist of a phospholipid bilayer and a variety of proteins that accomplish vital biological functions.
Electrochemical potential	In electrochemistry, the electrochemical potential, $\bar{\mu}$, sometimes abbreviated to ECP, is a thermodynamic measure that combines the concepts of energy stored in the form of chemical potential and electrostatics. Electrochemical potential is expressed in the unit of Jmol. Each chemical species (for example, 'water molecules', 'sodium ions', 'electrons', etc).
Polyribosome	Polyribosomes (or polysomes) also known as ergosomes are a cluster of ribosomes, bound to a mRNA molecule, first discovered and characterized by Jonathan Warner, Paul Knopf, and Alex Rich in 1963. Polyribosomes read one strand of mRNA simultaneously, helping to synthesize the same protein at different spots on the mRNA, mRNA being the 'messenger' in the process of protein synthesis. They may appear as clusters, linear arrays, or rosettes in routine: this is aided by the fact that mRNA is able to be twisted into a circular formation, creating a cycle of rapid ribosome recycling, and utilization of ribosomes.

Chapter 1. Part 1: The Microbial Cell

Ribosomal protein	Mitochondrial ribosomal protein L31 A ribosomal protein is any of the proteins that, in conjunction with rRNA, make up the ribosomal subunits involved in the cellular process of translation. A large part of the knowledge about these organic molecules has come from the study of E. coli ribosomes. Most ribosomic proteins have been isolated and specific anti-bodies have been produced.
Ribosome	The Ribosome is a large complex molecule which is responsible for catalyzing the formation of proteins from individual amino acids using messenger RNA as a template. This process is known as translation. Ribosomes are found in all living cells.
Fusion protein	Fusion proteins or chimeric proteins are proteins created through the joining of two or more genes which originally coded for separate proteins. Translation of this fusion gene results in a single polypeptide with functional properties derived from each of the original proteins. Recombinant fusion proteins are created artificially by recombinant DNA technology for use in biological research or therapeutics.
Lac operon	The lac operon is an operon required for the transport and metabolism of lactose in Escherichia coli and some other enteric bacteria. It consists of three adjacent structural genes, lacZ, lacY and lacA. The lac operon is regulated by several factors including the availability of glucose and of lactose. Gene regulation of the lac operon was the first complex genetic regulatory mechanism to be elucidated and is one of the foremost examples of prokaryotic gene regulation.
Protein A	Protein A is a 40-60 kDa MSCRAMM surface protein originally found in the cell wall of the bacterium Staphylococcus aureus. It is encoded by the spa gene and its regulation is controlled by DNA topology, cellular osmolarity, and a two-component system called ArlS-ArlR. It has found use in biochemical research because of its ability to bind immunoglobulins. It binds proteins from many of mammalian species, most notably IgGs.
Cytoskeleton	The cytoskeleton is a cellular 'scaffolding' or 'skeleton' contained within a cell's cytoplasm and is made out of protein. The cytoskeleton is present in all cells; it was once thought to be unique to eukaryotes, but recent research has identified the prokaryotic cytoskeleton. It has structures such as flagella, cilia and lamellipodia and plays important roles in both intracellular transport (the movement of vesicles and organelles, for example) and cellular division.
Phosphatidylethanola mine	Phosphatidylethanolamine is a lipid found in biological membranes. It is synthesized by the addition of CDP-ethanolamine to diglyceride, releasing CMP. S-adenosyl methionine can subsequently methylate the amine of phosphatidyl ethanolamine to yield phosphatidyl choline.

| Aquaporin | Aquaporins are proteins embedded in the cell membrane that regulate the flow of water. They are 'the plumbing system for cells.'

Aquaporins are integral membrane proteins from a larger family of major intrinsic proteins (MIP) that form pores in the membrane of biological cells.

Genetic defects involving aquaporin genes have been associated with several human diseases. |
|---|---|
| Aspirin | Aspirin also known as acetylsalicylic acid , is a salicylate drug, often used as an analgesic to relieve minor aches and pains, as an antipyretic to reduce fever, and as an anti-inflammatory medication.

Salicylic acid, the main metabolite of aspirin, is an integral part of human and animal metabolism. While much of it is attributable to diet, a substantial part is synthesized endogenously. |
| Brugia malayi | Brugia malayi is a nematode (roundworm), one of the three causative agents of lymphatic filariasis in humans. Lymphatic filariasis, also known as elephantiasis, is a condition characterized by swelling of the lower limbs. The two other filarial causes of lymphatic filariasis are Wuchereria bancrofti and Brugia timori, which differ from B. malayi morphologically, symptomatically, and in geographical extent. |
| Osmotic pressure | Osmotic pressure is the pressure which needs to be applied to a solution to prevent the inward flow of water across a semipermeable membrane. It is also defined as the minimum pressure needed to nullify osmosis.

The phenomenon of osmotic pressure arises from the tendency of a pure solvent to move through a semi-permeable membrane and into a solution containing a solute to which the membrane is impermeable. |
| Passive transport | Passive transport means moving biochemicals and other atomic or molecular substances across membranes. Unlike active transport, this process does not involve chemical energy, because, unlike in an active transport, the transport across membrane is always coupled with the growth of entropy of the system. So passive transport is dependent on the permeability of the cell membrane, which, in turn, is dependent on the organization and characteristics of the membrane lipids and proteins. |
| Active transport | Active transport is the movement of a substance against its concentration gradient (from low to high concentration). In all cells, this is usually concerned with accumulating high concentrations of molecules that the cell needs, such as ions, glucose and amino acids. |

Chapter 1. Part 1: The Microbial Cell

Virulence factor	Virulence factors are molecules expressed and secreted by pathogens (bacteria, viruses, fungi and protozoa) that enable them to achieve the following:•colonization of a niche in the host (this includes adhesion to cells)•Immunoevasion, evasion of the host's immune response•Immunosuppression, inhibition of the host's immune response•entry into and exit out of cells (if the pathogen is an intracellular one)•obtain nutrition from the host.
	Virulence factors are very often responsible for causing disease in the host as they inhibit certain host functions.
	Pathogens possess a wide array of virulence factors. Some are intrinsic to the bacteria (e.g. capsules and endotoxin) whereas others are obtained from plasmids (e.g. some toxins).
Weak acid	A weak acid is an acid that dissociates incompletely. It does not release all of its hydrogens in a solution, donating only a partial amount of its protons to the solution. These acids have higher pKa than strong acids, which release all of their hydrogen atoms when dissolved in water.
Weak base	In chemistry, a weak base is a chemical base that does not ionize fully in an aqueous solution. As Brønsted-Lowry bases are proton acceptors, a weak base may also be defined as a chemical base in which protonation is incomplete. This results in a relatively low pH compared to strong bases.
ATP hydrolysis	ATP hydrolysis is the reaction by which chemical energy that has been stored and transported in the high-energy phosphoanhydridic bonds in ATP (Adenosine triphosphate) is released, for example in the muscles, to produce work. The product is ADP (Adenosine diphosphate) and an inorganic phosphate, orthophosphate (Pi). ADP can be further hydrolyzed to give energy, AMP (Adenosine monophosphate), and another orthophosphate (Pi).
Energy carrier	According to ISO 13600, an energy carrier is either a substance (energy form) or a phenomenon (energy system) that can be used to produce mechanical work or heat or to operate chemical or physical processes.
	In the field of Energetics, however, an energy carrier corresponds only to an energy form (not an energy system) of energy input required by the various sectors of society to perform their functions.
	Examples of energy carriers include liquid fuel in a furnace, gasoline in a pump, electricity in a factory or a house, and hydrogen in a tank of a car.
Chemical reaction	A chemical reaction is a process that leads to the transformation of one set of chemical substances to another.

Chemical reactions can be either spontaneous, requiring no input of energy, or non-spontaneous, typically following the input of some type of energy, such as heat, light or electricity. Classically, chemical reactions encompass changes that strictly involve the motion of electrons in the forming and breaking of chemical bonds, although the general concept of a chemical reaction, in particular the notion of a chemical equation, is applicable to transformations of elementary particles (such as illustrated by Feynman diagrams), as well as nuclear reactions.

Cardiolipin	Cardiolipin (IUPAC name '1,3-bis(sn-3'-phosphatidyl)-sn-glycerol') is an important component of the inner mitochondrial membrane, where it constitutes about 20% of the total lipid composition. The only other place that cardiolipin can be found is in the membranes of most bacteria. The name 'cardiolipin' is derived from the fact that it was first found in animal hearts.
Membrane lipids	The three major classes of membrane lipids are phospholipids, glycolipids, and cholesterol. Phospholipids and glycolipids consist of two long, nonpolar (hydrophobic) hydrocarbon chains linked to a hydrophilic head group. The heads of phospholipids are phosphorylated and they consist of either:•Glycerol (and hence the name phosphoglycerides given to this group of lipids).•Sphingosine (with only one member - sphingomyelin).Glycolipids The heads of glycolipids contain a sphingosine with one or several sugar units attached to it.
Oleic acid	Oleic acid is a fatty acid that occurs naturally in various animal and vegetable fats and oils. It is an odorless, colourless oil, although commercial samples may be yellowish. In chemical terms, oleic acid is classified as a monounsaturated omega-9 fatty acid.
Phosphatidylglycerol	Phosphatidylglycerol is a glycerophospholipid found in pulmonary surfactant. The general structure of phosphatidylglycerol consists of a L-glycerol 3-phosphate backbone ester-bonded to either saturated or unsaturated fatty acids on carbons 1 and 2. The head group substituent glycerol is bonded through a phosphomonoester. It is the precursor of surfactant and its presence (>0.3) in the amniotic fluid of the newborn indicates fetal lung maturity.
Streptococcus pneumoniae	Streptococcus pneumoniae, is a Gram-positive, alpha-hemolytic, aerotolerant anaerobic member of the genus Streptococcus. A significant human pathogenic bacterium, S. pneumoniae was recognized as a major cause of pneumonia in the late 19th century, and is the subject of many humoral immunity studies.

Chapter 1. Part 1: The Microbial Cell

Trichophyton rubrum	Trichophyton rubrum is a fungus that is the most common cause of athlete's foot, jock itch and ringworm. This fungus was first described by Malmsten in 1845. The growth rate of Trichophyton colonies in the lab can be slow to rather quick. Their texture is waxy, smooth and even to cottony.
Fatty acid	In chemistry, especially biochemistry, a fatty acid is a carboxylic acid with a long aliphatic tail (chain), which is either saturated or unsaturated. Most naturally occurring fatty acids have a chain of an even number of carbon atoms, from 4 to 28. Fatty acids are usually derived from triglycerides or phospholipids. When they are not attached to other molecules, they are known as 'free' fatty acids.
Side chain	In organic chemistry and biochemistry, a side chain is a chemical group that is attached to a core part of the molecule called 'main chain' or backbone. The placeholder R is often used as a generic placeholder for alkyl (saturated hydrocarbon) group side chains in chemical structure diagrams. To indicate other non-carbon groups in structure diagrams, X, Y, or Z is often used.
Cholesterol	Cholesterol is an organic chemical substance classified as a waxy steroid of fat. It is an essential structural component of mammalian cell membranes and is required to establish proper membrane permeability and fluidity. In addition to its importance within cells, cholesterol is an important component in the hormonal systems of the body for the manufacture of bile acids, steroid hormones, and vitamin D. Cholesterol is the principal sterol synthesized by animals; in vertebrates it is formed predominantly in the liver.
Hyperthermophile	A hyperthermophile is an organism that thrives in extremely hot environments-- from 60 degrees C (140 degrees F) upwards. An optimal temperature for the existence of hyperthermophiles is above 80°C (176°F). Hyperthermophiles are a subset of extremophiles, micro-organisms within the domain Archaea, although some bacteria are able to tolerate temperatures of around 100°C (212° F), as well.
Terpenoid	The terpenoids , sometimes called isoprenoids, are a large and diverse class of naturally occurring organic chemicals similar to terpenes, derived from five-carbon isoprene units assembled and modified in thousands of ways. Most are multicyclic structures that differ from one another not only in functional groups but also in their basic carbon skeletons. These lipids can be found in all classes of living things, and are the largest group of natural products.
Hopanoids	Hopanoids are natural pentacyclic compounds (containing five rings) based on the chemical structure of hopane. Their primary function is to improve plasma membrane strength and rigidity in bacteria. In eukaryotes (including humans) cholesterol serves a similar function.
Glycan	The term glycan refers to a polysaccharide or oligosaccharide.

	Glycans usually consist solely of O-glycosidic linkages of monosaccharides. For example, cellulose is a glycan composed of beta-1,4-linked D-glucose, and chitin is a glycan composed of beta-1,4-linked N-acetyl-D-glucosamine.
N-Acetylglucosamine	N-Acetylglucosamine is a monosaccharide derivative of glucose. It is an amide between glucosamine and acetic acid. It has a molecular formula of $C_8H_{15}NO_6$, a molar mass of 221.21 g/mol, and it is significant in several biological systems.
N-Acetylmuramic acid	N-Acetylmuramic acid, is the ether of lactic acid and N-acetylglucosamine with a chemical formula of $C_{11}H_{19}NO_8$. It is part of a biopolymer in the bacterial cell wall, built from alternating units of N-acetylglucosamine (GlcNAc) and N-acetylmuramic acid cross-linked with oligopeptides at the lactic acid residue of MurNAc. This layered structure is called peptidoglycan.
Clostridium difficile	Clostridium difficile (from the Greek kloster , spindle, and Latin difficile, difficult), also known as 'CDF/cdf', or 'C. diff', is a species of Gram-positive bacteria of the genus Clostridium that causes severe diarrhea and other intestinal disease when competing bacteria in the gut flora have been wiped out by antibiotics.

Clostridia are anaerobic, spore-forming rods (bacilli). C. difficile is the most serious cause of antibiotic-associated diarrhea (AAD) and can lead to pseudomembranous colitis, a severe inflammation of the colon, often resulting from eradication of the normal gut flora by antibiotics. |
| Toxic shock syndrome | Toxic shock syndrome is a potentially fatal illness caused by a bacterial toxin. Different bacterial toxins may cause toxic shock syndrome, depending on the situation. The causative bacteria include Staphylococcus aureus and Streptococcus pyogenes. |
| S-layer | An S-layer is a part of the cell envelope commonly found in bacteria, as well as among archaea . It consists of a monomolecular layer composed of identical proteins or glycoproteins. This two-dimensional structure is built via self-assembly and encloses the whole cell surface. |
| Teichoic acid | Teichoic acids are bacterial polysaccharides of glycerol phosphate or ribitol phosphate linked via phosphodiester bonds.

Teichoic acids are found within the cell wall of Gram-positive bacteria such as species in the genera Staphylococcus, Streptococcus, Bacillus, Clostridium, Corynebacterium and Listeria, and appear to extend to the surface of the peptidoglycan layer. Teichoic acids are not found in Gram-negative bacteria. |
| Ethambutol | Ethambutol is a bacteriostatic antimycobacterial drug prescribed to treat tuberculosis. It is usually given in combination with other tuberculosis drugs, such as isoniazid, rifampicin and pyrazinamide. |

Chapter 1. Part 1: The Microbial Cell

Galactan	Galactan is a polysaccharide consisting of polymerized galactose. Galactan derived from Anogeissus latifolia is primarily a(1→6), but galactan from acacia trees is primarily a(1→3).
Mycoplasma	Mycoplasma refers to a genus of bacteria that lack a cell wall. Without a cell wall, they are unaffected by many common antibiotics such as penicillin or other beta-lactam antibiotics that target cell wall synthesis. They can be parasitic or saprotrophic.
Gram-negative bacteria	Gram-negative bacteria are bacteria that do not retain crystal violet dye in the Gram staining protocol. In a Gram stain test, a counterstain (commonly safranin) is added after the crystal violet, coloring all Gram-negative bacteria with a red or pink color. The test itself is useful in classifying two distinct types of bacteria based on the structural differences of their bacterial cell walls.
Lipoprotein	Lipoprotein(a) (also called Lp(a)) is a lipoprotein subclass. Genetic studies and numerous epidemiologic studies have identified Lp(a) as a risk factor for atherosclerotic diseases such as coronary heart disease and stroke. Lipoprotein(a) was discovered in 1963 by Kåre Berg and the human gene encoding apolipoprotein (a) was cloned in 1987.
Density gradient	Density gradient is a spatial variation in density over an area. The term is used in the natural sciences to describe varying density of matter, but can apply to any quantity whose density can be measured. In the study of supersonic flight, Schlieren photography observes the density gradient of air as it interacts with aircraft.
Glucosamine	Glucosamine is an amino sugar and a prominent precursor in the biochemical synthesis of glycosylated proteins and lipids. Glucosamine is part of the structure of the polysaccharides chitosan and chitin, which compose the exoskeletons of crustaceans and other arthropods, cell walls in fungi and many higher organisms. Glucosamine is one of the most abundant monosaccharides.
Porin	Porins are beta barrel proteins that cross a cellular membrane and act as a pore through which molecules can diffuse. Unlike other membrane transport proteins, porins are large enough to allow passive diffusion, i.e., they act as channels that are specific to different types of molecules.

Beta barrel	A beta barrel is a large beta-sheet that twists and coils to form a closed structure in which the first strand is hydrogen bonded to the last. Beta-strands in beta-barrels are typically arranged in an antiparallel fashion. Barrel structures are commonly found in porins and other proteins that span cell membranes and in proteins that bind hydrophobic ligands in the barrel center, as in lipocalins.
Endotoxin	The term endotoxin was coined by Richard Friedrich Johannes Pfeiffer, who distinguished between exotoxin, which he classified as a toxin that is released by bacteria into the environment, and endotoxin, which he considered to be a toxin kept 'within' the bacterial cell and to be released only after destruction of the bacterial cell wall. Today, the term 'endotoxin' is used synonymously to the term lipopolysaccharide, which is a major constituent of the outer cell wall of Gram-negative bacteria. Larger amounts of endotoxins can be mobilized if Gram-negative bacteria are killed or destroyed by detergents.
Chitin	Chitin is a long-chain polymer of a N-acetylglucosamine, a derivative of glucose, and is found in many places throughout the natural world. It is the main component of the cell walls of fungi, the exoskeletons of arthropods such as crustaceans (e.g., crabs, lobsters and shrimps) and insects, the radulas of mollusks, and the beaks of cephalopods, including squid and octopuses. In terms of structure, chitin may be compared to the polysaccharide cellulose and, in terms of function, to the protein keratin.
Osmotic shock	Osmotic shock is a sudden change in the solute concentration around a cell, causing a rapid change in the movement of water across its cell membrane. Under conditions of high concentrations of either salts, substrates or any solute in the supernatant, water is drawn out of the cells through osmosis. This also inhibits the transport of substrates and cofactors into the cell thus 'shocking' the cell.
Amp resistance	Amp resistance is a term for resistance to the antibiotic ampicillin. It is used as a selectable marker in bacterial transformation.
Coccidioides	Coccidioides is a genus of dimorphic ascomycete, cause of Coccidioidomycosis, also known as San Joaquin Valley Fever, an infectious fungal disease endemic in American deserts. The host acquires the disease via respiratory inhalation of spores disseminated in their natural habitat. The primary disease is auto-limited, although fewer than 1% of the cases develop complications, which result in high morbidity.
DNA gyrase	DNA gyrase, often referred to simply as gyrase, is an enzyme that relieves strain while double-stranded DNA is being unwound by helicase. This causes negative supercoiling of the DNA. Bacterial DNA gyrase is the target of many antibiotics, including nalidixic acid, novobiocin, and ciprofloxacin.

Chapter 1. Part 1: The Microbial Cell

DNA replication	DNA replication is a biological process that occurs in all living organisms and copies their DNA; it is the basis for biological inheritance. The process starts when one double-stranded DNA molecule produces two identical copies of the molecule. The cell cycle (mitosis) also pertains to the DNA replication/reproduction process.
Okazaki fragment	An Okazaki fragment is a relatively short fragment of DNA (with no RNA primer at the 5' terminus) created on the lagging strand during DNA replication. The lengths of Okazaki fragments are between 1,000 to 2,000 nucleotides long in E. coli and are generally between 100 to 200 nucleotides long in eukaryotes. It was originally discovered in 1968 by Reiji Okazaki, Tsuneko Okazaki, and their colleagues while studying replication of bacteriophage DNA in Escherichia coli.
Quinolone	The quinolones are a family of synthetic broad-spectrum antibiotics. The term quinolone(s) refers to potent synthetic chemotherapeutic antibacterials.
	The first generation of the quinolones begins with the introduction of nalidixic acid in 1962 for treatment of urinary tract infections in humans.
RNA polymerase	RNA polymerase also known as DNA-dependent RNA polymerase, is an enzyme that produces RNA. In cells, RNAP is necessary for constructing RNA chains using DNA genes as templates, a process called transcription. RNA polymerase enzymes are essential to life and are found in all organisms and many viruses. In chemical terms, RNAP is a nucleotidyl transferase that polymerizes ribonucleotides at the 3' end of an RNA transcript.
Southern blot	A Southern blot is a method routinely used in molecular biology for detection of a specific DNA sequence in DNA samples. Southern blotting combines transfer of electrophoresis-separated DNA fragments to a filter membrane and subsequent fragment detection by probe hybridization. ts inventor, the British biologist Edwin Southern.
Biocrystallization	Biocrystallization is the formation of crystals from organic macromolecules by living organisms. This may be a stress response, a normal part of metabolism such as processes that dispose of waste compounds, or a pathology. Template mediated crystallization is qualitatively different from in vitro crystallization.
Protein synthesis inhibitor	A protein synthesis inhibitor is a substance that stops or slows the growth or proliferation of cells by disrupting the processes that lead directly to the generation of new proteins.
	While a broad interpretation of this definition could be used to describe nearly any antibiotic, in practice, it usually refers to substances that act at the ribosome level, taking advantages of the major differences between prokaryotic and eukaryotic ribosome structures.

Cell division	Cell division is the process by which a parent cell divides into two or more daughter cells. Cell division is usually a small segment of a larger cell cycle. This type of cell division in eukaryotes is known as mitosis, and leaves the daughter cell capable of dividing again.
Endoplasmic reticulum	The endoplasmic reticulum is an organelle of cells in eukaryotic organisms that forms an interconnected network of tubules, vesicles, and cisternae. Rough endoplasmic reticula are involved in the synthesis of proteins and is also a membrane factory for the cell, while smooth endoplasmic reticula are involved in the synthesis of lipids, including oils, phospholipids and steroids, metabolism of carbohydrates, regulation of calcium concentration and detoxification of drugs and poisons. Sarcoplasmic reticula solely regulate calcium levels.
Replication fork	The replication fork is a structure that forms within the nucleus during DNA replication. It is created by helicases, which break the hydrogen bonds holding the two DNA strands together. The resulting structure has two branching 'prongs', each one made up of a single strand of DNA. These two strands serve as the template for the leading and lagging strands, which will be created as DNA polymerase matches complementary nucleotides to the templates.
Replisome	The replisome is a complex molecular machine that carries out replication of DNA. The replisome first unwinds double stranded DNA into two single strands. For each of the resulting single strands, a new complementary sequence of DNA is synthesized. The net result is formation of two new double stranded DNA sequences that are exact copies of the original double stranded DNA sequence.
Septum	In anatomy, a septum is a wall, dividing a cavity or structure into smaller ones. In human anatomy •Interatrial septum, the wall of tissue that is a sectional part of the left and right atria of the heart•Interventricular septum or median septum, the wall separating the left and right ventricles of the heart•Lingual septum, a vertical layer of fibrous tissue that separates the halves of the tongue•Nasal septum: the cartilage wall separating the nostrils of the nose•Alveolar septum: the thin wall which separates the alveoli from each other in the lungs•Orbital septum, a palpabral ligament in the upper and lower eyelids•Septum pellucidum or septum lucidum, a thin structure separating two fluid pockets in the brain•Medial septum, a cluster of neurons in close proximity to the septum pellucidum•Uterine septum, a malformation of the uterus•Vaginal septum, a lateral or transverse partition inside the vagina

Histological septa are seen throughout most tissues of the body, particularly where they are needed to stiffen soft cellular tissue, and they also provide planes of ingress for small blood vessels. Because the dense collagen fibres of a septum usually extend out into the softer adjacent tissues, microscopic fibrous septa are less clearly defined than the macroscopic types of septa listed above. |

Chapter 1. Part 1: The Microbial Cell

Caulobacter crescentus	Caulobacter crescentus is a Gram-negative, oligotrophic bacterium widely distributed in fresh water lakes and streams. It plays an important role in the carbon cycle. Caulobacter is an important model for studying the regulation of the cell cycle and cellular differentiation.
Protein Z	Protein Z is encoded by the PROZ gene. Protein Z is a member of the coagulation cascade, the group of blood proteins that leads to the formation of blood clots. It is vitamin K-dependent, and its functionality is therefore impaired in warfarin therapy.
Permissive temperature	The permissive temperature is the temperature at which a temperature sensitive mutant gene product takes on a normal, functional phenotype. When a temperature sensitive mutant is grown in a permissive condition, the mutated gene product behaves normally (meaning that the phenotype isn't observed), even if there is a mutant allele present. This results in the survival of the cell or organism, as if it were a wild type strain.
Verrucomicrobia	Verrucomicrobia is a recently described phylum of bacteria. This phylum contains only a few described species . The species identified have been isolated from fresh water and soil environments and human feces.
Thylakoid	A thylakoid is a membrane-bound compartment inside chloroplasts and cyanobacteria. They are the site of the light-dependent reactions of photosynthesis. Thylakoids consist of a thylakoid membrane surrounding a thylakoid lumen.
Carboxysome	Carboxysomes are bacterial microcompartments that contain enzymes involved in carbon fixation. Carboxysomes are made of polyhedral protein shells about 80 to 140 nanometres in diameter. These compartments are thought to concentrate carbon dioxide to overcome the inefficiency of RuBisCO - the predominant enzyme in carbon fixation and the rate limiting enzyme in the Calvin cycle.
Holdfast	A holdfast is a root-like structure that anchors aquatic sessile organisms, such as seaweed, other sessile algae, stalked crinoids, benthic cnidarians, and sponges, to the substrate. Holdfasts vary in shape and form depending on both the species and the substrate type. The holdfasts of organisms that live in muddy substrates often have complex tangles of root-like growths, while those of organisms that live in sandy substrates are bulb-like and very flexible, such as the holdfast of sea pens, allowing the organism(s) to pull the entire body into the substrate when the holdfast is contracted.

| Magnetotaxis | Logically, magnetotaxis describes an ability to sense a magnetic field and coordinate movement in response. It was applied to the behavior of certain motile, aquatic bacteria in 1975 by R. P. Blakemore.

However, it is now known that these bacteria orient to the Earth's magnetic field even when they are dead, just as a compass needle does. |
| --- | --- |
| Motility | Motility is a biological term which refers to the ability to move spontaneously and actively, consuming energy in the process. Most animals are motile but the term applies to single-celled and simple multicellular organisms, as well as to some mechanisms of fluid flow in multicellular organs, in addition to animal locomotion. Motile marine animals are commonly called free-swimming. |
| Periodontal disease | Periodontal disease is a type of disease that affects one or more of the periodontal tissues:•alveolar bone•periodontal ligament•cementum•gingiva

While many different diseases affect the tooth-supporting structures, plaque-induced inflammatory lesions make up the vast majority of periodontal diseases and have traditionally been divided into two categories:•gingivitis or•periodontitis.

While in some sites or individuals, gingivitis never progresses to periodontitis, data indicates that gingivitis always precedes periodontitis.

Diagnosis

In 1976, Page & Schroeder introduced an innovative new analysis of periodontal disease based on histopathologic and ultrastructural features of the diseased gingival tissue. Although this new classification does not correlate with clinical signs and symptoms and is admittedly 'somewhat arbitrary,' it permits a focus of attention on important pathologic aspects of the disease that were, until recently, not well understood. |
| Pilin | Pilin

Pilin refers to a class of fibrous proteins that are found in pilus structures in bacteria. Bacterial pili are used in the exchange of genetic material during bacterial conjugation, and a short pilus called a fimbrium is used as a cell adhesion mechanism. Although not all bacteria have pili or fimbriae, bacterial pathogens often use their fimbriae to attach to host cells. |
| Porphyromonas gingivalis | Porphyromonas gingivalis belongs to the phylum Bacteroidetes and is a non-motile, Gram-negative, rod-shaped, anaerobic pathogenic bacterium. It forms black colonies on blood agar. |

Chapter 1. Part 1: The Microbial Cell

Sulfur-reducing bacteria	Sulfur-reducing bacteria get their energy by reducing elemental sulfur to hydrogen sulfide. They couple this reaction with the oxidation of acetate, succinate or other organic compounds. Several types of bacteria and many non-methanogenic archaea can reduce sulfur.
Biomineralization	Biomineralization is the process by which living organisms produce minerals, often to harden or stiffen existing tissues. Such tissues are called mineralized tissues. It is an extremely widespread phenomenon; all six taxonomic kingdoms contain members that are able to form minerals, and over 60 different minerals have been identified in organisms.
Flagellin	Flagellin is a protein that arranges itself in a hollow cylinder to form the filament in bacterial flagellum. It has a mass of about 30,000 to 60,000 daltons. Flagellin is the principal substituent of bacterial flagellum, and is present in large amounts on nearly all flagellated bacteria.
Chemotaxis	Chemotaxis is the phenomenon whereby somatic cells, bacteria, and other single-cell or multicellular organisms direct their movements according to certain chemicals in their environment. This is important for bacteria to find food (for example, glucose) by swimming towards the highest concentration of food molecules, or to flee from poisons (for example, phenol). In multicellular organisms, chemotaxis is critical to early development (e.g. movement of sperm towards the egg during fertilization) and subsequent phases of development (e.g. migration of neurons or lymphocytes) as well as in normal function.
Myxobacteria	The myxobacteria are a group of bacteria that predominantly live in the soil. The myxobacteria have very large genomes, relative to other bacteria, e.g. 9-10 million nucleotides. Sorangium cellulosum has the largest known (as of 2008) bacterial genome, at 13.0 million nucleotides.
Lactic acid	Lactic acid, is a chemical compound that plays a role in various biochemical processes and was first isolated in 1780 by the Swedish chemist Carl Wilhelm Scheele. Lactic acid is a carboxylic acid with the chemical formula $C_3H_6O_3$. It has a hydroxyl group adjacent to the carboxyl group, making it an alpha hydroxy acid (AHA).
Marine habitats	Marine habitats can be divided into coastal and open ocean habitats. Coastal habitats are found in the area that extends from as far as the tide comes in on the shoreline out to the edge of the continental shelf. Most marine life is found in coastal habitats, even though the shelf area occupies only seven percent of the total ocean area.
Algal bloom	An algal bloom is a rapid increase or accumulation in the population of algae (typically microscopic) in an aquatic system. Algal blooms may occur in freshwater as well as marine environments.

Nutrition	Nutrition is the provision, to cells and organisms, of the materials necessary (in the form of food) to support life. Many common health problems can be prevented or alleviated with a healthy diet.
	The diet of an organism is what it eats, which is largely determined by the perceived palatability of foods.
Generation time	Generation time is a quantity used in population biology and demography to reflect the relative size of intervals of offspring production. Generation time usually expresses the average age of breeding females within a population. Suppose females begin breeding at age α and stop breeding at age ω, then the average age of first reproduction of a cohort of females is $$T = \frac{\sum_{x=\alpha}^{\omega} x l(x) m(x)}{\sum_{x=\alpha}^{\omega} l(x) m(x)}$$ where $l(x)$ is the hazard function and $m(x)$ is the fecundity of females aged x.
Legionella pneumophila	Legionella pneumophila is a thin, pleomorphic, flagellated Gram-negative bacterium of the genus Legionella. L. pneumophila is the primary human pathogenic bacterium in this group and is the causative agent of legionellosis or Legionnaires' disease.
	Characterization
	L. pneumophila is non-acid-fast, non-sporulating, and morphologically a non-capsulated rod-like bacteria.
Cobalt	Cobalt is a chemical element with symbol Co and atomic number 27. It is found naturally only in chemically combined form. The free element, produced by reductive smelting, is a hard, lustrous, silver-gray metal.
	Cobalt-based blue pigments have been used since ancient times for jewelry and paints, and to impart a distinctive blue tint to glass, but the color was later thought by alchemists to be due to the known metal bismuth.
Growth factor	A growth factor is a naturally occurring substance capable of stimulating cellular growth, proliferation and cellular differentiation. Usually it is a protein or a steroid hormone. Growth factors are important for regulating a variety of cellular processes.

Chapter 1. Part 1: The Microbial Cell

MHC restriction	MHC-restricted antigen recognition, or MHC restriction, refers to the fact that a given T cell will recognize a peptide antigen only when it is bound to a host body's own MHC molecule. Normally, as T cells are stimulated only in the presence of self-MHC molecules, antigen is recognized only as peptides bound to self-MHC molecules. MHC restriction is particularly important when primary lymphocytes are developing and differentiating in the thymus or bone marrow.
Micronutrient	Micronutrients are nutrients required by humans and other living things throughout life in small quantities to orchestrate a whole range of physiological functions, but which the organism itself cannot produce. For people, they include dietary trace minerals in amounts generally less than 100 milligrams/day - as opposed to macrominerals which are required in larger quantities. The microminerals or trace elements include at least iron, cobalt, chromium, copper, iodine, manganese, selenium, zinc and molybdenum.
Molybdenum	Molybdenum is a Group 6 chemical element with the symbol Mo and atomic number 42. The name is from Neo-Latin Molybdaenum, from Ancient Greek M?λυβδος molybdos, meaning lead, since its ores were confused with lead ores. Molybdenum minerals have been known into prehistory, but the element was 'discovered' (in the sense of differentiating it as a new entity from the mineral salts of other metals) in 1778 by Carl Wilhelm Scheele. The metal was first isolated in 1781 by Peter Jacob Hjelm.
Autotroph	An autotroph,(self-feeding) or producer, is an organism that produces complex organic compounds (such as carbohydrates, fats, and proteins) from simple substances present in its surroundings . For example using energy from light (by photosynthesis) or inorganic chemical reactions (chemosynthesis). They are the producers in a food chain, such as plants on land or algae in water.
Biogeochemical cycle	In geography and Earth science, a biogeochemical cycle is a pathway by which a chemical element or molecule moves through both biotic (biosphere) and abiotic (lithosphere, atmosphere, and hydrosphere) compartments of Earth. A cycle is a series of change which comes back to the starting point and which can be repeated. The term 'biogeochemical' tells us that biological; geological and chemical factors are all involved.
Carbon cycle	The carbon cycle is the biogeochemical cycle by which carbon is exchanged among the biosphere, pedosphere, geosphere, hydrosphere, and atmosphere of the Earth.

	It is one of the most important cycles of the Earth and allows for carbon to be recycled and reused throughout the biosphere and all of its organisms.
	The global carbon budget is the balance of the exchanges (incomes and losses) of carbon between the carbon reservoirs or between one specific loop (e.g., atmosphere ↔ biosphere) of the carbon cycle.
Q fever	Q fever is a disease caused by infection with Coxiella burnetii, a bacterium that affects humans and other animals. This organism is uncommon, but may be found in cattle, sheep, goats and other domestic mammals, including cats and dogs. The infection results from inhalation of a spore-like small cell variant, and from contact with the milk, urine, feces, vaginal mucus, or semen of infected animals.
Rickettsia prowazekii	Rickettsia prowazekii is a species of gram negative, Alpha Proteobacteria, obligate intracellular parasitic, aerobic bacteria that is the etiologic agent of epidemic typhus, transmitted in the faeces of lice. In North America, the main reservoir for R. prowazekii is the flying squirrel. R. prowazekii is often surrounded by a protein microcapsular layer and slime layer; the natural life cycle of the bacterium generally involves a vertebrate and an invertebrate host, usually an arthropod, typically the human body louse.
Heterotroph	A heterotroph is an organism that cannot fix carbon and uses organic carbon for growth. This contrasts with autotrophs, such as plants and algae, which can use energy from sunlight (photoautotrophs) or inorganic compounds (lithoautotrophs) to produce organic compounds such as carbohydrates, fats, and proteins from inorganic carbon dioxide. These reduced carbon compounds can be used as an energy source by the autotroph and provide the energy in food consumed by heterotrophs.
Parasitism	Parasitism is a type of non mutual relationship between organisms of different species where one organism, the parasite, benefits at the expense of the other, the host. Traditionally parasite referred to organisms with lifestages that needed more than one host (e.g. Taenia solium). These are now called macroparasites (typically protozoa and helminths).
Membrane potential	Membrane potential (also transmembrane potential is the difference in electrical potential between the interior and the exterior of a biological cell. All animal cells are surrounded by a plasma membrane composed of a lipid bilayer with a variety of types of proteins embedded in it. The membrane potential arises primarily from the interaction between the membrane and the actions of two types of transmembrane proteins embedded in the plasma membrane.
Microbial metabolism	Microbial metabolism is the means by which a microbe obtains the energy and nutrients (e.g. carbon) it needs to live and reproduce.

Chapter 1. Part 1: The Microbial Cell

Chapter 1. Part 1: The Microbial Cell

	Microbes use many different types of metabolic strategies and species can often be differentiated from each other based on metabolic characteristics. The specific metabolic properties of a microbe are the major factors in determining that microbe's ecological niche, and often allow for that microbe to be useful in industrial processes or responsible for biogeochemical cycles.
Bradyrhizobium	Bradyrhizobium is a genus of Gram-negative soil bacteria, many of which fix nitrogen. Nitrogen fixation is an important part of the nitrogen cycle. Plants cannot use atmospheric nitrogen (N_2) they must use nitrogen compounds such as nitrates.
Denitrification	Denitrification is a microbially facilitated process of nitrate reduction that may ultimately produce molecular nitrogen (N_2) through a series of intermediate gaseous nitrogen oxide products. This respiratory process reduces oxidized forms of nitrogen in response to the oxidation of an electron donor such as organic matter. The preferred nitrogen electron acceptors in order of most to least thermodynamically favorable include nitrate (NO_3^-), nitrite (NO_2^-), nitric oxide (NO), and nitrous oxide (N_2O)and dinitrigen [N2].
Denitrifying bacteria	Denitrifying bacteria form a necessary part of the process known as denitrification as part of the nitrogen cycle, their primary purpose being to metabolise nitrogenous compounds, with the assistance of the nitrate reductase enzyme, to turn oxides back to nitrogen gas or nitrous oxides for energy generation. This process takes place only in the absence of oxygen, as most denitrifying bacteria are facultative aerobes (prefers to use oxygen as their terminal electron acceptors), however, they could also utilize nitrate instead. Therefore, denitrification can only be performed under anaerobic conditions.
Nitrification	Nitrification is the biological oxidation of ammonia with oxygen into nitrite followed by the oxidation of these nitrites into nitrates. Degradation of ammonia to nitrite is usually the rate limiting step of nitrification. Nitrification is an important step in the nitrogen cycle in soil.
Rhizobium	Rhizobium is a genus of Gram-negative soil bacteria that fix nitrogen. Rhizobium forms an endosymbiotic nitrogen fixing association with roots of legumes and Parasponia. The bacteria colonize plant cells within root nodules; here the bacteria convert atmospheric nitrogen to ammonia and then provide organic nitrogenous compounds such as glutamine or ureides to the plant.
Sinorhizobium meliloti	Sinorhizobium meliloti is a Gram-negative nitrogen-fixing bacterium (rhizobium). It forms a symbiotic relationship with legumes from the genera Medicago, Melilotus and Trigonella, including the model legume Medicago truncatula.

Lignin	Lignin is a complex chemical compound most commonly derived from wood, and an integral part of the secondary cell walls of plants and some algae. The term was introduced in 1819 by de Candolle and is derived from the Latin word lignum, meaning wood. It is one of the most abundant organic polymers on Earth, exceeded only by cellulose, employing 30% of non-fossil organic carbon and constituting from a quarter to a third of the dry mass of wood.
Facilitated diffusion	Facilitated diffusion is a process of passive transport (as opposed to active transport), with this passive transport aided by integral membrane proteins. Facilitated diffusion is the spontaneous passage of molecules or ions across a biological membrane passing through specific transmembrane integral proteins. The facilitated diffusion may occur either across biological membranes or through aqueous compartments of an organism.
Permease	The permeases are membrane transport proteins, a class of multipass transmembrane proteins that facilitate the diffusion of a specific molecule in or out of the cell by passive transport. In contrast, active transporters couple molecule transmembrane transport with an energy source such as ATP or a favorable ion gradient.
Lactose permease	LacY proton/sugar symporter Lactose permease is a membrane protein which is a member of the major facilitator superfamily. Lactose permease can be classified as a symporter, which uses the gradient of H+ towards the cell to transport lactose in the same direction into the cell. The protein has twelve transmembrane helices and exhibits an internal two-fold symmetry, relating the N-terminal six helices onto the C-terminal helices.
Lactococcus	Lactococcus is a genus of lactic acid bacteria that were formerly included in the genus Streptococcus Group N1. They are known as homofermentors meaning that they produce a single product, lactic acid in this case, as the major or only product of glucose fermentation. Their homofermentative character can be altered by adjusting cultural conditions like pH, glucose concentration, and nutrient limitation. They are gram-positive, catalase negative, non-motile cocci that are found singly, in pairs, or in chains.
Neisseria gonorrhoeae	Neisseria gonorrhoeae, or gonococcus, is a species of Gram-negative coffee bean-shaped diplococci bacteria responsible for the sexually transmitted infection gonorrhea. N. gonorrhoea was first described by Albert Neisser in 1879. Microbiology

Chapter 1. Part 1: The Microbial Cell

Siderophore	Siderophores (compound from the Ancient Greek nouns síderos and phoros (φορος) meaning 'iron carrier') are small, high-affinity iron chelating compounds secreted by grasses and microorganisms such as bacteria and fungi. Siderophores are amongst the strongest soluble Fe^{3+} binding agents known.
	Iron is essential for almost all life, essential for processes such as respiration and DNA synthesis.
Ames test	The Ames test is a biological assay to assess the mutagenic potential of chemical compounds. A positive test indicates that the chemical is mutagenic and therefore may act as a carcinogen, since cancer is often linked to mutation. However, a number of false-positives and false-negatives are known.
Salmonella enterica	Salmonella enterica is a rod-shaped flagellated, facultative anaerobic, Gram-negative bacterium, and a member of the genus Salmonella.
	Most cases of salmonellosis are caused by food infected with S. enterica, which often infects cattle and poultry, though also other animals such as domestic cats and hamsters have also been shown to be sources of infection to humans. However, investigations of vacuum cleaner bags have shown that households can act as a reservoir of the bacterium; this is more likely if the household has contact with an infection source, for example members working with cattle or in a veterinary clinic.
Hemocytometer	The hemocytometer is a device originally designed for the counting of blood cells. It is now also used to count other types of cells as well as other microscopic particles.
	The hemocytometer was invented by Louis-Charles Malassez and consists of a thick glass microscope slide with a rectangular indentation that creates a chamber.
Propidium iodide	Propidium iodide is an intercalating agent and a fluorescent molecule with a molecular mass of 668.4 Da that can be used to stain cells. When excited with 488 nm wavelength light, it fluoresces red. Propidium iodide is used as a DNA stain for both flow cytometry to evaluate cell viability or DNA content in cell cycle analysis and microscopy to visualise the nucleus and other DNA containing organelles.
Cell counting	Cell counting is a general name for various methods for the quantification of cells in molecular biology and in medicine.
	Numerous procedures in biology and medicine require the counting of cells.

Cell growth	The term cell growth is used in the contexts of cell development and cell division (reproduction). When used in the context of cell division, it refers to growth of cell populations, where one cell (the 'mother cell') grows and divides to produce two 'daughter cells' (M phase). When used in the context of cell development, the term refers to increase in cytoplasmic and organelle volume (G1 phase), as well as increase in genetic material before replication (G2 phase).
Binary fission	Binary fission, or prokaryotic fission, is a form of asexual reproduction and cell division used by all prokaryotes, some protozoa, and some organelles within eukaryotic organisms. This process results in the reproduction of a living prokaryotic cell by division into two parts which each have the potential to grow to the size of the original cell.
Budding	Budding is a form of asexual reproduction in which a new organism develops from an outgrowth or bud on another one. The new organism remains attached as it grows, separating from the parent organism only when it is mature. Since the reproduction is asexual, the newly created organism is a clone and is genetically identical to the parent organism.
Deinococcus radiodurans	Deinococcus radiodurans is an extremophilic bacterium, one of the most radioresistant organisms known. It can survive cold, dehydration, vacuum, and acid, and is therefore known as a polyextremophile and has been listed as the world's toughest bacterium in The Guinness Book Of World Records. The name Deinococcus radiodurans derives from the Ancient Greek δειν?ς (deinos) and κ?κκος (kokkos) meaning 'terrible grain/berry' and the Latin radius and durare, meaning 'radiation surviving'.
Mitosis	Mitosis is the process by which a eukaryotic cell separates the chromosomes in its cell nucleus into two identical sets, in two separate nuclei. It is generally followed immediately by cytokinesis, which divides the nuclei, cytoplasm, organelles and cell membrane into two cells containing roughly equal shares of these cellular components. Mitosis and cytokinesis together define the mitotic (M) phase of the cell cycle--the division of the mother cell into two daughter cells, genetically identical to each other and to their parent cell.
DNA Research	DNA Research is an international, peer reviewed journal of genomics and DNA research.
Plasmodium falciparum	Plasmodium falciparum is a protozoan parasite, one of the species of Plasmodium that cause malaria in humans. It is transmitted by the female Anopheles mosquito. P. falciparum is the most dangerous of these infections as P. falciparum (or malignant) malaria has the highest rates of complications and mortality.
Reproduction	Reproduction is the biological process by which new 'offspring' individual organisms are produced from their 'parents'.

Chapter 1. Part 1: The Microbial Cell

	Reproduction is a fundamental feature of all known life; each individual organism exists as the result of reproduction. The known methods of reproduction are broadly grouped into two main types: sexual and asexual.
Quorum sensing	Quorum sensing is a system of stimulus and response correlated to population density. Many species of bacteria use quorum sensing to coordinate gene expression according to the density of their local population. In similar fashion, some social insects use quorum sensing to determine where to nest.
Turbidostat	A turbidostat is a continuous culture device, similar to a chemostat or an auxostat, which has feedback between the turbidity of the culture vessel and the dilution rate. The theoretical relationship between growth in a chemostat and growth in a turbidostat is somewhat complex, in part because it is similar. A chemostat technically has a fixed volume and flow rate - thus a fixed dilution rate.
Cystic fibrosis	Cystic fibrosis is an autosomal recessive genetic disorder affecting most critically the lungs, and also the pancreas, liver, and intestine. It is characterized by abnormal transport of chloride and sodium across an epithelium, leading to thick, viscous secretions. The name cystic fibrosis refers to the characteristic scarring (fibrosis) and cyst formation within the pancreas, first recognized in the 1930s.
Exopolysaccharide	Exopolysaccharides are high-molecular-weight polymers that are composed of sugar residues and are secreted by a microorganism into the surrounding environment. Microorganisms synthesize a wide spectrum of multifunctional polysaccharides including intracellular polysaccharides, structural polysaccharides and extracellular polysaccharides or exopolysaccharides (EPS). Exopolysaccharides generally consist of monosaccharides and some non-carbohydrate substituents (such as acetate, pyruvate, succinate, and phosphate).
Marine snow	In the deep ocean, marine snow is a continuous shower of mostly organic detritus falling from the upper layers of the water column. It is a significant means of exporting energy from the light-rich photic zone to the aphotic zone below. The term was first coined by the explorer William Beebe as he observed it from his bathysphere.
Aphotic zone	The aphotic zone is the portion of a lake or ocean where there is little or no sunlight. It is formally defined as the depths beyond which less than 1% of sunlight penetrates. Consequently, bioluminescence is essentially the only light found in this zone.

Dictyostelium discoideum	Dictyostelium discoideum is a species of soil-living amoeba belonging to the phylum Mycetozoa. D. discoideum, commonly referred to as slime mold, is a eukaryote that transitions from a collection of unicellular amoebae into a multicellular slug and then into a fruiting body within its lifetime. D. discoideum has a unique asexual lifecycle that consists of four stages: vegetative, aggregation, migration, and culmination.
Anabaena	Anabaena is a genus of filamentous cyanobacteria that exists as plankton. It is known for its nitrogen fixing abilities, and they form symbiotic relationships with certain plants, such as the mosquito fern. They are one of four genera of cyanobacteria that produce neurotoxins, which are harmful to local wildlife, as well as farm animals and pets.
Clostridium tetani	Clostridium tetani is a rod-shaped, anaerobic bacterium of the genus Clostridium. Like other Clostridium species, it is Gram-positive, and its appearance on a gram stain resembles tennis rackets or drumsticks. C. tetani is found as spores in soil or in the gastrointestinal tract of animals.
Dipicolinic acid	Dipicolinic acid is a chemical compound which composes 5% to 15% of the dry weight of bacterial spores. It is implicated as responsible for the heat resistance of the endospore. However, mutants resistant to heat but lacking dipicolinic acid have been isolated, suggesting other mechanisms contributing to heat resistance are at work.
Germination	Germination is the process in which a plant or fungus emerges from a seed or spore, respectively, and begins growth. The most common example of germination is the sprouting of a seedling from a seed of an angiosperm or gymnosperm. However the growth of a sporeling from a spore, for example the growth of hyphae from fungal spores, is also germination.
Heterocyst	Heterocysts are specialized nitrogen-fixing cells formed by some filamentous cyanobacteria, such as Nostoc punctiforme, Cylindrospermum stagnale and Anabaena sphaerica, during nitrogen starvation. They fix nitrogen from dinitrogen (N_2) in the air using the enzyme nitrogenase, in order to provide the cells in the filament with nitrogen for biosynthesis. Nitrogenase is inactivated by oxygen, so the heterocyst must create a microanaerobic environment.
Sporangium	A sporangium (modern Latin, from Greek σπ?ρος (sporos) 'spore' + αγγε?ον (angeion) 'vessel') is an enclosure in which spores are formed. It can be composed of a single cell or can be multicellular. All plants, fungi, and many other lineages form sporangia at some point in their life cycle.
Suspended animation	Suspended animation is the slowing of life processes by external means without termination.

	Breathing, heartbeat, and other involuntary functions may still occur, but they can only be detected by artificial means. Extreme cold can be used to precipitate the slowing of an individual's functions; use of this process has led to the developing science of cryonics.
Gliding motility	Gliding motility is a form of motility specific to apicomplexa that uses a large complex of proteins around the cell surface.
Myxococcus xanthus	Myxococcus xanthus colonies exist as a self-organized, predatory, saprotrophic, single-species biofilm called a swarm. Myxococcus xanthus, which can be found almost ubiquitously in soil, are thin rod shaped, gram-negative cells that exhibit self-organizing behavior as a response to environmental cues. The swarm, which has been compared to a 'wolf-pack,' modifies its environment through stigmergy.
Starvation	Starvation is a severe deficiency in caloric energy, nutrient and vitamin intake. It is the most extreme form of malnutrition. In humans, prolonged starvation can cause permanent organ damage and eventually, death.
Lactobacillus plantarum	Lactobacillus plantarum is a widespread member of the genus Lactobacillus, commonly found in many fermented food products as well as anaerobic plant matter. It is also present in saliva (from which it was first isolated). It has the ability to liquefy gelatin.
Extremophile	An extremophile (from Latin extremus meaning 'extreme' and Greek philia meaning 'love') is an organism that thrives in physically or geochemically extreme conditions that are detrimental to most life on Earth. In contrast, organisms that live in more moderate environments may be termed mesophiles or neutrophiles. The category name is unfortunate as it calls for subjective judgements of two issues - firstly, the degree of deviation from 'normal' justifying the use of 'extreme', and secondly, whether the organism prefers the environment or merely tolerates it.
Photosystem I	Photosystem I is the second photosystem in the photosynthetic light reactions of algae, plants, and some bacteria. Photosystem I is so named because it was discovered before photosystem II. Aspects of PS I were discovered in the 1950s, but the significances of these discoveries was not yet known. Louis Duysens first proposed the concepts of photosystems I and II in 1960, and, in the same year, a proposal by Fay Bendall and Robert Hill assembled earlier discoveries into a cohesive theory of serial photosynthetic reactions.
Photosystem II	Photosystem II is the first protein complex in the Light-dependent reactions. It is located in the thylakoid membrane of plants, algae, and cyanobacteria. The enzyme uses photons of light to energize electrons that are then transferred through a variety of coenzymes and cofactors to reduce plastoquinone to plastoquinol.

On Plants	On Plants is a work, sometimes attributed to Aristotle, but generally believed to have been written by Nicolaus of Damascus, which deals with a number of plant related topics.
	The work is divided in two parts. The first part discusses the nature of plant life, sex in plants, the parts of plants, the structure of plants, the classification of plants, the composition and products of plants, the methods of propagation and fertilization of plants, and the changes and variations of plants.
Sequence analysis	In bioinformatics, the term sequence analysis refers to the process of subjecting a DNA, RNA or peptide sequence to any of a wide range of analytical methods to understand its features, function, structure, or evolution. Methodologies used include sequence alignment, searches against biological databases, and others. Since the development of methods of high-throughput production of gene and protein sequences, the rate of addition of new sequences to the databases increased exponentially.
Arrhenius equation	The Arrhenius equation is a simple, but remarkably accurate, formula for the temperature dependence of the reaction rate constant, and therefore, rate of a chemical reaction. The equation was first proposed by the Dutch chemist J. H. van 't Hoff in 1884; five years later in 1889, the Swedish chemist Svante Arrhenius provided a physical justification and interpretation for it. Currently, it is best seen as an empirical relationship.
Listeria monocytogenes	Listeria monocytogenes, a facultative anaerobe, intracellular bacterium, is the causative agent of listeriosis. It is one of the most virulent foodborne pathogens, with 20 to 30 percent of clinical infections resulting in death. Responsible for approximately 2,500 illnesses and 500 deaths in the United States (U.S).
Mesophile	A mesophile is an organism that grows best in moderate temperature, neither too hot nor too cold, typically between 20 and 45 °C (68 and 113 °F). The term is mainly applied to microorganisms.
	The habitats of these organisms include especially cheese, yogurt, and mesophile organisms are often included in the process of beer and wine making.
Psychrophile	Psychrophiles or cryophiles (adj. cryophilic) are extremophilic organisms that are capable of growth and reproduction in cold temperatures, ranging from −15°C to +10°C. Temperatures as low as −15°C are found in pockets of very salty water (brine) surrounded by sea ice. They can be contrasted with thermophiles, which thrive at unusually hot temperatures.
Food processing	Food processing is the set of methods and techniques used to transform raw ingredients into food or to transform food into other forms for consumption by humans or animals either in the home or by the food processing industry.

Chapter 1. Part 1: The Microbial Cell

	Food processing typically takes clean, harvested crops or butchered animal products and uses these to produce attractive, marketable and often long shelf-life food products. Similar processes are used to produce animal feed.
Erwinia	Erwinia is a genus of Enterobacteriaceae bacteria containing mostly plant pathogenic species which was named for the first phytobacteriologist, Erwin Smith. It is a gram negative bacterium related to E. coli, Shigella, Salmonella and Yersinia. It is primarily a rod-shaped bacteria.
Pseudomonas syringae	Pseudomonas syringae is a rod shaped, Gram-negative bacterium with polar flagella. It is a plant pathogen which can infect a wide range of plant species, and exists as over 50 different pathovars, all of which are available to legitimate researches via international culture collections such as the NCPPB, ICMP, and others. Many of these pathovars were once considered to be individual species within the Pseudomonas genus, but molecular biology techniques such as DNA hybridization have shown these to in fact all be part of the P. syringae species.
Enzyme	Enzymes ( ) are biological molecules that catalyze (i.e., increase the rates of) chemical reactions. In enzymatic reactions, the molecules at the beginning of the process, called substrates, are converted into different molecules, called products. Almost all chemical reactions in a biological cell need enzymes in order to occur at rates sufficient for life.
Thermophile	A thermophile is an organism -- a type of extremophile -- that thrives at relatively high temperatures, between 45 and 80 °C (113 and 176 °F). Many thermophiles are archaea. It has been suggested that thermophilic eubacteria are among the earliest bacteria.
Piezophile	A piezophile is an organism which thrives at high pressures, such as deep sea bacteria or archaea. They are generally found on ocean floors, where pressure often exceeds 380 atm (38 MPa). Some have been found at the bottom of the Pacific Ocean where the maximum pressure is roughly 117 MPa.
Hypertonic	A hypertonic solution is a solution having a greater solute concentration than the cytosol. It contains a greater concentration of impermeable solutes on the external side of the membrane. When a cell's cytoplasm is bathed in a hypertonic solution the water will be drawn into the solution and out of the cell by osmosis. If water molecules continue to diffuse out of the cell, it will cause the cell to shrink, or crenate.A hypertonic solution is used in osmotherapy to treat cerebral hemorrhage.
Osmolarity	Osmolarity is the measure of solute concentration, defined as the number of osmoles (Osm) of solute per litre (L) of solution (osmol/L or Osm/L). The osmolarity of a solution is usually expressed as Osm/L, in the same way that the molarity of a solution is expressed as 'M' .

Water activity	Water activity or a_w was developed to account for the intensity with which water associates with various non-aqueous constituents and solids. Simply stated, it is a measure of the energy status of the water in a system. It is defined as the vapor pressure of a liquid divided by that of pure water at the same temperature; therefore, pure distilled water has a water activity of exactly one.
Halophile	Halophiles are extremophile organisms that thrive in environments with very high concentrations of salt. The name comes from the Greek for 'salt-loving'. While the term is perhaps most often applied to some halophiles classified into the Archaea domain, there are also bacterial halophiles and some eukaryota, such as the alga Dunaliella salina.
Hypotonic	A hypotonic solution is a solution having a lesser solute concentration than the cytosol. It contains a lesser concentration of impermeable solutes on the external side of the membrane. When a cell's cytoplasm is bathed in a hypotonic solution the water will be drawn out of the solution and into the cell by osmosis. If water molecules continue to diffuse into the cell, it will cause the cell to swell, up to the point that cytolysis (rupture) may occur. In plant cells, the cell will not always rupture. When placed in a hypotonic solution, the cell will have Turgor Pressure and proceed with its normal functions.
Yogurt	Yogurt, UK: /ˈjɒɡət/) is a dairy product produced by bacterial fermentation of milk. The bacteria used to make yogurt are known as 'yogurt cultures'. Fermentation of lactose by these bacteria produces lactic acid, which acts on milk protein to give yogurt its texture and its characteristic tang.
Alkaliphile	Alkaliphiles are microbes classified as extremophiles that thrive in alkaline environments with a pH of 9 to 11 such as playa lakes and carbonate-rich soils. To survive, alkaliphiles maintain a relatively low alkaline level of about 8 pH inside their cells by constantly pumping hydrogen ions (H^+) in the form of hydronium ions (H_3O^+) across their cell membranes into their cytoplasm. Examples include:•Geoalkalibacter ferrihydriticus•Bacillus okhensis •Alkalibacterium iburiense.
Cyclodextrin	Cyclodextrins (sometimes called cycloamyloses) are a family of compounds made up of sugar molecules bound together in a ring (cyclic oligosaccharides). Cyclodextrins are produced from starch by means of enzymatic conversion. They are used in food, pharmaceutical, drug delivery, and chemical industries, as well as agriculture and environmental engineering.
Halobacterium salinarum	Halobacterium salinarum is an extremely halophilic marine gram-negative obligate aerobic archaeon. Despite its name, this microorganism is not a bacterium, but rather a member of the Kingdom Archaea.

Chapter 1. Part 1: The Microbial Cell

Ammonia production	Because of its many uses, ammonia is one of the most highly-produced inorganic chemicals. There are numerous large-scale ammonia production plants worldwide, producing a total of 131,000,000 metric tons of ammonia in 2010. China produced 32.1% of the worldwide production, followed by India with 8.9%, Russia with 7.9%, and the United States with 6.3%. 80% or more of the ammonia produced is used for fertilizing agricultural crops.
Electron transport chain	An electron transport chain couples electron transfer between an electron donor (such as NADH) and an electron acceptor (such as O_2) with the transfer of H^+ ions (protons) across a membrane. The resulting electrochemical proton gradient is used to generate chemical energy in the form of adenosine triphosphate (ATP). Electron transport chains are the cellular mechanisms used for extracting energy from sunlight in photosynthesis and also from redox reactions, such as the oxidation of sugars (respiration).
Electrochemical gradient	An electrochemical gradient is a gradient of electrochemical potential, usually for an ion that can move across membrane. The gradient consist of two parts, the electrical potential and a difference in the chemical concentration across a membrane. The difference of electrochemical potentials can be interpreted as a type of potential energy available for work in a cell.
Electron acceptor	An electron acceptor is a chemical entity that accepts electrons transferred to it from another compound. It is an oxidizing agent that, by virtue of its accepting electrons, is itself reduced in the process. Typical oxidizing agents undergo permanent chemical alteration through covalent or ionic reaction chemistry, resulting in the complete and irreversible transfer of one or more electrons.
Anaerobic respiration	Anaerobic respiration is a form of respiration using electron acceptors other than oxygen. Although oxygen is not used as the final electron acceptor, the process still uses a respiratory electron transport chain; it is respiration without oxygen. In order for the electron transport chain to function, an exogenous final electron acceptor must be present to allow electrons to pass through the system.
Hydrogen peroxide	Hydrogen peroxide is the simplest peroxide (a compound with an oxygen-oxygen single bond) and an oxidizer. Hydrogen peroxide is a clear liquid, slightly more viscous than water. In dilute solution it appears colorless.
Hydroxyl radical	The hydroxyl radical, $^{\bullet}OH$, is the neutral form of the hydroxide ion (OH^-). Hydroxyl radicals are highly reactive and consequently short-lived; however, they form an important part of radical chemistry. Most notably hydroxyl radicals are produced from the decomposition of hydroperoxides (ROOH) or, in atmospheric chemistry, by the reaction of excited atomic oxygen with water.

Campylobacter jejuni	Campylobacter jejuni is a species of curved, helical shaped, non-spore forming, Gram-negative microaerophilic, bacteria commonly found in animal feces.Gorbach, Sherwood L., Falagas, Matthew (editors) (2001). The 5 minute infectious diseases consult (1st ed).. Lippincott Williams & Wilkins.
Oligotroph	An oligotroph is an organism that can live in an environment that offers very low levels of nutrients. They may be contrasted with copiotrophs, which prefer nutritionally rich environments. Oligotrophs are characterized by slow growth, low rates of metabolism, and generally low population density.
Eutrophication	Eutrophication, is the ecosystem response to the addition of artificial or natural substances, such as nitrates and phosphates, through fertilizers or sewage, to an aquatic system. One example is the 'bloom' or great increase of phytoplankton in a water body as a response to increased levels of nutrients. Negative environmental effects include hypoxia, the depletion of oxygen in the water, which induces reductions in specific fish and other animal populations.
Bacteriostatic agent	A bacteriostatic agent, abbreviated Bstatic, is a biological or chemical agent that stops bacteria from reproducing, while not necessarily harming them otherwise. Depending on their application, bacteriostatic antibiotics, disinfectants, antiseptics and preservatives can be distinguished. Upon removal of the bacteriostat, the bacteria usually start to grow again.
Fuel cell	A fuel cell is a device that converts the chemical energy from a fuel into electricity through a chemical reaction with oxygen or another oxidizing agent. Hydrogen is the most common fuel, but hydrocarbons such as natural gas and alcohols like methanol are sometimes used. Fuel cells are different from batteries in that they require a constant source of fuel and oxygen to run, but they can produce electricity continually for as long as these inputs are supplied.
Sanitation	Sanitation is the hygienic means of promoting health through prevention of human contact with the hazards of wastes. Hazards can be either physical, microbiological, biological or chemical agents of disease. Wastes that can cause health problems are human and animal feces, solid wastes, domestic wastewater (sewage, sullage, greywater), industrial wastes and agricultural wastes.
Botulinum toxin	Botulinum toxin is a protein and neurotoxin produced by the bacterium Clostridium botulinum. Botulinum toxin can cause botulism, a serious and life-threatening illness in humans and animals. When introduced intravenously in monkeys, type A (Botox Cosmetic) of the toxin exhibits an LD_{50} of 40-56 ng, type C1 around 32 ng, type D 3200 ng, and type E 88 ng; these are some of the most potent neurotoxins known.
Ethanol	Ethanol, pure alcohol, grain alcohol, or drinking alcohol, is a volatile, flammable, colorless liquid. It is a psychoactive drug and one of the oldest recreational drugs.

Chapter 1. Part 1: The Microbial Cell

Toxin	A toxin is a poisonous substance produced within living cells or organisms; man-made substances created by artificial processes are thus excluded. The term was first used by organic chemist Ludwig Brieger (1849-1919).
	For a toxic substance not produced within living organisms, 'toxicant' and 'toxics' are also sometimes used..
Laminar flow	Laminar flow, occurs when a fluid flows in parallel layers, with no disruption between the layers. At low velocities the fluid tends to flow without lateral mixing, and adjacent layers slide past one another like playing cards. There are no cross currents perpendicular to the direction of flow, nor eddies or swirls of fluids.
Pasteurization	Pasteurization is a process of heating a food, usually a liquid, to a specific temperature for a definite length of time and then cooling it immediately. This process slows spoilage due to microbial growth in the food.
	Unlike sterilization, pasteurization is not intended to kill all micro-organisms in the food.
Botulism	Botulism is metabolic waste produced under anaerobic conditions by the bacterium Clostridium botulinum, and affecting a wide range of mammals, birds and fish.
	The toxin enters the human body in one of three ways: by colonization of the digestive tract by the bacterium in children (infant botulism) or adults (adult intestinal toxemia), by ingestion of toxin from foods or by contamination of a wound by the bacterium (wound botulism). Person to person transmission of botulism does not occur.
Yersinia enterocolitica	Yersinia enterocolitica is a species of gram-negative coccobacillus-shaped bacterium, belonging to the family Enterobacteriaceae. Primarily a zoonotic disease (cattle, deer, pigs, and birds), animals that recover frequently become asymptomatic carriers of the disease.
	Signs and symptoms
	Acute Y. enterocolitica infections produce severe diarrhea in humans, along with Peyer's patch necrosis, chronic lymphadenopathy, and hepatic or splenic abscesses.
Irradiation	Irradiation is the process by which an item is exposed to radiation. The exposure can originate from any of various sources, including those occurring naturally, or as part of a mechanical process, or otherwise. In common usage the term refers specifically to ionizing radiation, and to a level of radiation that will serve that specific purpose, rather than radiation exposure to normal levels of background radiation or abnormal levels of radiation due to accidental exposure.

Phenol coefficient	Phenol coefficient is a measure of the bactericidal activity of a chemical compound in relation to phenol. When listed numerically, the figure expressing the disinfecting power of a substance by relating it to the disinfecting power of phenol may be a function of the standardized test performed. For example, the Rideal-Walker method gives a Rideal-Walker coefficient and the U.S. Department of Agriculture method gives a U.S. Department of Agriculture coefficient.
Aldehyde	An aldehyde is an organic compound containing a formyl group. This functional group, with the structure R-CHO, consists of a carbonyl centre bonded to hydrogen and an R group. The group without R is called the aldehyde group or formyl group.
Ethylene oxide	Ethylene oxide, is the organic compound with the formula C_2H_4O. It is a cyclic ether. This means that it is composed of 2 alkyl groups attached to an oxygen atom in a cyclic shape (circular). This colorless flammable gas with a faintly sweet odor is the simplest epoxide, a three-membered ring consisting of two carbon and one oxygen atom.
Molecular structure	The molecular structure of a substance is described by the combination of nuclei and electrons that comprise its constitute molecules. This includes the molecular geometry (essentially the arrangement, in space, of the equilibrium positions of the constituent atoms -- in reality, these are in a state of constant vibration, at temperatures above absolute zero), the electronic properties of the bonds, and further molecular properties. The determination of molecular structure uses a multitude of experimental methods, that include X-ray diffraction, electron diffraction, many kinds of optical spectroscopy, nuclear magnetic resonance, electron spin resonance, and mass spectrometry.
Streptomycin	Streptomycin is an antibiotic drug, the first of a class of drugs called aminoglycosides to be discovered, and was the first antibiotic remedy for tuberculosis. It is derived from the actinobacterium Streptomyces griseus. Streptomycin is a bactericidal antibiotic.
Phage therapy	Phage therapy is the therapeutic use of bacteriophages to treat pathogenic bacterial infections. Although extensively used and developed mainly in former Soviet Union countries circa 1920, the treatment is not approved in countries other than Georgia. Phage therapy has many potential applications in human medicine as well as dentistry, veterinary science, and agriculture.
Phytophthora cinnamomi	Phytophthora cinnamomi is a soil-borne water mould that produces an infection which causes a condition in plants called root rot or dieback. The plant pathogen is one of the world's most invasive species and is present in over 70 countries from around the world. P. cinnamomi lives in the soil and in plant tissues, can take different shapes and can move in water.

Chapter 1. Part 1: The Microbial Cell

Probiotic	Probiotic are live microorganisms thought to be beneficial to the host organism. According to the currently adopted definition by FAOWHO, probiotics are: 'Live microorganisms which when administered in adequate amounts confer a health benefit on the host'. Lactic acid bacteria (LAB) and bifidobacteria are the most common types of microbes used as probiotics; but certain yeasts and bacilli may also be used.
Rous sarcoma virus	Rous sarcoma virus is a retrovirus and is the first oncovirus to have been described: it causes sarcoma in chickens. As with all retroviruses, it reverse transcribes its RNA genome into cDNA before integration into the host DNA.
Pandemic	A pandemic is an epidemic of infectious disease that has spread through human populations across a large region; for instance multiple continents, or even worldwide. A widespread endemic disease that is stable in terms of how many people are getting sick from it is not a pandemic. Further, flu pandemics generally exclude recurrences of seasonal flu.
Bacteriophage	A bacteriophage is any one of a number of viruses that infect bacteria. They do this by injecting genetic material, which they carry enclosed in an outer protein capsid. The genetic material can be ssRNA, dsRNA, ssDNA, or dsDNA ('ss-' or 'ds-' prefix denotes single-strand or double-strand) along with either circular or linear arrangement.
Capsid	A capsid is the protein shell of a virus. It consists of several oligomeric structural subunits made of protein called protomers. The observable 3-dimensional morphological subunits, which may or may not correspond to individual proteins, are called capsomeres.
Viral envelope	Many viruses (e.g. influenza and many animal viruses) have viral envelopes covering their protein capsids. The envelopes typically are derived from portions of the host cell membranes (phospholipids and proteins), but include some viral glycoproteins. Functionally, viral envelopes are used to help viruses enter host cells.
West Nile virus	West Nile virus is a virus of the family Flaviviridae. Part of the Japanese encephalitis (JE) antigenic complex of viruses, it is found in both tropical and temperate regions. It mainly infects birds, but is known to infect humans, horses, dogs, cats, bats, chipmunks, skunks, squirrels, domestic rabbits, crows, robins, crocodiles and alligators.
Zidovudine	Zidovudine or azidothymidine (AZT) (also called ZDV) is a nucleoside analog reverse transcriptase inhibitor (NRTI), a type of antiretroviral drug used for the treatment of HIV/AIDS. It is an analog of thymidine. AZT was the first approved treatment for HIV, sold under the names Retrovir and Retrovis.

Cloning vector	A cloning vector is a small piece of DNA into which a foreign DNA fragment can be inserted. The insertion of the fragment into the cloning vector is carried out by treating the vehicle and the foreign DNA with a restriction enzyme that creates the same overhang, then ligating the fragments together. There are many types of cloning vectors.
Protease inhibitor	Protease inhibitors are a class of drugs used to treat or prevent infection by viruses, including HIV and Hepatitis C. Protease inhibitors prevent viral replication by inhibiting the activity of proteases, e.g.HIV-1 protease, enzymes used by the viruses to cleave nascent proteins for final assembly of new virions.

Protease inhibitors have been developed or are presently undergoing testing for treating various viruses:•HIV/AIDS: antiretroviral protease inhibitors (saquinavir, ritonavir, indinavir, nelfinavir, amprenavir etc).•Hepatitis C: Boceprevir•Hepatitis C: Telaprevir

Given the specificity of the target of these drugs there is the risk, as in antibiotics, of the development of drug-resistant mutated viruses. To reduce this risk it is common to use several different drugs together that are each aimed at different targets. |
Reading frame	In biology, a reading frame is a way of breaking a sequence of nucleotides in DNA or RNA into three letter codons which can be translated in amino acids. There are 3 possible reading frames in an mRNA strand: each reading frame corresponds to a different starting alignment. Double stranded DNA has six different reading frames per molecule due to the two strands from which transcription is possible--three of them reading forward and three of them reading backwards.
Rice dwarf virus	Rice dwarf virus is a plant pathogenic virus of the family Reoviridae.
Inhibitor protein	The inhibitor protein is situated in the mitochondrial matrix and protects the cell against rapid ATP hydrolysis during momentary ischaemia. In oxygen absence, the pH of the matrix drops. This causes IP to become protonated and change its conformation to one that can bind to the F1Fo synthetase and stops it thereby preventing it from moving in a backwards direction and hydrolyze ATP instead of make it.
Scrapie	Scrapie is a fatal, degenerative disease that affects the nervous systems of sheep and goats. It is one of several transmissible spongiform encephalopathies (TSEs), which are related to bovine spongiform encephalopathy (BSE or 'mad cow disease') and chronic wasting disease of deer. Like other spongiform encephalopathies, scrapie is caused by a prion.
Serum albumin	Serum albumin, often referred to simply as albumin is a protein that in humans is encoded by the ALB gene.

	Serum albumin is the most abundant plasma protein in mammals. Albumin is essential for maintaining the osmotic pressure needed for proper distribution of body fluids between intravascular compartments and body tissues.
Filamentous phage	A filamentous phage is a type of bacteriophage shaped like a rod filament. Filamentous phages usually contain a genome of single-stranded DNA and infect Gram-negative bacteria. •Ff phages - these infect E. coli that carry the F episome •M13 bacteriophage•f1 phage•fd phage•Ike phage•N1 phage.
Propionibacterium freudenreichii	Propionibacterium freudenreichii is a Gram-positive, nonmotile bacterium that plays an important role in the creation of Emmental cheese, and to some extent, Leerdammer. Its concentration in Swiss-type cheeses is higher than in any other cheese. Propionibacteria are commonly found in milk and dairy products, though they have also been extracted from soil.
Tegument	Tegument /'t?gj?m?nt/ is a terminology in helminthology for the name of the outer body covering among members of the phylum Platyhelminthes. The name is derived from a Latin word tegumentum or tegere, meaning 'to cover'. It is characteristic of all flatworms including the broad groups of tapeworms and flukes.
International Committee on Taxonomy of Viruses	The International Committee on Taxonomy of Viruses is a committee which authorizes and organizes the taxonomic classification of viruses. They have developed a universal taxonomic scheme for viruses and aim to describe all the viruses of living organisms. Members of the committee are considered to be world experts on viruses.
Rabies virus	The rabies virus is neurotropic virus that causes fatal disease in human and animals. Rabies transmission can occur through the saliva of animals.
	The rabies virus has a cylindrical morphology and is the type species of the Lyssavirus genus of the Rhabdoviridae family.
Reverse transcriptase	In the fields of molecular biology and biochemistry, a reverse transcriptase, is a DNA polymerase enzyme that transcribes single-stranded RNA into single-stranded DNA. It also is a DNA-dependent DNA polymerase which synthesizes a second strand of DNA complementary to the reverse-transcribed single-stranded cDNA after degrading the original mRNA with its RNaseH activity. Normal transcription involves the synthesis of RNA from DNA; hence, reverse transcription is the reverse of this.

	Well studied reverse transcriptases include:•HIV-1 reverse transcriptase from human immunodeficiency virus type 1 (PDB 1HMV)•M-MLV reverse transcriptase from the Moloney murine leukemia virus•AMV reverse transcriptase from the avian myeloblastosis virus•Telomerase reverse transcriptase that maintains the telomeres of eukaryotic chromosomesHistory Reverse transcriptase was discovered by Howard Temin at the University of Wisconsin-Madison, and independently by David Baltimore in 1970 at MIT. The two shared the 1975 Nobel Prize in Physiology or Medicine with Renato Dulbecco for their discovery.
Virus classification	Virus classification is the process of naming viruses and placing them into a taxonomic system. Similar to the classification systems used for cellular organisms, virus classification is the subject of ongoing debate and proposals. This is mainly due to the pseudo-living nature of viruses, which is to say they are non-living particles with some chemical characteristics similar to those of life.
Hepatitis B	Hepatitis B is an infectious inflammatory illness of the liver caused by the hepatitis B virus (HBV) that affects hominoidea, including humans. Originally known as 'serum hepatitis', the disease has caused epidemics in parts of Asia and Africa, and it is endemic in China. About a third of the world population has been infected at one point in their lives, including 350 million who are chronic carriers.
Inverted repeat	An inverted repeat is a sequence of nucleotides that is the reversed complement of another sequence further downstream. For example, 5'---GACTGC....GCAGTC---3'. When no nucleotides intervene between the sequence and its downstream complement, it is called a palindrome.
RNA-dependent RNA polymerase	RNA-directed RNA polymerase, flaviviral RNA-dependent RNA polymerase (RDR), or RNA replicase, is an enzyme (EC 2.7.7.48) that catalyzes the replication of RNA from an RNA template. This is in contrast to a typical DNA-dependent RNA polymerase, which catalyzes the transcription of RNA from a DNA template. RNA-dependent RNA polymerase is an essential protein encoded in the genomes of all RNA-containing viruses with no DNA stage that have sense negative RNA. It catalyses synthesis of the RNA strand complementary to a given RNA template.
Cauliflower mosaic virus	Cauliflower mosaic virus is the type member of the caulimoviruses, one of the six genera in the Caulimoviridae family, pararetroviruses that infect plants (Pringle, 1999).

Pararetroviruses replicate through reverse transcription just like retroviruses, but the viral particles contain DNA instead of RNA (Rothnie et al., 1994).

Structure

The CaMV particle is an icosahedron with a diameter of 52 nm built from 420 capsid protein (CP) subunits arranged with a triangulation T = 7, which surrounds a solvent-filled central cavity (Cheng et al., 1992).

Evolution	Evolution is any change across successive generations in the inherited characteristics of biological populations. Evolutionary processes give rise to diversity at every level of biological organisation, including species, individual organisms and molecules such as DNA and proteins.

Life on Earth originated and then evolved from a universal common ancestor approximately 3.7 billion years ago.

Molecular evolution	Molecular evolution is in part a process of evolution at the scale of DNA, RNA, and proteins. Molecular evolution emerged as a scientific field in the 1960s as researchers from molecular biology, evolutionary biology and population genetics sought to understand recent discoveries on the structure and function of nucleic acids and protein. Some of the key topics that spurred development of the field have been the evolution of enzyme function, the use of nucleic acid divergence as a 'molecular clock' to study species divergence, and the origin of noncoding DNA.

Recent advances in genomics, including whole-genome sequencing, high-throughput protein characterization, and bioinformatics have led to a dramatic increase in studies on the topic.

Plant cell	Plant cells are eukaryotic cells that differ in several key respects from the cells of other eukaryotic organisms. Their distinctive features include:•A large central vacuole, a water-filled volume enclosed by a membrane known as the tonoplast maintains the cell's turgor, controls movement of molecules between the cytosol and sap, stores useful material and digests waste proteins and organelles.•A cell wall composed of cellulose and hemicellulose, pectin and in many cases lignin, is secreted by the protoplast on the outside of the cell membrane. This contrasts with the cell walls of fungi (which are made of chitin), and of bacteria, which are made of peptidoglycan.•Specialised cell-cell communication pathways known as plasmodesmata, pores in the primary cell wall through which the plasmalemma and endoplasmic reticulum of adjacent cells are continuous.•Plastids, the most notable being the chloroplasts, which contain chlorophyll a green coloured pigment which is used for absorbing sunlight and is used by a plant to make its own food in the process is known as photosynthesis.

Corynebacterium diphtheriae	Corynebacterium diphtheriae is a pathogenic bacterium that causes diphtheria. It is also known as the Klebs-Löffler bacillus, because it was discovered in 1884 by German bacteriologists Edwin Klebs (1834 - 1912) and Friedrich Löffler (1852 - 1915). Classification Four subspecies are recognized: C. diphtheriae mitis, C. diphtheriae intermedius, C. diphtheriae gravis, and C. diphtheriae belfanti.
Hin recombinase	Hin recombinase is a 21kD protein composed of 198 amino acids that is found in the bacteria Salmonella. Hin belongs to the serine recombinase family of DNA invertases in which it relies on the active site serine to initiate DNA cleavage and recombination. The related protein, gamma-delta resolvase shares high similarity to Hin, of which much structural work has been done, including structures bound to DNA and reaction intermediates.
Lysogeny	Lysogeny, is one of two methods of viral reproduction (the lytic cycle is the other). Lysogeny is characterized by integration of the bacteriophage nucleic acid into the host bacterium's genome. The newly integrated genetic material, called a prophage can be transmitted to daughter cells at each subsequent cell division, and a later event (such as UV radiation) can release it, causing proliferation of new phages via the lytic cycle.
Prophage	A prophage is a phage (viral) genome inserted and integrated into the circular bacterial DNA chromosome. A prophage, also known as a temperate phage, is any virus in the lysogenic cycle; it is integrated into the host chromosome or exists as an extrachromosomal plasmid. Technically, a virus may be called a prophage only while the viral DNA remains incorporated in the host DNA. This is a latent form of a bacteriophage, in which the viral genes are incorporated into the bacterial chromosome without causing disruption of the bacterial cell.
Shiga toxin	Shiga toxins are a family of related toxins with two major groups, Stx1 and Stx2, whose genes are considered to be part of the genome of lambdoid prophages. The toxins are named for Kiyoshi Shiga, who first described the bacterial origin of dysentery caused by Shigella dysenteriae. The most common sources for Shiga toxin are the bacteria S. dysenteriae and the Shigatoxigenic group of Escherichia coli (STEC), which includes serotypes O157:H7, O104:H4, and other enterohemorrhagic E. coli (EHEC).
Shigella	Shigella is a genus of Gram-negative, nonspore forming, non-motile, rod-shaped bacteria closely related to Escherichia coli and Salmonella. The causative agent of human shigellosis, Shigella causes disease in primates, but not in other mammals. It is only naturally found in humans and apes.

Chapter 1. Part 1: The Microbial Cell

Site-specific recombination	Site-specific recombination, is a type of genetic recombination in which DNA strand exchange takes place between segments possessing only a limited degree of sequence homology. Site-specific recombinases perform rearrangements of DNA segments by recognizing and binding to short DNA sequences (sites), at which they cleave the DNA backbone, exchange the two DNA helices involved and rejoin the DNA strands. While in some site-specific recombination systems just a recombinase enzyme and the recombination sites is enough to perform all these reactions, in other systems a number of accessory proteins and/or accessory sites are also needed.
Lysis	Lysis refers to the breaking down of a cell, often by viral, enzymic, or osmotic mechanisms that compromise its integrity. A fluid containing the contents of lysed cells is called a 'lysate'. Many species of bacteria are subject to lysis by the enzyme lysozyme, found in animal saliva, egg white, and other secretions.
Lytic cycle	The lytic cycle is one of the two cycles of viral reproduction, the other being the lysogenic cycle. The lytic cycle is typically considered the main method of viral replication, since it results in the destruction of the infected cell. A key difference between the lytic and lysogenic phage cycles is that in the lytic phage, the viral DNA exists as a separate molecule within the bacterial cell, and replicates separately from the host bacterial DNA. The location of viral DNA in the lysogenic phage cycle is within the host DNA, therefore in both cases the virus/phage replicates using the host DNA machinery, but in the lytic phage cycle, the phage is a free floating separate molecule to the host DNA. Viruses of the lytic cycle are called virulent viruses.
Human genome	The human genome is stored on 23 chromosome pairs and in the small mitochondrial DNA. Twenty-two of the 23 chromosomes belong to autosomal chromosome pairs, while the remaining pair is sex determinative. The haploid human genome occupies a total of just over three billion DNA base pairs. The Human Genome Project (HGP) produced a reference sequence of the euchromatic human genome and which is used worldwide in the biomedical sciences.
Ebola virus	Ebola virus causes severe disease in humans and in nonhuman primates in the form of viral hemorrhagic fever. EBOV is a select agent, World Health Organization Risk Group 4 Pathogen (requiring Biosafety Level 4-equivalent containment), National Institutes of HealthNational Institute of Allergy and Infectious Diseases Category A Priority Pathogen, Centers for Disease Control and Prevention Category A Bioterrorism Agent, and listed as a Biological Agent for Export Control by the Australia Group. Ebola virus was first described in 1976.

Endocytosis	Endocytosis is a process by which cells absorb molecules (such as proteins) by engulfing them. It is used by all cells of the body because most substances important to them are large polar molecules that cannot pass through the hydrophobic plasma or cell membrane. The process which is the opposite to endocytosis is exocytosis.
Hepatitis C	Hepatitis C is an infectious disease affecting the liver, caused by the hepatitis C virus (HCV). The infection is often asymptomatic, but once established, chronic infection can progress to scarring of the liver (fibrosis), and advanced scarring (cirrhosis) which is generally apparent after many years. In some cases, those with cirrhosis will go on to develop liver failure or other complications of cirrhosis, including liver cancer or life threatening esophageal varices and gastric varices.
Tropism	A tropism is a biological phenomenon, indicating growth or turning movement of a biological organism, usually a plant, in response to an environmental stimulus. In tropisms, this response is dependent on the direction of the stimulus (as opposed to nastic movements which are non-directional responses). Viruses and other pathogens also affect what is called 'host tropism' or 'cell tropism' in which case tropism refers to the way in which different viruses/pathogens have evolved to preferentially target specific host species, or specific cell types within those species.
Sialic acid	Sialic acid is a generic term for the N- or O-substituted derivatives of neuraminic acid, a monosaccharide with a nine-carbon backbone. It is also the name for the most common member of this group, N-acetylneuraminic acid (Neu5Ac or NANA). Sialic acids are found widely distributed in animal tissues and to a lesser extent in other species ranging from plants and fungi to yeasts and bacteria, mostly in glycoproteins and gangliosides.
Tissue tropism	Tissue tropism is a term most often used in virology to define the cells and tissues of a host which support growth of a particular virus. Bacteria and other parasites may also be referred to as having a tissue tropism. Some viruses have a broad tissue tropism and can infect many types of cells and tissues.
Wart	A wart is generally a small, rough growth, typically on hands and feet but often other locations, that can resemble a cauliflower or a solid blister. They are caused by a viral infection, specifically by human papillomavirus 2 and 7. There are as many as 10 varieties of warts, the most common considered to be mostly harmless. It is possible to get warts from others; they are contagious and usually enter the body in an area of broken skin.
Keratinocyte	Keratinocyte is the predominant cell type in the epidermis, the outermost layer of the skin, constituting 95% of the cells found there. Those keratinocytes found in the basal layer (Stratum germinativum) of the skin are sometimes referred to as 'basal cells' or 'basal keratinocytes'.

Chapter 1. Part 1: The Microbial Cell

Oncogene	An oncogene is a gene that has the potential to cause cancer. In tumor cells, they are often mutated or expressed at high levels.
	Most normal cells undergo a programmed form of death (apoptosis).
Plant virus	Plant viruses are viruses that affect plants. Like all other viruses, plant viruses are obligate intracellular parasites that do not have the molecular machinery to replicate without a host. Plant viruses are pathogenic to higher plants.
Latent period	Latent period is the time elapsed from virus entry into the cell until the first progeny are released.
Virulence	Virulence is by MeSH definition the degree of pathogenicity within a group or species of parasites as indicated by case fatality rates and/or the ability of the organism to invade the tissues of the host. The pathogenicity of an organism - its ability to cause disease - is determined by its virulence factors. The noun virulence derives from the adjective virulent.
Tissue culture	Tissue culture is the growth of tissues or cells separate from the organism. This is typically facilitated via use of a liquid, semi-solid, or solid growth medium, such as broth or agar. Tissue culture commonly refers to the culture of animal cells and tissues, while the more specific term plant tissue culture is being named for the plants.
Plaque forming unit	A plaque forming unit is a measure of the number of particles capable of forming plaques per unit volume, such as virus particles. It is a functional measurement rather than a measurement of the absolute quantity of particles: viral particles that are defective or which fail to infect their target cell will not produce a plaque and thus will not be counted. For example, a solution of Tick-borne encephalitis virus with a concentration of 1,000 Plaque forming unit/µl indicates that there are 1,000 infectious virus particles in one microliter of solution.
Reservoir	A reservoir is used to store water. Reservoirs may be created in river valleys by the construction of a dam or may be built by excavation in the ground or by conventional construction techniques such a brickwork or cast concrete.
	The term reservoir may also be used to describe underground reservoirs such as an oil or water well.
Population density	Population density is a measurement of population per unit area or unit volume. It is frequently applied to living organisms, and particularly to humans. It is a key geographic term.

1. In the fields of medicine, biotechnology and pharmacology, _____ is the process by which drugs are discovered or designed.

   In the past most drugs have been discovered either by identifying the active ingredient from traditional remedies or by serendipitous discovery. As our understanding of disease has increased to the extent that we know how disease and infection are controlled at the molecular and physiological level, scientists are now able to try to find compounds that specifically modulate those molecules, for instance via high throughput screening.

   a. Drug eruption
   b. Drug holiday
   c. Drug Identification Number
   d. Drug discovery

2. The _____ is the initial patient in the population of an epidemiological investigation. The _____ may indicate the source of the disease, the possible spread, and which reservoir holds the disease in between outbreaks. The _____ is the first patient that indicates the existence of an outbreak.

   a. Indicator bacteria
   b. Indices of deprivation 2004
   c. Indices of deprivation 2007
   d. Index case

3. _____ is a sexually transmitted infection caused by the spirochete bacterium Treponema pallidum subspecies pallidum. The primary route of transmission is through sexual contact; however, it may also be transmitted from mother to fetus during pregnancy or at birth, resulting in congenital _____. Other human diseases caused by related Treponema pallidum include yaws (subspecies pertenue), pinta (subspecies carateum) and bejel (subspecies endemicum).

   a. Syphilis
   b. Tetanus
   c. Vancomycin-resistant Staphylococcus aureus
   d. Vertebral osteomyelitis

4. _____ is the growth of tissues or cells separate from the organism. This is typically facilitated via use of a liquid, semi-solid, or solid growth medium, such as broth or agar. _____ commonly refers to the culture of animal cells and tissues, while the more specific term plant _____ is being named for the plants.

   a. Tissue culture
   b. Trichrome stain
   c. Tubule
   d. Variegation

5. . _____ is an American molecular biologist, geneticist, and zoologist, best known as one of the discoverers of the structure of DNA in 1953 with Francis Crick and Rosalind Franklin.

Chapter 1. Part 1: The Microbial Cell

Watson, Crick, and Maurice Wilkins were awarded the 1962 Nobel Prize in Physiology or Medicine 'for their discoveries concerning the molecular structure of nucleic acids and its significance for information transfer in living material'.

a. Brett Abrahams
b. Twelvefold way
c. Vexillary permutation
d. James D. Watson

ANSWER KEY
Chapter 1. Part 1: The Microbial Cell

1. d
2. d
3. a
4. a
5. d

You can take the complete Chapter Practice Test

for Chapter 1. Part 1: The Microbial Cell
on all key terms, persons, places, and concepts.

Online 99 Cents

http://www.epub13.5.20451.1.cram101.com/

Use www.Cram101.com for all your study needs

including Cram101's online interactive problem solving labs in

chemistry, statistics, mathematics, and more.

_____ | Caulobacter crescentus

_____ | Deinococcus radiodurans

_____ | Cell division

_____ | Cytoskeleton

_____ | DNA replication

_____ | Intermediate filament

_____ | Cell cycle

_____ | Metagenomics

_____ | Genome

_____ | Q fever

_____ | Sinorhizobium meliloti

_____ | Protein Z

_____ | Adenine

_____ | DNA Research

_____ | Genomic island

_____ | Guanine

_____ | Staphylococcus aureus

_____ | Horizontal transmission

_____ | Structural gene

Toxic shock syndrome

Agrobacterium tumefaciens

Ames test

Borrelia burgdorferi

Mitosis

Mycoplasma genitalium

Salmonella enterica

Chromosome

Confocal microscopy

Genome size

Transfer DNA

Molecular model

Noncoding DNA

Intron

Regulon

Reporter gene

Lac operon

Messenger RNA

Lactose permease

Chapter 2. Part 2: Genes and Genomes

CHAPTER OUTLINE: KEY TERMS, PEOPLE, PLACES, CONCEPTS

- Phosphodiester bond
- Recombinant DNA
- Chemical structure
- Purine
- Pyrimidine
- Southern blot
- Uracil
- Histone
- DNA gyrase
- RNA polymerase
- Topoisomerase
- Quinolone
- DNA polymerase
- Okazaki fragment
- Replication fork
- Semiconservative replication
- Deoxyribonucleic acid
- Primase
- Inhibitor protein

Methylation

Origin of replication

Replication timing

DNA-binding protein

DNA ligase

DNA polymerase I

Exonuclease

Replisome

Bacillus subtilis

Catenane

Antibiotic resistance

Drug resistance

Plasmid

Ti plasmid

Mycoplasma pneumoniae

Saccharomyces cerevisiae

Telomerase

DNA sequencing

Sequence analysis

Chapter 2. Part 2: Genes and Genomes

CHAPTER OUTLINE: KEY TERMS, PEOPLE, PLACES, CONCEPTS

- Restriction site
- Cloning
- Polymerase chain reaction
- Shuttle vector
- Taq polymerase
- Thermus aquaticus
- Chain reaction
- Dideoxynucleotide
- Gattaca
- Luciferase
- Pyrophosphate
- Pyrosequencing
- Contig
- DNA sequencer
- Human microbiome
- Microbiome
- Molecular clock
- Prochlorococcus
- Ribosomal RNA

_____ | Rifamycin

_____ | Charles Robert Darwin

_____ | Genetic code

_____ | Sigma factor

_____ | Consensus sequence

_____ | Sense strand

_____ | Heteroduplex

_____ | Neisseria meningitidis

_____ | Catabolite repression

_____ | Codon

_____ | Shine-Dalgarno sequence

_____ | Candida albicans

_____ | Inosine

_____ | Mycoplasma

_____ | Pseudouridine

_____ | Saccharomyces

_____ | Star formation

_____ | Start codon

_____ | Stop codon

Chapter 2. Part 2: Genes and Genomes

CHAPTER OUTLINE: KEY TERMS, PEOPLE, PLACES, CONCEPTS

- _______ Transfer RNA
- _______ Cistron
- _______ Northern blot
- _______ Ribosome
- _______ Ribosomal protein
- _______ T cell
- _______ Reading frame
- _______ Methanobrevibacter smithii
- _______ Elongation factor
- _______ Initiation factor
- _______ Aminoglycoside
- _______ Enterococcus faecalis
- _______ Release factor
- _______ Protein synthesis inhibitor
- _______ Brugia malayi
- _______ Fusidic acid
- _______ Isocitrate dehydrogenase
- _______ Macrolide
- _______ Molecular mimicry

Chapter 2. Part 2: Genes and Genomes

CHAPTER OUTLINE: KEY TERMS, PEOPLE, PLACES, CONCEPTS

- ______ Puromycin
- ______ Streptococcus pneumoniae
- ______ Biosynthesis
- ______ Protease
- ______ Glutamine synthetase
- ______ GroEL
- ______ Hemolysin
- ______ Secretion
- ______ Membrane protein
- ______ Cell membrane
- ______ Degron
- ______ Phase variation
- ______ Proteasome
- ______ Ubiquitin
- ______ Bioinformatics
- ______ Dihydrolipoamide dehydrogenase
- ______ Exon
- ______ Sequence alignment
- ______ ExPASy

CHAPTER OUTLINE: KEY TERMS, PEOPLE, PLACES, CONCEPTS

Functional genomics

Helicobacter pylori

Joint Genome Institute

KEGG

Multiple sequence alignment

Rickettsia prowazekii

Genomics

Fertility factor

Pseudomonas aeruginosa

Pyrococcus furiosus

Biofilm

Haemophilus

Colin Munro MacLeod

Electroporation

Horizontal gene transfer

Neisseria gonorrhoeae

Quorum sensing

Gene mapping

Bacillus thuringiensis

Permease

Rhizosphere

Nematode

Pathogenicity island

Prophage

Site-specific recombination

Holliday junction

RecA

Beta-galactosidase

Deletion

Point mutation

Silent mutation

Transversion

Frameshift mutation

Nonsense mutation

Missense mutation

Synonymous substitution

Mutagen

Mutation rate

Chapter 2. Part 2: Genes and Genomes

CHAPTER OUTLINE: KEY TERMS, PEOPLE, PLACES, CONCEPTS

______ Hydrogen peroxide

______ Hydroxyl radical

______ Essential nutrient

______ Pyrimidine dimers

______ Mutation frequency

______ Frequency

______ AP site

______ DNA repair

______ Excision repair

______ Base excision repair

______ SOS response

______ Transcription-coupled repair

______ AP endonuclease

______ Hypoxanthine

______ Cockayne syndrome

______ Lyme disease

______ Transposable element

______ Insertion sequence

______ Inverted repeat

Transposon

Replicative transposition

Transposase

Composite transposon

Genome evolution

Integron

Evolution

Human genome

Integrase

Mobile genetic elements

Salmonella

Bordetella pertussis

Bradyrhizobium

Gene duplication

Divergent evolution

Genetic variation

Myxococcus xanthus

Blood-brain barrier

Corepressor

	Vibrio fischeri
	Derepression
	Repressor
	TATA-binding protein
	Signal transduction
	ATP hydrolysis
	Energy carrier
	Active transport
	Chemical reaction
	Galactose
	Amino acid
	Diauxic growth
	Stringent response
	Eukaryote
	Anti-sigma factors
	Clostridium
	Antisense RNA
	Direct repeat
	Hin recombinase

Tandem repeat

Flagellin

Gene family

Chemotaxis

Trypanosoma brucei

Methyl-accepting chemotaxis protein

Nitrogen cycle

Nitrogen fixation

Autoinducer

Vibrio harveyi

Pathogen

Virulence factor

Enteromorpha

Vibrio anguillarum

Isoelectric focusing

Isoelectric point

Proteome

Transcriptome

Complementary DNA

Gel electrophoresis

Polyacrylamide gel

Proteomics

Metronidazole

RNA virus

Life cycle

Molecular biology

Chemokine receptor

Stem cell

Capsid

Lysogeny

Nonsense suppressor

Polio vaccine

Poliovirus

Post-polio syndrome

Gut-associated lymphoid tissue

Viremia

Palmaria palmata

RNA-dependent RNA polymerase

Chapter 2. Part 2: Genes and Genomes

CHAPTER OUTLINE: KEY TERMS, PEOPLE, PLACES, CONCEPTS

- Hemagglutinin
- Pandemic
- Haemophilus influenzae
- Influenza
- Neuraminidase
- Vaccine
- Sialic acid
- Zidovudine
- Oncogene
- Rous sarcoma virus
- Gene therapy
- Herpes simplex
- Protein A
- Protein structure
- Chemokine
- Langerhans cell
- Microglia
- Infectious disease
- Opportunistic infection

CHAPTER OUTLINE: KEY TERMS, PEOPLE, PLACES, CONCEPTS

Reverse transcriptase

Long terminal repeat

Protease inhibitor

Epstein-Barr virus

Herpes simplex virus

Human papillomavirus

Retrotransposon

Helicobacter

Molecular evolution

Viroid

Protein homology

Tegument

Bubonic plague

Exocytosis

Listeria monocytogenes

Yersinia pestis

Pseudotyping

Severe combined immunodeficiency

Adenoviruses

Chapter 2. Part 2: Genes and Genomes

Combined immunodeficiencies

Viral vector

Nerve growth factor

Transfection

Transgene

Fusion protein

Growth factor

Breast cancer

Glioma

Biotechnology

Genetic analysis

Phytophthora infestans

Western blot

Infectious dose

Electrophoretic mobility shift assay

Glutamate decarboxylase

Affinity chromatography

Primer extension

Deoxyribonuclease I

CHAPTER OUTLINE: KEY TERMS, PEOPLE, PLACES, CONCEPTS

_____ Immunoprecipitation

_____ Transcription factor

_____ Cell physiology

_____ Interaction network

_____ ATP synthase

_____ Gene gun

_____ Insecticide

_____ Parasporal body

_____ Systemic disease

_____ Cystic fibrosis

_____ Biopanning

_____ Hapten

_____ Phage display

_____ Directed evolution

_____ Bacillus cereus

Chapter 2. Part 2: Genes and Genomes

Caulobacter crescentus	Caulobacter crescentus is a Gram-negative, oligotrophic bacterium widely distributed in fresh water lakes and streams. It plays an important role in the carbon cycle. Caulobacter is an important model for studying the regulation of the cell cycle and cellular differentiation.
Deinococcus radiodurans	Deinococcus radiodurans is an extremophilic bacterium, one of the most radioresistant organisms known. It can survive cold, dehydration, vacuum, and acid, and is therefore known as a polyextremophile and has been listed as the world's toughest bacterium in The Guinness Book Of World Records. The name Deinococcus radiodurans derives from the Ancient Greek δειν?ς (deinos) and κ?κκος (kokkos) meaning 'terrible grain/berry' and the Latin radius and durare, meaning 'radiation surviving'.
Cell division	Cell division is the process by which a parent cell divides into two or more daughter cells. Cell division is usually a small segment of a larger cell cycle. This type of cell division in eukaryotes is known as mitosis, and leaves the daughter cell capable of dividing again.
Cytoskeleton	The cytoskeleton is a cellular 'scaffolding' or 'skeleton' contained within a cell's cytoplasm and is made out of protein. The cytoskeleton is present in all cells; it was once thought to be unique to eukaryotes, but recent research has identified the prokaryotic cytoskeleton. It has structures such as flagella, cilia and lamellipodia and plays important roles in both intracellular transport (the movement of vesicles and organelles, for example) and cellular division.
DNA replication	DNA replication is a biological process that occurs in all living organisms and copies their DNA; it is the basis for biological inheritance. The process starts when one double-stranded DNA molecule produces two identical copies of the molecule. The cell cycle (mitosis) also pertains to the DNA replication/reproduction process.
Intermediate filament	Intermediate filaments are a family of related proteins that share common structural and sequence features. Intermediate filaments have an average diameter of 10 nanometers, which is between that of actin (microfilaments) and microtubules, although they were initially designated 'intermediate' because their average diameter is between those of narrower microfilaments (actin) and wider myosin filaments. Most types of intermediate filaments are cytoplasmic, but one type, the lamins, are nuclear.
Cell cycle	The cell cycle, is the series of events that take place in a cell leading to its division and duplication (replication). In cells without a nucleus (prokaryotic), the cell cycle occurs via a process termed binary fission.

Metagenomics	Metagenomics is the study of metagenomes, genetic material recovered directly from environmental samples. The broad field may also be referred to as environmental genomics, ecogenomics or community genomics. While traditional microbiology and microbial genome sequencing and genomics rely upon cultivated clonal cultures, early environmental gene sequencing cloned specific genes (often the 16S rRNA gene) to produce a profile of diversity in a natural sample.
Genome	In modern molecular biology and genetics, the genome is the entirety of an organism's hereditary information. It is encoded either in DNA or, for many types of virus, in RNA. The genome includes both the genes and the non-coding sequences of the DNA/RNA. The term was adapted in 1920 by Hans Winkler, Professor of Botany at the University of Hamburg, Germany. The Oxford English Dictionary suggests the name to be a blend of the words gene and chromosome.
Q fever	Q fever is a disease caused by infection with Coxiella burnetii, a bacterium that affects humans and other animals. This organism is uncommon, but may be found in cattle, sheep, goats and other domestic mammals, including cats and dogs. The infection results from inhalation of a spore-like small cell variant, and from contact with the milk, urine, feces, vaginal mucus, or semen of infected animals.
Sinorhizobium meliloti	Sinorhizobium meliloti is a Gram-negative nitrogen-fixing bacterium (rhizobium). It forms a symbiotic relationship with legumes from the genera Medicago, Melilotus and Trigonella, including the model legume Medicago truncatula. This symbiosis results in a new plant organ termed a root nodule.
Protein Z	Protein Z is encoded by the PROZ gene. Protein Z is a member of the coagulation cascade, the group of blood proteins that leads to the formation of blood clots. It is vitamin K-dependent, and its functionality is therefore impaired in warfarin therapy.
Adenine	Adenine is a nucleobase (a purine derivative) with a variety of roles in biochemistry including cellular respiration, in the form of both the energy-rich adenosine triphosphate (ATP) and the cofactors nicotinamide adenine dinucleotide (NAD) and flavin adenine dinucleotide (FAD), and protein synthesis, as a chemical component of DNA and RNA. The shape of adenine is complementary to either thymine in DNA or uracil in RNA. Adenine forms several tautomers, compounds that can be rapidly interconverted and are often considered equivalent. However, in isolated conditions, i.e. in an inert gas matrix and in the gas phase, mainly the 9H-adenine tautomer is found. Biosynthesis

Chapter 2. Part 2: Genes and Genomes

DNA Research	DNA Research is an international, peer reviewed journal of genomics and DNA research.
Genomic island	A Genomic island is part of a genome that has evidence of horizontal origins. The term is usually used in microbiology, especially with regard to bacteria. A GI can code for many functions, can be involved in symbiosis or pathogenesis, and may help an organism's adaptation.
Guanine	Guanine is one of the four main nucleobases found in the nucleic acids DNA and RNA, the others being adenine, cytosine, and thymine (uracil in RNA). In DNA, guanine is paired with cytosine. With the formula $C_5H_5N_5O$, guanine is a derivative of purine, consisting of a fused pyrimidine-imidazole ring system with conjugated double bonds.
Staphylococcus aureus	Staphylococcus aureus is a bacterial species named from Greek σταφυλ?κοκκος meaning the 'golden grape-cluster berry'. Also known as 'golden staph' and Oro staphira, it is a facultative anaerobic Gram-positive coccal bacterium. It is frequently found as part of the normal skin flora on the skin and nasal passages.
Horizontal transmission	Horizontal disease transmission is the transmission of an infectious agent, such as bacterial, fungal, or viral infection, between members of the same species that are not in a parent-child relationship. Horizontal transmission tends to evolve virulence. It is therefore a critical concept for evolutionary medicine.
Structural gene	A structural gene is a gene that codes for any RNA or protein product other than a regulatory factor (i.e. regulatory protein). It may code for a structural protein, an enzyme, or an RNA molecule not involved in regulation. Structural genes represent an enormous variety of protein structures and functions, including structural proteins, enzymes with catalytic activities and so on.
Toxic shock syndrome	Toxic shock syndrome is a potentially fatal illness caused by a bacterial toxin. Different bacterial toxins may cause toxic shock syndrome, depending on the situation. The causative bacteria include Staphylococcus aureus and Streptococcus pyogenes.
Agrobacterium tumefaciens	Agrobacterium tumefaciens is the causal agent of crown gall disease (the formation of tumours) in over 140 species of dicot. It is a rod shaped, Gram negative soil bacterium (Smith et al., 1907). Symptoms are caused by the insertion of a small segment of DNA (known as the T-DNA, for 'transfer DNA'), from a plasmid, into the plant cell, which is incorporated at a semi-random location into the plant genome.
Ames test	The Ames test is a biological assay to assess the mutagenic potential of chemical compounds.

A positive test indicates that the chemical is mutagenic and therefore may act as a carcinogen, since cancer is often linked to mutation. However, a number of false-positives and false-negatives are known.

Borrelia burgdorferi	Borrelia burgdorferi is a species of Gram negative bacteria of the spirochete class of the genus Borrelia. B. burgdorferi is predominant in North America, but also exists in Europe, and is the agent of Lyme disease. It is a zoonotic, vector-borne disease transmitted by ticks and is named after the researcher Willy Burgdorfer who first isolated the bacterium in 1982. B. burgdorferi is one of the few pathogenic bacteria that can survive without iron, having replaced all of its iron-sulfur cluster enzymes with enzymes that use manganese, thus avoiding the problem many pathogenic bacteria face in acquiring iron.
Mitosis	Mitosis is the process by which a eukaryotic cell separates the chromosomes in its cell nucleus into two identical sets, in two separate nuclei. It is generally followed immediately by cytokinesis, which divides the nuclei, cytoplasm, organelles and cell membrane into two cells containing roughly equal shares of these cellular components. Mitosis and cytokinesis together define the mitotic (M) phase of the cell cycle--the division of the mother cell into two daughter cells, genetically identical to each other and to their parent cell.
Mycoplasma genitalium	Mycoplasma genitalium is a small parasitic bacterium that lives on the ciliated epithelial cells of the primate genital and respiratory tracts. M. genitalium is the smallest known genome that can constitute a cell, and the second-smallest bacterium after the endosymbiont Carsonella ruddii. Until the discovery of Nanoarchaeum in 2002, M. genitalium was also considered to be the organism with the smallest genome.
Salmonella enterica	Salmonella enterica is a rod-shaped flagellated, facultative anaerobic, Gram-negative bacterium, and a member of the genus Salmonella. Most cases of salmonellosis are caused by food infected with S. enterica, which often infects cattle and poultry, though also other animals such as domestic cats and hamsters have also been shown to be sources of infection to humans. However, investigations of vacuum cleaner bags have shown that households can act as a reservoir of the bacterium; this is more likely if the household has contact with an infection source, for example members working with cattle or in a veterinary clinic.
Chromosome	In genetic algorithms, a chromosome (also sometimes called a genome) is a set of parameters which define a proposed solution to the problem that the genetic algorithm is trying to solve.

Chapter 2. Part 2: Genes and Genomes

Confocal microscopy	Confocal microscopy is an optical imaging technique used to increase optical resolution and contrast of a micrograph by using point illumination and a spatial pinhole to eliminate out-of-focus light in specimens that are thicker than the focal plane. It enables the reconstruction of three-dimensional structures from the obtained images. This technique has gained popularity in the scientific and industrial communities and typical applications are in life sciences, semiconductor inspection and materials science.
Genome size	Genome size is the total amount of DNA contained within one copy of a single genome. It is typically measured in terms of mass in picograms (trillionths (10^{-12}) of a gram, abbreviated pg) or less frequently in Daltons or as the total number of nucleotide base pairs typically in megabases . One picogram equals 978 megabases.
Transfer DNA	The transfer DNA is the transferred DNA of the tumor-inducing (Ti) plasmid of some species of bacteria such as Agrobacterium tumefaciens and Agrobacterium rhizogenes. It derives its name from the fact that the bacterium transfers this DNA fragment into the host plant's nuclear DNA genome. The T-DNA is bordered by 25-base-pair repeats on each end.
Molecular model	A molecular model, in this article, is a physical model that represents molecules and their processes. The creation of mathematical models of molecular properties and behaviour is molecular modelling, and their graphical depiction is molecular graphics, but these topics are closely linked and each uses techniques from the others. In this article, 'molecular model' will primarily refer to systems containing more than one atom and where nuclear structure is neglected.
Noncoding DNA	In genetics, noncoding DNA describes components of an organism's DNA sequences that do not encode for protein sequences. In many eukaryotes, a large percentage of an organism's total genome size is noncoding DNA, although the amount of noncoding DNA, and the proportion of coding versus noncoding DNA varies greatly between species. Much of this DNA has no known biological function and is sometimes referred to as 'junk DNA'.
Intron	An intron is any nucleotide sequence within a gene that is removed by RNA splicing while the final mature RNA product of a gene is being generated. The term intron refers to both the DNA sequence within a gene, and the corresponding sequence in RNA transcripts. Sequences that are joined together in the final mature RNA after RNA splicing are exons.
Regulon	In cell biology and genetics, a regulon is a collection of genes or operons under regulation by the same regulatory protein. This term is generally used for prokaryotic systems, for example quorum sensing in bacteria. It is a group of operons/genes spread around the chromosome but controlled by a common factor or stimulus.

Reporter gene	In molecular biology, a reporter gene is a gene that researchers attach to a regulatory sequence of another gene of interest in bacteria, cell culture, animals or plants. Certain genes are chosen as reporters because the characteristics they confer on oanisms expressing them are easily identified and measured, or because they are selectable markers. Reporter genes are often used as an indication of whether a certain gene has been taken up by or expressed in the cell or oanism population.
Lac operon	The lac operon is an operon required for the transport and metabolism of lactose in Escherichia coli and some other enteric bacteria. It consists of three adjacent structural genes, lacZ, lacY and lacA. The lac operon is regulated by several factors including the availability of glucose and of lactose. Gene regulation of the lac operon was the first complex genetic regulatory mechanism to be elucidated and is one of the foremost examples of prokaryotic gene regulation.
Messenger RNA	Messenger RNA is a molecule of RNA that encodes a chemical 'blueprint' for a protein product. mRNA is transcribed from a DNA template, and carries coding information to the sites of protein synthesis, the ribosomes. In the ribosomes, the mRNA is translated into a polymer of amino acids: a protein.
Lactose permease	LacY proton/sugar symporter Lactose permease is a membrane protein which is a member of the major facilitator superfamily. Lactose permease can be classified as a symporter, which uses the gradient of H+ towards the cell to transport lactose in the same direction into the cell. The protein has twelve transmembrane helices and exhibits an internal two-fold symmetry, relating the N-terminal six helices onto the C-terminal helices.
Phosphodiester bond	A phosphodiester bond is a group of strong covalent bonds between a phosphate group and two 5-carbon ring carbohydrates (pentoses) over two ester bonds. Phosphodiester bonds are central to all known life, as they make up the backbone of each helical strand of DNA. In DNA and RNA, the phosphodiester bond is the linkage between the 3' carbon atom of one sugar molecule and the 5' carbon atom of another; the sugar molecules being deoxyribose in DNA and ribose in RNA. The phosphate groups in the phosphodiester bond are negatively-charged. Because the phosphate groups have a pK_a near 0, they are negatively-charged at pH 7. This repulsion forces the phosphates to take opposite sides of the DNA strands and is neutralized by proteins (histones), metal ions such as magnesium, and polyamines.

Chapter 2. Part 2: Genes and Genomes

Recombinant DNA	Recombinant DNA molecules are DNA sequences that result from the use of laboratory methods (molecular cloning) to bring together genetic material from multiple sources, creating sequences that would not otherwise be found in biological organisms. Recombinant DNA is possible because DNA molecules from all organisms share the same chemical structure; they differ only in the sequence of nucleotides within that identical overall structure. Consequently, when DNA from a foreign source is linked to host sequences that can drive DNA replication and then introduced into a host organism, the foreign DNA is replicated along with the host DNA.
	Recombinant DNA molecules are sometimes called chimeric DNA, because they are usually made of material from two different species, like the mythical chimera.
Chemical structure	A chemical structure includes molecular geometry, electronic structure and crystal structure of molecules. Molecular geometry refers to the spatial arrangement of atoms in a molecule and the chemical bonds that hold the atoms together. Molecular geometry can range from the very simple, such as diatomic oxygen or nitrogen molecules, to the very complex, such as protein or DNA molecules.
Purine	A purine is a heterocyclic aromatic organic compound, consisting of a pyrimidine ring fused to an imidazole ring. Purines, including substituted purines and their tautomers, are the most widely distributed kind of nitrogen-containing heterocycle in nature.
	Purines and pyrimidines make up the two groups of nitrogenous bases, including the two groups of nucleotide bases.
Pyrimidine	Pyrimidine is a heterocyclic aromatic organic compound similar to benzene and pyridine, containing two nitrogen atoms at positions 1 and 3 of the six-member ring. It is isomeric with two other forms of diazine: Pyridazine, with the nitrogen atoms in positions 1 and 2; and Pyrazine, with the nitrogen atoms in positions 1 and 4.
	A pyrimidine has many properties in common with pyridine, as the number of nitrogen atoms in the ring increases the ring pi electrons become less energetic and electrophilic aromatic substitution gets more difficult while nucleophilic aromatic substitution gets easier.
Southern blot	A Southern blot is a method routinely used in molecular biology for detection of a specific DNA sequence in DNA samples. Southern blotting combines transfer of electrophoresis-separated DNA fragments to a filter membrane and subsequent fragment detection by probe hybridization. ts inventor, the British biologist Edwin Southern.
Uracil	Uracil is one of the four nucleobases in the nucleic acid of RNA that are represented by the letters A, C, G, and U. The others are adenine, cytosine, and guanine. In RNA, uracil binds to adenine (A) via two hydrogen bonds.

Histone	In biology, histones are highly alkaline proteins found in eukaryotic cell nuclei that package and order the DNA into structural units called nucleosomes. They are the chief protein components of chromatin, acting as spools around which DNA winds, and play a role in gene regulation. Without histones, the unwound DNA in chromosomes would be very long (a length to width ratio of more than 10 million to one in human DNA).
DNA gyrase	DNA gyrase, often referred to simply as gyrase, is an enzyme that relieves strain while double-stranded DNA is being unwound by helicase. This causes negative supercoiling of the DNA. Bacterial DNA gyrase is the target of many antibiotics, including nalidixic acid, novobiocin, and ciprofloxacin. DNA gyrase is a type II topoisomerase (EC 5.99.1.3) that introduces negative supercoils into DNA by looping the template so as to form a crossing, then cutting one of the double helices and passing the other through it before releasing the break, changing the linking number by two in each enzymatic step.
RNA polymerase	RNA polymerase also known as DNA-dependent RNA polymerase, is an enzyme that produces RNA. In cells, RNAP is necessary for constructing RNA chains using DNA genes as templates, a process called transcription. RNA polymerase enzymes are essential to life and are found in all organisms and many viruses. In chemical terms, RNAP is a nucleotidyl transferase that polymerizes ribonucleotides at the 3' end of an RNA transcript.
Topoisomerase	Topoisomerases (type I: EC 5.99.1.2, type II: EC 5.99.1.3) are enzymes that regulate the overwinding or underwinding of DNA. The winding problem of DNA arises due to the intertwined nature of its double helical structure. For example, during DNA replication, DNA becomes overwound ahead of a replication fork. If left unabated, this tension would eventually grind replication to a halt (a similar event happens during transcription).
Quinolone	The quinolones are a family of synthetic broad-spectrum antibiotics. The term quinolone(s) refers to potent synthetic chemotherapeutic antibacterials. The first generation of the quinolones begins with the introduction of nalidixic acid in 1962 for treatment of urinary tract infections in humans.
DNA polymerase	A DNA polymerase is an enzyme (the suffix -ase is used to identify enzymes) that helps catalyze the polymerization of deoxyribonucleotides into a DNA strand. DNA polymerases are best known for their feedback role in DNA replication, in which the polymerase 'reads' an intact DNA strand as a template and uses it to synthesize the new strand. This process copies a piece of DNA. The newly polymerized molecule is complementary to the template strand and identical to the template's original partner strand.

Chapter 2. Part 2: Genes and Genomes

Okazaki fragment	An Okazaki fragment is a relatively short fragment of DNA (with no RNA primer at the 5' terminus) created on the lagging strand during DNA replication. The lengths of Okazaki fragments are between 1,000 to 2,000 nucleotides long in E. coli and are generally between 100 to 200 nucleotides long in eukaryotes. It was originally discovered in 1968 by Reiji Okazaki, Tsuneko Okazaki, and their colleagues while studying replication of bacteriophage DNA in Escherichia coli.
Replication fork	The replication fork is a structure that forms within the nucleus during DNA replication. It is created by helicases, which break the hydrogen bonds holding the two DNA strands together. The resulting structure has two branching 'prongs', each one made up of a single strand of DNA. These two strands serve as the template for the leading and lagging strands, which will be created as DNA polymerase matches complementary nucleotides to the templates.
Semiconservative replication	Semiconservative replication describes the mechanism by which DNA is replicated in all known cells. This mechanism of replication was one of three models originally proposed for DNA replication:•Semiconservative replication would produce two copies that each contained one of the original strands and one new strand.•Conservative replication would leave the two original template DNA strands together in a double helix and would produce a copy composed of two new strands containing all of the new DNA base pairs.•Dispersive replication would produce two copies of the DNA, both containing distinct regions of DNA composed of either both original strands or both new strands.

The deciphering of the structure of DNA by Watson and Crick in 1953 suggested that each strand of the double helix would serve as a template for synthesis of a new strand. However, there was no way of knowing how the newly synthesized strands might combine with the template strands to form two double helical DNA molecules. |
| Deoxyribonucleic acid | Deoxyribonucleic acid is a nucleic acid that contains the genetic instructions used in the development and functioning of all known living organisms (with the exception of RNA viruses). The main role of DNA molecules is the long-term storage of information. DNA is often compared to a set of blueprints, like a recipe or a code, since it contains the instructions needed to construct other components of cells, such as proteins and RNA molecules. |
| Primase | DNA primase is an enzyme involved in the replication of DNA.

Primase catalyzes the synthesis of a short RNA (or DNA in some organisms) segment called a primer complementary to a ssDNA template. Primase is of key importance in DNA replication because no known DNA polymerases can initiate the synthesis of a DNA strand without an initial RNA or DNA primer (for temporary DNA elongation). Function |

Inhibitor protein	The inhibitor protein is situated in the mitochondrial matrix and protects the cell against rapid ATP hydrolysis during momentary ischaemia. In oxygen absence, the pH of the matrix drops. This causes IP to become protonated and change its conformation to one that can bind to the F1Fo synthetase and stops it thereby preventing it from moving in a backwards direction and hydrolyze ATP instead of make it.
Methylation	In the chemical sciences, methylation denotes the addition of a methyl group to a substrate or the substitution of an atom or group by a methyl group. Methylation is a form of alkylation with, to be specific, a methyl group, rather than a larger carbon chain, replacing a hydrogen atom. These terms are commonly used in chemistry, biochemistry, soil science, and the biological sciences.
Origin of replication	The origin of replication is a particular sequence in a genome at which replication is initiated. This can either involve the replication of DNA in living organisms such as prokaryotes and eukaryotes, or that of DNA or RNA in viruses, such as double-stranded RNA viruses. DNA replication may proceed from this point bidirectionally or unidirectionally.
Replication timing	Replication Timing refers to the order in which segments of DNA along the length of a chromosome are duplicated. In eukaryotic cells (cells that package their DNA within a nucleus), chromosomes consist of very long linear double-stranded DNA molecules. During the S-phase of each cell cycle, all of the DNA in a cell is duplicated in order to provide one copy to each of the daughter cells after the next cell division.
DNA-binding protein	DNA-binding proteins are proteins that are composed of DNA-binding domains and thus have a specific or general affinity for either single or double stranded DNA. Sequence-specific DNA-binding proteins generally interact with the major groove of B-DNA, because it exposes more functional groups that identify a base pair. However there are some known minor groove DNA-binding ligands such as Netropsin, Distamycin, Hoechst 33258, Pentamidine and others. DNA-binding proteins include transcription factors which modulate the process of transcription, various polymerases, nucleases which cleave DNA molecules, and histones which are involved in chromosome packaging and transcription in the cell nucleus.
DNA ligase	In molecular biology, DNA ligase is a specific type of enzyme, a ligase, (EC 6.5.1.1) that facilitates the joining of DNA strands together by catalyzing the formation of a phosphodiester bond. It plays a role in repairing single-strand breaks in duplex DNA in living organisms, but some forms (such as DNA ligase IV) may specifically repair double-strand breaks (i.e. a break in both complementary strands of DNA).

	Single-strand breaks are repaired by DNA ligase using the complementary strand of the double helix as a template with DNA ligase creating the final phosphodiester bond to fully repair the DNA. DNA ligase has applications in both DNA repair and DNA replication .
DNA polymerase I	DNA Polymerase I is an enzyme that participates in the process of DNA replication and is exclusively found in prokaryotes. It is composed of 928 amino acids, and is an example of a processive enzyme - it can sequentially catalyze multiple polymerisations. Discovered by Arthur Kornberg in 1956, it was the first known DNA polymerase (and, indeed, the first known of any kind of polymerase).
Exonuclease	Exonucleases are enzymes that work by cleaving nucleotides one at a time from the end (exo) of a polynucleotide chain. A hydrolyzing reaction that breaks phosphodiester bonds at either the 3' or the 5' end occurs. Its close relative is the endonuclease, which cleaves phosphodiester bonds in the middle (endo) of a polynucleotide chain.
Replisome	The replisome is a complex molecular machine that carries out replication of DNA. The replisome first unwinds double stranded DNA into two single strands. For each of the resulting single strands, a new complementary sequence of DNA is synthesized. The net result is formation of two new double stranded DNA sequences that are exact copies of the original double stranded DNA sequence.
Bacillus subtilis	Bacillus subtilis, known also as the hay bacillus or grass bacillus, is a Gram-positive, catalase-positive bacterium commonly found in soil. A member of the genus Bacillus, B. subtilis is rod-shaped, and has the ability to form a tough, protective endospore, allowing the organism to tolerate extreme environmental conditions. Unlike several other well-known species, B. subtilis has historically been classified as an obligate aerobe, though recent research has demonstrated that this is not strictly correct.
Catenane	A catenane is a mechanically-interlocked molecular architecture consisting of two or more interlocked macrocycles. The interlocked rings cannot be separated without breaking the covalent bonds of the macrocycles. Catenane is derived from the Latin catena meaning 'chain'.
Antibiotic resistance	Antibiotic resistance is a type of drug resistance where a microorganism is able to survive exposure to an antibiotic. While a spontaneous or induced genetic mutation in bacteria may confer resistance to antimicrobial drugs, genes that confer resistance can be transferred between bacteria in a horizontal fashion by conjugation, transduction, or transformation. Thus, a gene for antibiotic resistance that evolves via natural selection may be shared.

Drug resistance	Drug resistance is the reduction in effectiveness of a drug such as an antimicrobial or an antineoplastic in curing a disease or condition. When the drug is not intended to kill or inhibit a pathogen, then the term is equivalent to dosage failure or drug tolerance. More commonly, the term is used in the context of resistance acquired by pathogens.
Plasmid	In microbiology and genetics, a plasmid is a DNA molecule that is separate from, and can replicate independently of, the chromosomal DNA. They are double-stranded and, in many cases, circular. Plasmids usually occur naturally in bacteria, but are sometimes found in eukaryotic organisms (e.g., the 2-micrometre ring in Saccharomyces cerevisiae). Plasmid sizes vary from 1 to over 1,000 kbp.
Ti plasmid	Ti plasmid is a circular plasmid that often, but not always, is a part of the genetic equipment that Agrobacterium tumefaciens and Agrobacterium rhizogenes use to transduce its genetic material to plants. Ti stands for tumor inducing. The Ti plasmid is lost when Agrobacterium is grown above 28° C. Such cured bacteria do not induce crown galls, i.e. they become avirulent.
Mycoplasma pneumoniae	Mycoplasma pneumoniae is a very small bacterium in the class Mollicutes. It causes the disease mycoplasma pneumonia, a form of atypical bacterial pneumonia, and is related to cold agglutinin disease.
Saccharomyces cerevisiae	Saccharomyces cerevisiae is a species of yeast. It is perhaps the most useful yeast, having been instrumental to baking and brewing since ancient times. It is believed that it was originally isolated from the skin of grapes .
Telomerase	Telomerase is a ribonucleoprotein that is an enzyme which adds DNA sequence repeats ('TTAGGG' in all vertebrates) to the 3' end of DNA strands in the telomere regions, which are found at the ends of eukaryotic chromosomes. This region of repeated nucleotide called telomeres contains noncoding DNA and hinders the loss of important DNA from chromosome ends. As a result, every time the chromosome is copied only 100-200 nucleotides are lost, which causes no damage to the organism's DNA. Telomerase is a reverse transcriptase that carries its own RNA molecule, which is used as a template when it elongates telomeres, which are shortened after each replication cycle.
DNA sequencing	DNA sequencing includes several methods and technologies that are used for determining the order of the nucleotide bases--adenine, guanine, cytosine, and thymine--in a molecule of DNA. Knowledge of DNA sequences has become indispensable for basic biological research, other research branches utilizing DNA sequencing, and in numerous applied fields such as diagnostic, biotechnology, forensic biology and biological systematics. The advent of DNA sequencing has significantly accelerated biological research and discovery.

Chapter 2. Part 2: Genes and Genomes

Sequence analysis	In bioinformatics, the term sequence analysis refers to the process of subjecting a DNA, RNA or peptide sequence to any of a wide range of analytical methods to understand its features, function, structure, or evolution. Methodologies used include sequence alignment, searches against biological databases, and others. Since the development of methods of high-throughput production of gene and protein sequences, the rate of addition of new sequences to the databases increased exponentially.
Restriction site	Restriction sites, or restriction recognition sites, are locations on a DNA molecule containing specific sequences of nucleotides, which are recognized by restriction enzymes. These are generally palindromic sequences (because restriction enzymes usually bind as homodimers), and a particular restriction enzyme may cut the sequence between two nucleotides within its recognition site, or somewhere nearby. For example, the common restriction enzyme EcoRI recognizes the palindromic sequence GAATTC and cuts between the G and the A on both the top and bottom strands, leaving an overhang (an end-portion of a DNA strand with no attached complement) on each end, of AATT. This overhang can then be used to ligate in a piece of DNA with a complementary overhang (another EcoRI-cut piece, for example).
Cloning	Cloning in biology is the process of producing similar populations of genetically identical individuals that occurs in nature when organisms such as bacteria, insects or plants reproduce asexually. Cloning in biotechnology refers to processes used to create copies of DNA fragments (molecular cloning), cells (cell cloning), or organisms. The term also refers to the production of multiple copies of a product such as digital media or software.
Polymerase chain reaction	The polymerase chain reaction is a scientific technique in molecular biology to amplify a single or a few copies of a piece of DNA across several orders of magnitude, generating thousands to millions of copies of a particular DNA sequence.
	Developed in 1983 by Kary Mullis, PCR is now a common and often indispensable technique used in medical and biological research labs for a variety of applications. These include DNA cloning for sequencing, DNA-based phylogeny, or functional analysis of genes; the diagnosis of hereditary diseases; the identification of genetic fingerprints (used in forensic sciences and paternity testing); and the detection and diagnosis of infectious diseases.
Shuttle vector	A shuttle vector is a vector (usually a plasmid) constructed so that it can propagate in two different host species . Therefore, DNA inserted into a shuttle vector can be tested or manipulated in two different cell types. The main advantage of these vectors is they can be manipulated in E. coli then used in a system which is more difficult or slower to use (e.g. yeast, other bacteria).
Taq polymerase	Taq polymerase, exonuclease

Taq polymerase is a thermostable DNA polymerase named after the thermophilic bacterium Thermus aquaticus from which it was originally isolated by Thomas D. Brock in 1965. It is often abbreviated to 'Taq Pol' (or simply 'Taq'), and is frequently used in polymerase chain reaction (PCR), a method for greatly amplifying short segments of DNA.

T. aquaticus is a bacterium that lives in hot springs and hydrothermal vents, and Taq polymerase was identified as an enzyme able to withstand the protein-denaturing conditions (high temperature) required during PCR. Therefore it replaced the DNA polymerase from E. coli originally used in PCR. Taq's optimum temperature for activity is 75-80°C, with a half-life of greater than 2 hours at 92.5°C, 40 minutes at 95°C and 9 minutes at 97.5°C, and can replicate a 1000 base pair strand of DNA in less than 10 seconds at 72°C.

One of Taq's drawbacks is its relatively low replication fidelity. It lacks a 3' to 5' exonuclease proofreading activity, and has an error rate measured at about 1 in 9,000 nucleotides. The remaining two domains however may act in coordination, via coupled domain motion.

Thermus aquaticus	Thermus aquaticus is a species of bacterium that can tolerate high temperatures, one of several thermophilic bacteria that belong to the Deinococcus-Thermus group. It is the source of the heat-resistant enzyme Taq DNA polymerase, one of the most important enzymes in molecular biology because of its use in the polymerase chain reaction (PCR) DNA amplification technique.

When studies of biological organisms in hot springs began in the 1960s, scientists thought that the life of thermophilic bacteria could not be sustained in temperatures above about 55° Celsius (131° Fahrenheit).

Chain reaction	A chain reaction is a sequence of reactions where a reactive product or by-product causes additional reactions to take place. In a chain reaction, positive feedback leads to a self-amplifying chain of events.

Chain reactions are one way in which systems which are in thermodynamic non-equilibrium can release energy or increase entropy in order to reach a state of higher entropy.

Dideoxynucleotide	Dideoxynucleotides, or ddNTPs, are nucleotides lacking a 3'-hydroxyl (-OH) group on their deoxyribose sugar. Since deoxyribose already lacks a 2'-OH, dideoxyribose lacks hydroxyl groups at both its 2' and 3' carbons.

Chapter 2. Part 2: Genes and Genomes

Gattaca	Gattaca is a 1997 American science fiction film written and directed by Andrew Niccol. It stars Ethan Hawke, Uma Thurman and Jude Law with supporting roles played by Loren Dean, Ernest Borgnine, Gore Vidal and Alan Arkin. The film was a 1997 nominee for the Academy Award for Best Art Direction -- Set Decoration.
Luciferase	Luciferase is a generic term for the class of oxidative enzymes used in bioluminescence and is distinct from a photoprotein. One famous example is the firefly luciferase from the firefly Photinus pyralis. 'Firefly luciferase' as a laboratory reagent usually refers to P. pyralis luciferase although recombinant luciferases from several other species of fireflies are also commercially available.
Pyrophosphate	In chemistry, the anion, the salts, and the esters of pyrophosphoric acid are called pyrophosphates. Any salt or ester containing two phosphate groups is called a diphosphate. As a food additive, diphosphates are known as E450.
Pyrosequencing	Pyrosequencing is a method of DNA sequencing (determining the order of nucleotides in DNA) based on the 'sequencing by synthesis' principle. It differs from Sanger sequencing, in that it relies on the detection of pyrophosphate release on nucleotide incorporation, rather than chain termination with dideoxynucleotides. The technique was developed by Pål Nyrén and Mostafa Ronaghi at the Royal Institute of Technology in Stockholm in 1996. The desired DNA sequence is able to be determined by light emitted upon incorporation of the next complementary nucleotide by the fact that only one out of four of the possible A/T/C/G nucleotides are added and available at a time so that only one letter can be incorporated on the single stranded template (which is the sequence to be determined).
Contig	A contig is a set of overlapping DNA segments that together represent a consensus region of DNA. In bottom-up sequencing projects, a contig refers to overlapping sequence data (reads); in top-down sequencing projects, contig refers to the overlapping clones that form a physical map of the genome that is used to guide sequencing and assembly. Contigs can thus refer both to overlapping DNA sequence and to overlapping physical segments (fragments) contained in clones depending on the context. A sequence contig is a contiguous, overlapping sequence read resulting from the reassembly of the small DNA fragments generated by bottom-up sequencing strategies.
DNA sequencer	A DNA sequencer is a scientific instrument used to automate the DNA sequencing process. It can be also considered an optical instrument as it generally analyzes light signals originating from fluorochromes attached to nucleotides. Modern automated DNA sequencing instruments (called DNA sequencers) are able to sequence multiple samples in a batch (run) and perform as many as 24 runs a day.

Human microbiome	The human microbiome is the aggregate of microorganisms that reside on the surface and in deep layers of skin, in the saliva and oral mucosa, in the conjunctiva, and in the gastrointestinal tracts. They include bacteria, fungi, and archaea. Some of these organisms perform tasks that are useful for the human host.
Microbiome	A microbiome is the totality of microbes, their genetic elements (genomes), and environmental interactions in a particular environment. The term 'microbiome' was coined by Joshua Lederberg, who argued that microorganisms inhabiting the human body should be included as part of the human genome, because of their influence on human physiology. The human body contains over 10 times more microbial cells than human cells.
Molecular clock	The molecular clock (based on the molecular clock hypothesis (MCH)) is a technique in molecular evolution that uses fossil constraints and rates of molecular change to deduce the time in geologic history when two species or other taxa diverged. It is used to estimate the time of occurrence of events called speciation or radiation. The molecular data used for such calculations is usually nucleotide sequences for DNA or amino acid sequences for proteins.
Prochlorococcus	Prochlorococcus is a genus of very small (0.6 μm) marine cyanobacteria with an unusual pigmentation (chlorophyll b). These bacteria belong to the photosynthetic picoplankton and are probably the most abundant photosynthetic organism on Earth. Microbes of the genus Prochlorococcus are among the major primary producers in the ocean, responsible for at least 20% of atmospheric oxygen.
Ribosomal RNA	Ribosomal RNA is the RNA component of the ribosome, the enzyme that is the site of protein synthesis in all living cells. Ribosomal RNA provides a mechanism for decoding mRNA into amino acids and interacts with tRNAs during translation by providing peptidyl transferase activity. The tRNAs bring the necessary amino acids corresponding to the appropriate mRNA codon.
Rifamycin	The rifamycins are a group of antibiotics that are synthesized either naturally by the bacterium Amycolatopsis mediterranei or artificially. They are a subclass of the larger family Ansamycin. Rifamycins are particularly effective against mycobacteria, and are therefore used to treat tuberculosis, leprosy, and mycobacterium avium complex (MAC) infections.
Charles Robert Darwin	Charles Robert Darwin FRS (12 February 1809 - 19 April 1882) was an English naturalist. He established that all species of life have descended over time from common ancestry, and proposed the scientific theory that this branching pattern of evolution resulted from a process that he called natural selection. He published his theory with compelling evidence for evolution in his 1859 book On the Origin of Species.

Chapter 2. Part 2: Genes and Genomes

Genetic code	The genetic code is the set of rules by which information encoded in genetic material (DNA or mRNA sequences) is translated into proteins (amino acid sequences) by living cells. The code defines how sequences of three nucleotides, called codons, specify which amino acid will be added next during protein synthesis. With some exceptions, a three-nucleotide codon in a nucleic acid sequence specifies a single amino acid.
Sigma factor	A sigma factor is a protein needed only for initiation of RNA synthesis. It is a bacterial transcription initiation factor that enables specific binding of RNA polymerase to gene promoters. The specific sigma factor used to initiate transcription of a given gene will vary, depending on the gene and on the environmental signals needed to initiate transcription of that gene.
Consensus sequence	In molecular biology and bioinformatics, the consensus sequence is the calculated order of most frequent residues, either nucleotide or amino acid, found at each position in a sequence alignment. It represents the results of a multiple sequence alignment in which related sequences are compared to each other and similar sequence motifs are calculated. Such information is important when considering sequence dependent enzymes such as RNA polymerase.
Sense strand	In genetics, a sense strand is complementary to the antisense strand or template strand. The sense strand is the strand of DNA that has the same sequence as the mRNA, which takes the antisense strand as its template during transcription, and eventually undergoes (typically, not always) translation into a protein. The immediate product of this transcription is a resultant initial RNA transcript, which contains sequencing of nucleotides that are identical to the sense strand.
Heteroduplex	A heteroduplex is a double-stranded (duplex) molecule of nucleic acid originated through the genetic recombination of single complementary strands derived from different sources, such as from different homologous chromosomes or even from different organisms. One such example is the heteroduplex DNA strand formed in hybridization processes, usually for biochemistry-based phylogenetical analyses. Another example is the heteroduplexes formed when non-natural analogs of nucleic acids are used to bind with nucleic acids; these heteroduplexes result from performing antisense techniques using single-stranded peptide nucleic acid, 2'-O-methyl phosphorothioate or Morpholino oligos to bind with RNA. In meiosis, the process of crossing-over occurs between non-sister chromatids, which results in new allelic combinations of a population.

Neisseria meningitidis	Neisseria meningitidis is a heterotrophic gram-negative diplococcal bacterium best known for its role in meningitis and other forms of meningococcal disease such as meningococcemia. N. meningitidis is a major cause of morbidity and mortality during childhood in industrialized countries and is responsible for epidemics in Africa and in Asia. Approximately 2500 to 3500 cases of N meningitidis infection occur annually in the United States, with a case rate of about 1 in 100,000. Children younger than 5 years are at greatest risk, followed by teenagers of high school age.
Catabolite repression	Carbon catabolite repression, is an important part of global control system of various bacteria and other micro-organisms. Catabolite repression allows bacteria to adapt quickly to a preferred (rapidly metabolisable) carbon and energy source first. This is usually achieved through inhibition of synthesis of enzymes involved in catabolism of carbon sources other than the preferred one.
Codon	Codons are the basic unit of the genetic code. The fact that codons consist of three DNA bases was first demonstrated in the Crick, Brenner et al. experiment. The first elucidation of a codon was done by Marshall Nirenberg and Heinrich J. Matthaei in 1961 at the National Institutes of Health. The genome of an organism is inscribed in DNA, or in the case of some viruses, RNA. The portion of the genome that codes for a protein or an RNA is referred to as a gene. Those genes that code for proteins are composed of tri-nucleotide units called codons, each coding for a single amino acid.
Shine-Dalgarno sequence	The Shine-Dalgarno sequence proposed by Australian scientists John Shine (b.1946) and Lynn Dalgarno (b.1935), is a ribosomal binding site in the mRNA, generally located 8 basepairs upstream of the start codon AUG. The Shine-Dalgarno sequence exists both in bacteria and archaea, being also present in some chloroplastic and mitochondial transcripts. The six-base consensus sequence is AGGAGG; in E. coli, for example, the sequence is AGGAGGU. This sequence helps recruit the ribosome to the mRNA to initiate protein synthesis by aligning it with the start codon. The complementary sequence (CCUCCU), is called the anti-Shine-Dalgarno sequence and is located at the 3' end of the 16S rRNA in the ribosome.
Candida albicans	Candida albicans is a diploid fungus that grows both as yeast and filamentous cells and a causal agent of opportunistic oral and genital infections in humans. Systemic fungal infections (fungemias) including those by C. albicans have emerged as important causes of morbidity and mortality in immunocompromised patients (e.g., AIDS, cancer chemotherapy, organ or bone marrow transplantation). C. albicans biofilms may form on the surface of implantable medical devices.

Chapter 2. Part 2: Genes and Genomes

Inosine	Inosine is a nucleoside that is formed when hypoxanthine is attached to a ribose ring (also known as a ribofuranose) via a β-N_9-glycosidic bond. Inosine is commonly found in tRNAs and is essential for proper translation of the genetic code in wobble base pairs. Knowledge of inosine metabolism has led to advances in immunotherapy in recent decades.
Mycoplasma	Mycoplasma refers to a genus of bacteria that lack a cell wall. Without a cell wall, they are unaffected by many common antibiotics such as penicillin or other beta-lactam antibiotics that target cell wall synthesis. They can be parasitic or saprotrophic.
Pseudouridine	Pseudouridine is the C-glycoside isomer of the nucleoside uridine, and it is the most prevalent of the over one hundred different modified nucleosides found in RNA. Ψ is found in all species and in many classes of RNA except mRNA. Ψ is formed by enzymes called Ψ synthases, which post-transcriptionally isomerize specific uridine residues in RNA in a process termed pseudouridylation. Recent studies suggest it may offer protection from radiation. It is commonly found in tRNA, associated with thymidine and cytosine in the TΨC arm and is one of the invariant regions of tRNA. The function of it is not very clear, but it is expected to play a role in association with aminoacyl transferases during their interaction with tRNA, and hence in the initiation of translation.
Saccharomyces	Saccharomyces is a genus in the kingdom of fungi that includes many species of yeast. Saccharomyces is from Greek σ?κχαρ (sugar) and μ?κης (mushroom) and means sugar fungus. Many members of this genus are considered very important in food production.
Star formation	Star formation is the process by which dense regions within molecular clouds, commonly referred to as 'stellar nurseries', collapse into spheres of plasma to form stars. As a branch of astronomy star formation includes the study of the interstellar medium and giant molecular clouds (GMC) as precursors to the star formation process and the study of young stellar objects and planet formation as its immediate products. Star formation theory, as well as accounting for the formation of a single star, must also account for the statistics of binary stars and the initial mass function.
Start codon	The start codon is generally defined as the point, sequence, at which a ribosome begins to translate a sequence of RNA into amino acids.

	When an RNA transcript is 'read' from the 5' nucleotide to the 3' nucleotide by the ribosome the start codon is the first codon on which the tRNA bound to Met, methionine, and ribosomal subunits attach. AUG denotes sequences of RNA that are the start codon or initiation codon encoding the amino acid methionine (Met) in eukaryotes and a modified Met (fMet) in prokaryotes.
Stop codon	In the genetic code, a stop codon is a nucleotide triplet within messenger RNA that signals a termination of translation. Proteins are based on polypeptides, which are unique sequences of amino acids. Most codons in messenger RNA correspond to the addition of an amino acid to a growing polypeptide chain, which may ultimately become a protein.
Transfer RNA	Transfer RNA is an adaptor molecule composed of RNA, typically 73 to 93 nucleotides in length, that is used in biology to bridge the four-letter genetic code (ACGU) in messenger RNA (mRNA) with the twenty-letter code of amino acids in proteins. The role of tRNA as an adaptor is best understood by considering its three-dimensional structure. One end of the tRNA carries the genetic code in a three-nucleotide sequence called the anticodon.
Cistron	A cistron is a gene. The term cistron is used to emphasize that genes exhibit a specific behavior in a cis-trans test; distinct positions within a genome are cistronic (i.e., within the same gene) when mutations at the loci exhibit the same simple Mendelian inheritance as would mutations at a single locus. For example, suppose a mutation at a chromosome position x is responsible for a recessive trait in a diploid organism (where chromosomes come in pairs).
Northern blot	The northern blot is a technique used in molecular biology research to study gene expression by detection of RNA in a sample. With northern blotting it is possible to observe cellular control over structure and function by determining the particular gene expression levels during differentiation, morphogenesis, as well as abnormal or diseased conditions. Northern blotting involves the use of electrophoresis to separate RNA samples by size and detection with a hybridization probe complementary to part of or the entire target sequence.
Ribosome	The Ribosome is a large complex molecule which is responsible for catalyzing the formation of proteins from individual amino acids using messenger RNA as a template. This process is known as translation. Ribosomes are found in all living cells.
Ribosomal protein	Mitochondrial ribosomal protein L31

Chapter 2. Part 2: Genes and Genomes

A ribosomal protein is any of the proteins that, in conjunction with rRNA, make up the ribosomal subunits involved in the cellular process of translation. A large part of the knowledge about these organic molecules has come from the study of E. coli ribosomes. Most ribosomic proteins have been isolated and specific anti-bodies have been produced.

T cell	T cells or T lymphocytes belong to a group of white blood cells known as lymphocytes, and play a central role in cell-mediated immunity. They can be distinguished from other lymphocytes, such as B cells and natural killer cells (NK cells), by the presence of a T cell receptor (TCR) on the cell surface. They are called T cells because they mature in the thymus.

Reading frame	In biology, a reading frame is a way of breaking a sequence of nucleotides in DNA or RNA into three letter codons which can be translated in amino acids. There are 3 possible reading frames in an mRNA strand: each reading frame corresponds to a different starting alignment. Double stranded DNA has six different reading frames per molecule due to the two strands from which transcription is possible--three of them reading forward and three of them reading backwards.

Methanobrevibacter smithii	Methanobrevibacter smithii is the dominant archaeon in the human gut. It is important for the efficient digestion of polysaccharides (complex sugars) because it consumes end products of bacterial fermentation. Methanobrevibacter smithii is a single-celled micro-organism from the Archaea domain.

| Elongation factor | Elongation factors are a set of proteins that facilitate the events of translational elongation, the steps in protein synthesis from the formation of the first peptide bond to the formation of the last one.

Elongation is the most rapid step in translation:•in prokaryotes it proceeds at a rate of 15 to 20 amino acids added per second (about 60 nucleotides per second)•in eukaryotes the rate is about two amino acids per second.

Elongation factors play a role in orchestrating the events of this process, and in ensuring the 99.99% accuracy of translation at this speed. |
|---|---|

| Initiation factor | Initiation factors are proteins that bind to the small subunit of the ribosome during the initiation of translation, a part of protein biosynthesis.

They are divided into three major groups:•Prokaryotic initiation factors•Archaeal initiation factors•Eukaryotic initiation factors. |
|---|---|

Aminoglycoside	An aminoglycoside is a molecule or a portion of a molecule composed of amino-modified sugars.

Several aminoglycosides function as antibiotics that are effective against certain types of bacteria. They include amikacin, arbekacin, gentamicin, kanamycin, neomycin, netilmicin, paromomycin, rhodostreptomycin, streptomycin, tobramycin, and apramycin . |
| Enterococcus faecalis | Enterococcus faecalis - formerly classified as part of the Group D Streptococcus system - is a Gram-positive commensal bacterium inhabiting the gastrointestinal tracts of humans and other mammals. It is among the main constituents of some probiotic food supplements. A commensal organism like other species in the genus Enterococcus, E. faecalis can cause life-threatening infections in humans, especially in the nosocomial (hospital) environment, where the naturally high levels of antibiotic resistance found in E. faecalis contribute to its pathogenicity. |
| Release factor | A release factor is a protein that allows for the termination of translation by recognizing the termination codon or stop codon in a mRNA sequence.

During translation of mRNA, most codons are recognized by 'charged' tRNA molecules, called aminoacyl-tRNAs because they are adhered to specific amino acids corresponding to each tRNA's anticodon. |
| Protein synthesis inhibitor | A protein synthesis inhibitor is a substance that stops or slows the growth or proliferation of cells by disrupting the processes that lead directly to the generation of new proteins.

While a broad interpretation of this definition could be used to describe nearly any antibiotic, in practice, it usually refers to substances that act at the ribosome level, taking advantages of the major differences between prokaryotic and eukaryotic ribosome structures.

Toxins such as ricin also function via protein synthesis inhibition. |
| Brugia malayi | Brugia malayi is a nematode (roundworm), one of the three causative agents of lymphatic filariasis in humans. Lymphatic filariasis, also known as elephantiasis, is a condition characterized by swelling of the lower limbs. The two other filarial causes of lymphatic filariasis are Wuchereria bancrofti and Brugia timori, which differ from B. malayi morphologically, symptomatically, and in geographical extent. |
| Fusidic acid | Fusidic acid is a bacteriostatic antibiotic that is often used topically in creams and eyedrops, but may also be given systemically as tablets or injections. The global problem of advancing antimicrobial resistance has led to a renewed interest in its use recently. |

Chapter 2. Part 2: Genes and Genomes

Isocitrate dehydrogenase	Isocitrate dehydrogenase and (EC 1.1.1.41), also known as IDH, is an enzyme that participates in the citric acid cycle. It catalyzes the third step of the cycle: the oxidative decarboxylation of isocitrate, producing alpha-ketoglutarate (α-ketoglutarate) and CO_2 while converting NAD^+ to NADH. This is a two-step process, which involves oxidation of isocitrate (a secondary alcohol) to oxalosuccinate (a ketone), followed by the decarboxylation of the carboxyl group beta to the ketone, forming alpha-ketoglutarate. Another isoform of the enzyme catalyzes the same reaction, however this reaction is unrelated to the citric acid cycle, is carried out in the cytosol as well as the mitochondrion and peroxisome and uses $NADP^+$ as a cofactor instead of NAD^+.
Macrolide	The macrolides are a group of drugs (typically antibiotics) whose activity stems from the presence of a macrolide ring, a large macrocyclic lactone ring to which one or more deoxy sugars, usually cladinose and desosamine, may be attached. The lactone rings are usually 14-, 15-, or 16-membered. Macrolides belong to the polyketide class of natural products.
Molecular mimicry	Molecular mimicry is defined as the theoretical possibility that sequence similarities between foreign and self-peptides are sufficient to result in the cross-activation of autoreactive T or B cells by pathogen-derived peptides. Despite the promiscuity of several peptide sequences which can be both foreign and self in nature, a single antibody or TCR (T cell receptor) can be activated by even a few crucial residues which stresses the importance of structural homology in the theory of molecular mimicry. Upon the activation of B or T cells, it is believed that these 'peptide mimic' specific T or B cells can cross-react with self-epitopes, thus leading to tissue pathology (autoimmunity).
Puromycin	Puromycin is an antibiotic that is a protein synthesis inhibitor by inhibiting translation.

Puromycin is an aminonucleoside antibiotic, derived from the Streptomyces alboniger bacterium, that causes premature chain termination during translation taking place in the ribosome. Part of the molecule resembles the 3' end of the aminoacylated tRNA. It enters the A site and transfers to the growing chain, causing the formation of a puromycylated nascent chain and premature chain release . |
| Streptococcus pneumoniae | Streptococcus pneumoniae, is a Gram-positive, alpha-hemolytic, aerotolerant anaerobic member of the genus Streptococcus. A significant human pathogenic bacterium, S. pneumoniae was recognized as a major cause of pneumonia in the late 19th century, and is the subject of many humoral immunity studies.

Despite the name, the organism causes many types of pneumococcal infections other than pneumonia. |
| Biosynthesis | Biosynthesis is an enzyme-catalyzed process in cells of living organisms by which substrates are converted to more complex products. |

The biosynthesis process often consists of several enzymatic steps in which the product of one step is used as substrate in the following step. Examples for such multi-step biosynthetic pathways are those for the production of amino acids, fatty acids, and natural products.

Protease	A protease is any enzyme that conducts proteolysis, that is, begins protein catabolism by hydrolysis of the peptide bonds that link amino acids together in the polypeptide chain forming the protein. Standard

Proteases are currently classified into six broad groups:•Serine proteases•Threonine proteases•Cysteine proteases•Aspartate proteases•Metalloproteases•Glutamic acid proteases

The threonine and glutamic-acid proteases were not described until 1995 and 2004, respectively. The mechanism used to cleave a peptide bond involves making an amino acid residue that has the cysteine and threonine (proteases) or a water molecule (aspartic acid, metallo- and glutamic acid proteases) nucleophilic so that it can attack the peptide carboxyl group.

Glutamine synthetase

Glutamine synthetase (EC 6.3.1.2) is an enzyme that plays an essential role in the metabolism of nitrogen by catalyzing the condensation of glutamate and ammonia to form glutamine:

Glutamate + ATP + NH_3 → Glutamine + ADP + phosphate + H_2O

Glutamine Synthetase uses ammonia produced by nitrate reduction, amino acid degradation, and photorespiration. The amide group of glutamate is a nitrogen source for the synthesis of glutamine pathway metabolites.

Other reactions may take place via Glutamine synthetase. Competition between ammonium ion and water, their binding affinities, and the concentration of ammonium ion, influences glutamine synthesis and glutamine hydrolysis.

GroEL

GroEL belongs to the chaperonin family of molecular chaperones, and is found in a large number of bacteria. It is required for the proper folding of many proteins. To function properly, GroEL requires the lid-like cochaperonin protein complex GroES. In eukaryotes the proteins Hsp60 and Hsp10 are structurally and functionally nearly identical to GroEL and GroES, respectively.

Hemolysin

Hemolytic toxin N terminal

Hemolysins are exotoxins produced by bacteria that cause lysis of red blood cells in vitro. Visualization of hemolysis of red blood cells in agar plates facilitates the categorization of some pathogenic bacteria such as Streptococcus and Staphylococcus.

Chapter 2. Part 2: Genes and Genomes

Secretion	Secretion is the process of elaborating, releasing, and oozing chemicals, or a secreted chemical substance from a cell or gland. In contrast to excretion, the substance may have a certain function, rather than being a waste product. Many cells contain this such as glucoma cells.
Membrane protein	A membrane protein is a protein molecule that is attached to, or associated with the membrane of a cell or an organelle. More than half of all proteins interact with membranes. Biological membranes consist of a phospholipid bilayer and a variety of proteins that accomplish vital biological functions.
Cell membrane	The cell membrane is a biological membrane that separates the interior of all cells from the outside environment. The cell membrane is selectively permeable to ions and organic molecules and controls the movement of substances in and out of cells. It basically protects the cell from outside forces.
Degron	A degron is a specific sequence of amino acids in a protein that directs the starting place of degradation. A degron sequence can occur at either the N or C-terminal region, these are called N-Degrons or C-degrons respectively. A temperature sensitive degron takes advantage of the N-end rule pathway, in which a destabilizing N-terminal residue dramatically decreases the in vivo half-life of a protein.
Phase variation	Phase variation is a method for dealing with rapidly varying environments without requiring random mutation employed by various types of bacteria, including Salmonella species. It involves the variation of protein expression, frequently in an on-off fashion, within different parts of a bacterial population. Although it has been most commonly studied in the context of immune evasion, it is observed in many other areas as well.
Proteasome	Proteasomes are very large protein complexes inside all eukaryotes and archaea, and in some bacteria. In eukaryotes, they are located in the nucleus and the cytoplasm. The main function of the proteasome is to degrade unneeded or damaged proteins by proteolysis, a chemical reaction that breaks peptide bonds.
Ubiquitin	Ubiquitin is a small regulatory protein that has been found in almost all tissues (ubiquitously) of eukaryotic organisms. Among other functions, it directs protein recycling. Ubiquitin can be attached to proteins and label them for destruction.
Bioinformatics	Bioinformatics is the application of computer science and information technology to the field of biology and medicine.

	Bioinformatics deals with algorithms, databases and information systems, web technologies, artificial intelligence and soft computing, information and computation theory, software engineering, data mining, image processing, modeling and simulation, signal processing, discrete mathematics, control and system theory, circuit theory, and statistics. Bioinformatics generates new knowledge as well as the computational tools to create that knowledge.
Dihydrolipoamide dehydrogenase	Dihydrolipoamide dehydrogenase also known as dihydrolipoyl dehydrogenase, mitochondrial, is an enzyme that in humans is encoded by the DLD gene. DLD is a flavoprotein enzyme that degrades dihydrolipoamide, and produces lipoamide. This gene encodes the L protein of the mitochondrial glycine cleavage system.
Exon	An exon is a nucleic acid sequence that is represented in the mature form of an RNA molecule either after portions of a precursor RNA (introns) have been removed by cis-splicing or when two or more precursor RNA molecules have been ligated by trans-splicing. The mature RNA molecule can be a messenger RNA or a functional form of a non-coding RNA such as rRNA or tRNA. Depending on the context, exon can refer to the sequence in the DNA or its RNA transcript. The term exon derives from expressed region and was coined by American biochemist Walter Gilbert in 1978: 'The notion of the cistron... must be replaced by that of a transcription unit containing regions which will be lost from the mature messenger - which I suggest we call introns (for intragenic regions) - alternating with regions which will be expressed - exons.' This definition was originally made for protein-coding transcripts that are spliced before being translated.
Sequence alignment	In bioinformatics, a sequence alignment is a way of arranging the sequences of DNA, RNA, or protein to identify regions of similarity that may be a consequence of functional, structural, or evolutionary relationships between the sequences. Aligned sequences of nucleotide or amino acid residues are typically represented as rows within a matrix. Gaps are inserted between the residues so that identical or similar characters are aligned in successive columns.
ExPASy	ExPASy is a bioinformatics resource portal operated by the Swiss Institute of Bioinformatics (SIB) and in particular the SIB Web Team. It is an extensible and integrative portal accessing many scientific resources, databases and software tools in different areas of life sciences. Scientists can henceforth access seamlessly a wide range of resources in many different domains, such as proteomics, genomics, phylogeny/evolution, systems biology, population genetics, transcriptomics, etc.

Chapter 2. Part 2: Genes and Genomes

Functional genomics	Functional genomics is a field of molecular biology that attempts to make use of the vast wealth of data produced by genomic projects (such as genome sequencing projects) to describe gene (and protein) functions and interactions. Unlike genomics, functional genomics focuses on the dynamic aspects such as gene transcription, translation, and protein-protein interactions, as opposed to the static aspects of the genomic information such as DNA sequence or structures. Functional genomics attempts to answer questions about the function of DNA at the levels of genes, RNA transcripts, and protein products.
Helicobacter pylori	Helicobacter pylori previously named Campylobacter pyloridis, is a Gram-negative, microaerophilic bacterium found in the stomach. It was identified in 1982 by Barry Marshall and Robin Warren, who found that it was present in patients with chronic gastritis and gastric ulcers, conditions that were not previously believed to have a microbial cause. It is also linked to the development of duodenal ulcers and stomach cancer.
Joint Genome Institute	The U.S. Department of Energy (DOE) Joint Genome Institute was created in 1997 to unite the expertise and resources in genome mapping, DNA sequencing, technology development, and information sciences pioneered at the DOE genome centers at Lawrence Berkeley National Laboratory (LBNL), Lawrence Livermore National Laboratory (LLNL) and Los Alamos National Laboratory (LANL). The DOE JGI also collaborates with other national labs such as Oak Ridge National Laboratory (ORNL) and Pacific Northwest National Laboratory (PNNL). In 1999, the University of California, which manages the three national labs for the DOE, leased laboratory and office space in a light industrial park in Walnut Creek, California to consolidate genome research activities.
KEGG	KEGG is a collection of online databases dealing with genomes, enzymatic pathways, and biological chemicals. The PATHWAY database records networks of molecular interactions in the cells, and variants of them specific to particular organisms. As of July 2011, KEGG has switched to a subscription model and access via FTP is no longer free.
Multiple sequence alignment	A multiple sequence alignment is a sequence alignment of three or more biological sequences, generally protein, DNA, or RNA. In many cases, the input set of query sequences are assumed to have an evolutionary relationship by which they share a lineage and are descended from a common ancestor. From the resulting MSA, sequence homology can be inferred and phylogenetic analysis can be conducted to assess the sequences' shared evolutionary origins. Visual depictions of the alignment as in the image at right illustrate mutation events such as point mutations (single amino acid or nucleotide changes) that appear as differing characters in a single alignment column, and insertion or deletion mutations (indels or gaps) that appear as hyphens in one or more of the sequences in the alignment.

Rickettsia prowazekii	Rickettsia prowazekii is a species of gram negative, Alpha Proteobacteria, obligate intracellular parasitic, aerobic bacteria that is the etiologic agent of epidemic typhus, transmitted in the faeces of lice. In North America, the main reservoir for R. prowazekii is the flying squirrel. R. prowazekii is often surrounded by a protein microcapsular layer and slime layer; the natural life cycle of the bacterium generally involves a vertebrate and an invertebrate host, usually an arthropod, typically the human body louse.
Genomics	Genomics is a discipline in genetics concerned with the study of the genomes of organisms. The field includes efforts to determine the entire DNA sequence of organisms and fine-scale genetic mapping. The field also includes studies of intragenomic phenomena such as heterosis, epistasis, pleiotropy and other interactions between loci and alleles within the genome.
Fertility factor	The Fertility factor was the first episome discovered (viral DNA that is separate from bacterial DNA). 'F' is often confused with sexual conjugation. Sexual conjugation is a bacterial DNA sequence that allows a bacterium to produce a sex pilus necessary for conjugation.
Pseudomonas aeruginosa	Pseudomonas aeruginosa is a common bacterium which can cause disease in animals, including humans. It is found in soil, water, skin flora, and most man-made environments throughout the world. It thrives not only in normal atmospheres, but also in hypoxic atmospheres, and has thus colonized many natural and artificial environments.
Pyrococcus furiosus	Pyrococcus furiosus is an extremophilic species of Archaea. It can be classified as a hyperthermophile because it thrives best under extremely high temperatures--higher than those preferred of a thermophile. It is notable for having an optimum growth temperature of 100°C (a temperature that would destroy most living organisms), and for being one of the few organisms identified as possessing enzymes containing tungsten, an element rarely found in biological molecules.
Biofilm	A biofilm is an aggregate of microorganisms in which cells adhere to each other on a surface. These adherent cells are frequently embedded within a self-produced matrix of extracellular polymeric substance (EPS). Biofilm EPS, which is also referred to as slime (although not everything described as slime is a biofilm), is a polymeric conglomeration generally composed of extracellular DNA, proteins, and polysaccharides.
Haemophilus	Haemophilus is a genus of Gram-negative, pleomorphic, coccobacilli bacteria belonging to the Pasteurellaceae family. While Haemophilus bacteria are typically small coccobacilli, they are categorized as pleomorphic bacteria because of the wide range of shapes they occasionally assume. The genus includes commensal organisms along with some significant pathogenic species such as H. influenzae--a cause of sepsis and bacterial meningitis in young children--and H.

Chapter 2. Part 2: Genes and Genomes

Colin Munro MacLeod	Colin Munro MacLeod was a Canadian-American geneticist.
	Born in Port Hastings, Nova Scotia, Canada MacLeod entered McGill University at the age of 16 (having skipped three grades in primary school), and completed his medical studies by age 23.
	In his early years as a research scientist, MacLeod, together with Oswald Avery and Maclyn McCarty, demonstrated that DNA is the active component responsible for bacterial transformation-- and in retrospect, the physical basis of the gene.
Electroporation	Electroporation, is a significant increase in the electrical conductivity and permeability of the cell plasma membrane caused by an externally applied electrical field. It is usually used in molecular biology as a way of introducing some substance into a cell, such as loading it with a molecular probe, a drug that can change the cell's function, or a piece of coding DNA.
	Electroporation is a dynamic phenomenon that depends on the local transmembrane voltage at each point on the cell membrane. It is generally accepted that for a given pulse duration and shape, a specific transmembrane voltage threshold exists for the manifestation of the electroporation phenomenon (from 0.5 V to 1 V).
Horizontal gene transfer	Horizontal gene transfer also lateral gene transfer (LGT) or transposition refers to the transfer of genetic material between organisms other than vertical gene transfer. Vertical transfer occurs when there is gene exchange from the parental generation to the offspring. LGT is then a mechanism of gene exchange that happens independently of reproduction.
Neisseria gonorrhoeae	Neisseria gonorrhoeae, or gonococcus, is a species of Gram-negative coffee bean-shaped diplococci bacteria responsible for the sexually transmitted infection gonorrhea.
	N. gonorrhoea was first described by Albert Neisser in 1879.
	Microbiology
	Neisseria are fastidious Gram-negative cocci that require nutrient supplementation to grow in laboratory cultures.
Quorum sensing	Quorum sensing is a system of stimulus and response correlated to population density. Many species of bacteria use quorum sensing to coordinate gene expression according to the density of their local population. In similar fashion, some social insects use quorum sensing to determine where to nest.
Gene mapping	Gene mapping, is the creation of a genetic map assigning DNA fragments to chromosomes.

When a genome is first investigated, this map is nonexistent. The map improves with the scientific progress and is perfect when the genomic DNA sequencing of the species has been completed.

Bacillus thuringiensis	Bacillus thuringiensis is a Gram-positive, soil-dwelling bacterium, commonly used as a biological pesticide; alternatively, the Cry toxin may be extracted and used as a pesticide. B. thuringiensis also occurs naturally in the gut of caterpillars of various types of moths and butterflies, as well as on the dark surface of plants. During sporulation many Bt strains produce crystal proteins (proteinaceous inclusions), called δ-endotoxins, that have insecticidal action.
Permease	The permeases are membrane transport proteins, a class of multipass transmembrane proteins that facilitate the diffusion of a specific molecule in or out of the cell by passive transport. In contrast, active transporters couple molecule transmembrane transport with an energy source such as ATP or a favorable ion gradient.
Rhizosphere	The rhizosphere is the narrow region of soil that is directly influenced by root secretions and associated soil microorganisms. Soil which is not part of the rhizosphere is known as bulk soil. The rhizosphere contains many bacteria that feed on sloughed-off plant cells, termed rhizodeposition, and the proteins and sugars released by roots.
Nematode	The nematodes ( ) or roundworms (phylum Nematoda) are the most diverse phylum of pseudocoelomates, and one of the most diverse of all animals. Nematode species are very difficult to distinguish; over 28,000 have been described, of which over 16,000 are parasitic. The total number of nematode species has been estimated to be about 1,000,000. Unlike cnidarians or flatworms, roundworms have tubular digestive systems with openings at both ends.
Pathogenicity island	Pathogenicity islands (PAIs) are a distinct class of genomic islands acquired by microorganisms through horizontal gene transfer. They are incorporated in the genome of pathogenic organisms, but are usually absent from those nonpathogenic organisms of the same or closely related species. These mobile genetic elements may range from 10-200 kb and encode genes which contribute to the virulence of the respective pathogen.
Prophage	A prophage is a phage (viral) genome inserted and integrated into the circular bacterial DNA chromosome. A prophage, also known as a temperate phage, is any virus in the lysogenic cycle; it is integrated into the host chromosome or exists as an extrachromosomal plasmid. Technically, a virus may be called a prophage only while the viral DNA remains incorporated in the host DNA.

Chapter 2. Part 2: Genes and Genomes

Site-specific recombination	Site-specific recombination, is a type of genetic recombination in which DNA strand exchange takes place between segments possessing only a limited degree of sequence homology. Site-specific recombinases perform rearrangements of DNA segments by recognizing and binding to short DNA sequences (sites), at which they cleave the DNA backbone, exchange the two DNA helices involved and rejoin the DNA strands. While in some site-specific recombination systems just a recombinase enzyme and the recombination sites is enough to perform all these reactions, in other systems a number of accessory proteins and/or accessory sites are also needed.
Holliday junction	A Holliday junction is a mobile junction between four strands of DNA. The structure is named after Robin Holliday, who proposed it in 1964 to account for a particular type of exchange of genetic information he observed in yeast known as homologous recombination. Holliday junctions are highly conserved structures, from prokaryotes to mammals.
	Because these junctions are between homologous sequences, they can slide up and down the DNA. In bacteria, this sliding is facilitated by the RuvABC complex or RecG protein, molecular motors that use the energy of ATP hydrolysis to push the junction around.
RecA	RecA bacterial DNA recombination protein
	RecA is a 38 kilodalton Escherichia coli protein essential for the repair and maintenance of DNA. A RecA structural and functional homolog has been found in every species in which one has been seriously sought and serves as an archetype for this class of homologous DNA repair proteins. The homologous protein in Homo sapiens is called RAD51.
	RecA has multiple activities, all related to DNA repair.
Beta-galactosidase	β-galactosidase, also called beta-gal or β-gal, is a hydrolase enzyme that catalyzes the hydrolysis of β-galactosides into monosaccharides. Substrates of different β-galactosidases include ganglioside GM1, lactosylceramides, lactose, and various glycoproteins. Lactase is often confused as an alternative name for β-galactosidase, but it is actually simply a sub-class of β-galactosidase.
	β-galactosidase is an exoglycosidase which hydrolyzes the β-glycosidic bond formed between a galactose and its organic moiety. It may also cleave fucosides and arabinosides but with much lower efficiency. It is an essential enzyme in the human body, deficiencies in the protein can result in galactosialidosis or Morquio B syndrome. In E. coli, the gene of β-galactosidase, the lacZ gene, is present as part of the inducible system lac operon which is activated in the presence of lactose when glucose level is low.
	It is commonly used in molecular biology as a reporter marker to monitor gene expression.

It also exhibits a phenomenon called α-complementation which forms the basis for the blue/white screening of recombinant clones. This enzyme can be split in two peptides, LacZα and LacZΩ, neither of which is active by itself but when both are present together, spontaneously reassemble into a functional enzyme. This property is exploited in many cloning vectors where the presence of the lacZα gene in a plasmid can complement in trans another mutant gene encoding the LacZΩ in specific laboratory strains of E. coli. However, when DNA fragments are inserted in the vector, the production of LacZα is disrupted, the cells therefore show no β-galactosidase activity. The presence or absence of an active β-galactosidase may be detected by X-gal, which produces a characteristic blue dye when cleaved by β-galactosidase, thereby providing an easy means of distinguishing the presence or absence of cloned product in a plasmid.

In 1995, Dimri et al. proposed a new isoform for beta-galactosidase with optimum activity at pH 6.0 (Senescence Associated beta-gal or SA-beta-gal) which would be specifically expressed in senescence (The irreversible growth arrest of cells).

Deletion

In genetics, a deletion (also called gene deletion, deficiency, or deletion mutation) (sign: Δ) is a mutation (a genetic aberration) in which a part of a chromosome or a sequence of DNA is missing. Deletion is the loss of genetic material. Any number of nucleotides can be deleted, from a single base to an entire piece of chromosome.

Point mutation

A point mutation, is a type of mutation that causes the replacement of a single base nucleotide with another nucleotide of the genetic material, DNA or RNA. The term point mutation also includes insertions or deletions of a single base pair.

A point mutant is an individual that is affected by a point mutation.

Repeat induced point mutations are recurring point mutations, discussed below.

Silent mutation

Silent mutations are DNA mutations that do not result in a change to the amino acid sequence of a protein. They may occur in a non-coding region (outside of a gene or within an intron), or they may occur within an exon in a manner that does not alter the final amino acid sequence. The phrase silent mutation is often used interchangeably with the phrase synonymous mutation; however, synonymous mutations are a subcategory of the former, occurring only within exons.

Transversion

In molecular biology, transversion refers to the substitution of a purine for a pyrimidine or vice versa. It can only be reverted by a spontaneous reversion. Because this type of mutation changes the chemical structure dramatically, the consequences of this change tend to be more drastic than those of transitions.

Chapter 2. Part 2: Genes and Genomes

Frameshift mutation	A frameshift mutation is a genetic mutation caused by indels (insertions or deletions) of a number of nucleotides that is not evenly divisible by three from a DNA sequence. Due to the triplet nature of gene expression by codons, the insertion or deletion can change the reading frame (the grouping of the codons), resulting in a completely different translation from the original. The earlier in the sequence the deletion or insertion occurs, the more altered the protein produced is.
Nonsense mutation	In genetics, a nonsense mutation is a point mutation in a sequence of DNA that results in a premature stop codon, or a nonsense codon in the transcribed mRNA, and in a truncated, incomplete, and usually nonfunctional protein product. It differs from a missense mutation, which is a point mutation where a double nucleotide is changed to cause substitution of a different amino acid. Some genetic disorders, such as thalassemia and DMD, result from nonsense mutations.
Missense mutation	In genetics, a missense mutation is a point mutation in which a single nucleotide is changed, resulting in a codon that codes for a different amino acid (mutations that change an amino acid to a stop codon are considered nonsense mutations, rather than missense mutations). This can render the resulting protein nonfunctional. Such mutations are responsible for diseases such as Epidermolysis bullosa, sickle-cell disease, and SOD1 mediated ALS (Boillée 2006, p. 39).
Synonymous substitution	A synonymous substitution is the evolutionary substitution of one base for another in an exon of a gene coding for a protein, such that the produced amino acid sequence is not modified. Synonymous substitutions and mutations affecting noncoding DNA are collectively known as silent mutations. A non-synonymous substitution results in a change in amino acid that may be arbitrarily further classified as conservative (change to an amino acid with similar physiochemical properties), semi-conservative (e.g. negative to positively charged amino acid), or radical (vastly different amino acid).
Mutagen	In genetics, a mutagen is a physical or chemical agent that changes the genetic material, usually DNA, of an organism and thus increases the frequency of mutations above the natural background level. As many mutations cause cancer, mutagens are therefore also likely to be carcinogens. Not all mutations are caused by mutagens: so-called 'spontaneous mutations' occur due to spontaneous hydrolysis, errors in DNA replication, repair and recombination.
Mutation rate	In genetics, the mutation rate is a measure of the rate at which various types of mutations occur during some unit of time. Mutation rates are typically given for a specific class of mutation, for instance point mutations, small or large scale insertions or deletions. The rate of substitutions can be further subdivided into a mutation spectrum which describes the influence of genetic context on the mutation rate.

Hydrogen peroxide	Hydrogen peroxide is the simplest peroxide (a compound with an oxygen-oxygen single bond) and an oxidizer. Hydrogen peroxide is a clear liquid, slightly more viscous than water. In dilute solution it appears colorless.
Hydroxyl radical	The hydroxyl radical, ${}^{\bullet}OH$, is the neutral form of the hydroxide ion (OH^-). Hydroxyl radicals are highly reactive and consequently short-lived; however, they form an important part of radical chemistry. Most notably hydroxyl radicals are produced from the decomposition of hydroperoxides (ROOH) or, in atmospheric chemistry, by the reaction of excited atomic oxygen with water.
Essential nutrient	An essential nutrient is a nutrient required for normal body functioning that either cannot be synthesized by the body at all, or cannot be synthesized in amounts adequate for good health (e.g. niacin, choline), and thus must be obtained from a dietary source. Essential nutrients are also defined by the collective physiological evidence for their importance in the diet, as represented in e.g. US government approved tables for Dietary Reference Intake. Some categories of essential nutrients include vitamins, dietary minerals, essential fatty acids, and essential amino acids.
Pyrimidine dimers	Pyrimidine dimers are molecular lesions formed from thymine or cytosine bases in DNA via photochemical reactions. Ultraviolet light induces the formation of covalent linkages by reactions localized on the C=C double bonds. In dsRNA, uracil dimers may also accumulate as a result of UV radiation.
Mutation frequency	Mutation frequency and mutation rates are highly correlated to each other. Mutation frequencies test are cost effective in laboratories however; these two concepts provide vital information in reference to accounting for the emergence of mutations on any given germ line. There are several test utilized in measuring the chances of mutation frequency and rates occurring in a particular gene pool.
Frequency	Frequency is the number of occurrences of a repeating event per unit time. It is also referred to as temporal frequency. The period is the duration of one cycle in a repeating event, so the period is the reciprocal of the frequency.
AP site	In biochemistry and molecular genetics, an AP site also known as an abasic site, is a location in DNA that has neither a purine nor a pyrimidine base, either spontaneously or due to DNA damage. It has been estimated that under physiological condition 10,000 apurinic sites may be generated in a cell daily.

Chapter 2. Part 2: Genes and Genomes

DNA repair	DNA repair refers to a collection of processes by which a cell identifies and corrects damage to the DNA molecules that encode its genome. In human cells, both normal metabolic activities and environmental factors such as UV light and radiation can cause DNA damage, resulting in as many as 1 million individual molecular lesions per cell per day. Many of these lesions cause structural damage to the DNA molecule and can alter or eliminate the cell's ability to transcribe the gene that the affected DNA encodes.
Excision repair	Excision repair is a term applied to several DNA repair mechanisms. They remove the damaged nucleotides and are able to determine the correct sequence from the complementary strand of DNA.

Specific mechanisms include:•Base excision repair which repairs damage due to a single nucleotide caused by oxidation, alkylation, hydrolysis, or deamination;•Nucleotide excision repair which repairs damage affecting 2−30 nucleotide-length strands. These include bulky, helix distorting damage, such as thymine dimerization and other types of cyclobutyl dimerization caused by UV light as well as single-strand breaks. |
Base excision repair	In biochemistry and genetics, base excision repair is a cellular mechanism that repairs damaged DNA throughout the cell cycle. It is responsible primarily for removing small, non-helix-distorting base lesions from the genome. The related nucleotide excision repair pathway repairs bulky helix-distorting lesions.
SOS response	The SOS response is a global response to DNA damage in which the cell cycle is arrested and DNA repair and mutagenesis are induced. The system involves the RecA protein (Rad51 in eukaryotes). The RecA protein, stimulated by single-stranded DNA, is involved in the inactivation of the LexA repressor thereby inducing the response.
Transcription-coupled repair	Transcription-coupled repair is a DNA repair mechanism which operates in tandem with transcription. The activity of TCR has been known for 20 years, but its mechanism of action is an area of current research. Failure of the transcription-coupled repair is the cause of Cockayne syndrome, an extreme form of accelerated aging that is fatal early in life.
AP endonuclease	Apurinic/apyrimidinic (AP) endonuclease (BRENDA = 4.2.99.18) is an enzyme that is involved in the DNA base excision repair pathway (BER). Its main role in the repair of damaged or mismatched nucleotides in DNA is to create a nick in the phosphodiester backbone of the AP site created when DNA glycosylase removes the damaged base. There are four types of AP endonucleases that have been classified according to their sites of incision.
Hypoxanthine	Hypoxanthine is a naturally occurring purine derivative. It is occasionally found as a constituent of nucleic acids, where it is present in the anticodon of tRNA in the form of its nucleoside inosine.

Chapter 2. Part 2: Genes and Genomes

Cockayne syndrome	Cockayne syndrome (also called Weber-Cockayne syndrome is a rare autosomal recessive, congenital disorder characterized by growth failure, impaired development of the nervous system, abnormal sensitivity to sunlight (photosensitivity), and premature aging. Hearing loss and eye abnormalities (pigmentary retinopathy) are other common features, but problems with any or all of the internal organs are possible. It is associated with a group of disorders called leukodystrophies.
Lyme disease	Lyme disease, is an emerging infectious disease caused by at least three species of bacteria belonging to the genus Borrelia. Borrelia burgdorferi sensu stricto is the main cause of Lyme disease in the United States, whereas Borrelia afzelii and Borrelia garinii cause most European cases. he town of Lyme, Connecticut, USA, where a number of cases were identified in 1975. Although Allen Steere realized that Lyme disease was a tick-borne disease in 1978, the cause of the disease remained a mystery until 1981, when B. burgdorferi was identified by Willy Burgdorfer.
Transposable element	A transposable element is a DNA sequence that can change its relative position (self-transpose) within the genome of a single cell. The mechanism of transposition can be either 'copy and paste' or 'cut and paste'. Transposition can create phenotypically significant mutations and alter the cell's genome size.
Insertion sequence	An insertion sequence is a short DNA sequence that acts as a simple transposable element. Insertion sequences have two major characteristics: they are small relative to other transposable elements (generally around 700 to 2500 bp in length) and only code for proteins implicated in the transposition activity (they are thus different from other transposons, which also carry accessory genes such as antibiotic resistance genes). These proteins are usually the transposase which catalyses the enzymatic reaction allowing the IS to move, and also one regulatory protein which either stimulates or inhibits the transposition activity.
Inverted repeat	An inverted repeat is a sequence of nucleotides that is the reversed complement of another sequence further downstream. For example, 5'---GACTGC....GCAGTC---3'. When no nucleotides intervene between the sequence and its downstream complement, it is called a palindrome.
Transposon	Transposons are sequences of DNA that can move or transpose themselves to new positions within the genome of a single cell. The mechanism of transposition can be either 'copy and paste' or 'cut and paste'. Transposition can create phenotypically significant mutations and alter the cell's genome size.
Replicative transposition	Replicative transposition is a mechanism of transposition in molecular biology, proposed by James A.

Shapiro in 1979, in which the transposable element is duplicated during the reaction, so that the transposing entity is a copy of the original element. In this mechanism, the donor and receptor DNA sequences form a characteristic intermediate 'theta' configuration, sometimes called a 'Shapiro intermediate'. Replicative transposition is characteristic to retrotransposons and occurs from time to time in class II transposons..

Transposase	Transposase is an enzyme that binds to the ends of a transposon and catalyzes the movement of the transposon to another part of the genome by a cut and paste mechanism or a replicative transposition mechanism. The word 'transposase' was first coined by the individuals who cloned the enzyme required for tranposition of the Tn3 transposon. The existence of transposons was postulated in the late 1940s by Barbara McClintock, who was studying the inheritance of maize, but the actual molecular basis for transposition was described by later groups.
Composite transposon	A composite transposon is similar in function to simple transposons and Insertion Sequence (IS) elements in that it has protein coding DNA segments flanked by inverted, repeated sequences that can be recognized by transposase enzymes. A composite transposon, however, is flanked by two separate IS elements which may or may not be exact replicas. Instead of each IS element moving separately, the entire length of DNA spanning from one IS element to the other is transposed as one complete unit.
Genome evolution	The genome evolution is a set of phenomena involved in the changing of the structure of a genome throughout the evolution. The study of genome evolution involves multiple fields such as structural analysis of the genome (evolution of its size, gene content etc)., the study of genomic parasites, gene and ancient genome duplications, polypoidy, comparative genomics. In multicellular organisms, there is a paradox observed, namely that there are similar genes and largely similar mechanisms operating, often with redundant function.
Integron	An integron is a two component gene capture and dissemination system, initially discovered in relation to antibiotic resistance, and which is found in plasmids, chromosomes and transposons. The first component consists of a gene encoding a site specific recombinase along with a specific site for recombination, while the second component comprises fragments of DNA called gene cassettes which can be incorporated or shuffled. A cassette may encode genes for antibiotic resistance, although most genes in integrons are uncharacterized.
Evolution	Evolution is any change across successive generations in the inherited characteristics of biological populations.

| | Evolutionary processes give rise to diversity at every level of biological organisation, including species, individual organisms and molecules such as DNA and proteins.

Life on Earth originated and then evolved from a universal common ancestor approximately 3.7 billion years ago. |
| --- | --- |
| Human genome | The human genome is stored on 23 chromosome pairs and in the small mitochondrial DNA. Twenty-two of the 23 chromosomes belong to autosomal chromosome pairs, while the remaining pair is sex determinative. The haploid human genome occupies a total of just over three billion DNA base pairs. The Human Genome Project (HGP) produced a reference sequence of the euchromatic human genome and which is used worldwide in the biomedical sciences. |
| Integrase | Integrase Zinc binding domain

Retroviral integrase is an enzyme produced by a retrovirus (such as HIV) that enables its genetic material to be integrated into the DNA of the infected cell. Retroviral INs are not to be confused with phage integrases, such as λ phage integrase .

IN is a key component in the retroviral pre-integration complex (PIC). |
| Mobile genetic elements | Mobile genetic elements are a type of DNA that can move around within the genome. They include:•Transposons (also called transposable elements) •Retrotransposons•DNA transposons•Insertion sequences•Plasmids•Bacteriophage elements, like Mu, which integrates randomly into the genome•Group II introns

The total of all mobile genetic elements in a genome may be referred to as the mobilome.

Barbara McClintock was awarded the 1983 Nobel Prize in Physiology or Medicine 'for her discovery of mobile genetic elements'. |
| Salmonella | Salmonella is a genus of rod-shaped, Gram-negative, non-spore-forming, predominantly motile enterobacteria with diameters around 0.7 to 1.5 μm, lengths from 2 to 5 μm, and flagella which grade in all directions (i.e. peritrichous). They are chemoorganotrophs, obtaining their energy from oxidation and reduction reactions using organic sources, and are facultative anaerobes. Most species produce hydrogen sulfide, which can readily be detected by growing them on media containing ferrous sulfate, such as TSI. Most isolates exist in two phases: a motile phase I and a nonmotile phase II. Cultures that are nonmotile upon primary culture may be switched to the motile phase using a Cragie tube. |

Chapter 2. Part 2: Genes and Genomes

Bordetella pertussis	Bordetella pertussis is a Gram-negative, aerobic coccobacillus capsulate of the genus Bordetella, and the causative agent of pertussis or whooping cough. Unlike B. bronchiseptica, B. pertussis is nonmotile. Its virulence factors include pertussis toxin, filamentous hæmagglutinin, pertactin, fimbria, and tracheal cytotoxin.
Bradyrhizobium	Bradyrhizobium is a genus of Gram-negative soil bacteria, many of which fix nitrogen. Nitrogen fixation is an important part of the nitrogen cycle. Plants cannot use atmospheric nitrogen (N_2) they must use nitrogen compounds such as nitrates.
Gene duplication	Gene duplication is any duplication of a region of DNA that contains a gene; it may occur as an error in homologous recombination, a retrotransposition event, or duplication of an entire chromosome. The second copy of the gene is often free from selective pressure -- that is, mutations of it have no deleterious effects to its host organism. Thus it accumulates mutations faster than a functional single-copy gene, over generations of organisms.
Divergent evolution	Divergent evolution is the accumulation of differences between groups which can lead to the formation of new species, usually a result of diffusion of the same species to different and isolated environments which blocks the gene flow among the distinct populations allowing differentiated fixation of characteristics through genetic drift and natural selection. Primarily diffusion is the basis of molecular division can be seen in some higher-level characters of structure and function that are readily observable in organisms. For example, the vertebrate limb is one example of divergent evolution.
Genetic variation	Genetic variation, variation in alleles of genes, occurs both within and among populations. Genetic variation is important because it provides the 'raw material' for natural selection. Genetic variation is brought about by mutation, which is a change in the chemical structure of a gene.
Myxococcus xanthus	Myxococcus xanthus colonies exist as a self-organized, predatory, saprotrophic, single-species biofilm called a swarm. Myxococcus xanthus, which can be found almost ubiquitously in soil, are thin rod shaped, gram-negative cells that exhibit self-organizing behavior as a response to environmental cues. The swarm, which has been compared to a 'wolf-pack,' modifies its environment through stigmergy.
Blood-brain barrier	The blood-brain barrier is a separation of circulating blood and cerebrospinal fluid (CSF) in the central nervous system (CNS). It occurs along all capillaries and consists of tight junctions around the capillaries that do not exist in normal circulation. Endothelial cells restrict the diffusion of microscopic objects (e.g. bacteria) and large or hydrophilic molecules into the CSF, while allowing the diffusion of small hydrophobic molecules (O_2, hormones, CO_2).
Corepressor	In the field of molecular biology, a corepressor is a substance that inhibits the expression of genes.

In prokaryotes, corepressors are small molecules whereas in eukaryotes, corepressors are proteins. A corepressor does not directly bind to DNA, but instead indirectly regulates gene expression by binding to repressors.

Vibrio fischeri	Vibrio fischeri is a Gram-negative, rod-shaped bacterium found globally in marine environments. V. fischeri has bioluminescent properties, and is found predominantly in symbiosis with various marine animals, such as the bobtail squid. It is heterotrophic and moves by means of flagella.
Derepression	In genetics and biochemistry, a repressor gene inhibits the activity of an operator gene. By inactivating the repressor, the operator gene becomes active again. This effect is called derepression.
Repressor	In molecular genetics, a repressor is a DNA-binding protein that regulates the expression of one or more genes by binding to the operator and blocking the attachment of RNA polymerase to the promoter, thus preventing transcription of the genes. This blocking of expression is called repression. Repressor proteins are coded for by regulator genes.
TATA-binding protein	The TATA-binding protein is a general transcription factor that binds specifically to a DNA sequence called the TATA box. This DNA sequence is found about 25 base pairs upstream of the transcription start site in some eukaryotic gene promoters. TBP, along with a variety of TBP-associated factors, make up the TFIID, a general transcription factor that in turn makes up part of the RNA polymerase II preinitiation complex.
Signal transduction	Signal transduction occurs when an extracellular signaling molecule activates a cell surface receptor. In turn, this receptor alters intracellular molecules creating a response. There are two stages in this process:•A signaling molecule activates a specific receptor protein on the cell membrane.•A second messenger transmits the signal into the cell, eliciting a physiological response. In either step, the signal can be amplified.
ATP hydrolysis	ATP hydrolysis is the reaction by which chemical energy that has been stored and transported in the high-energy phosphoanhydridic bonds in ATP (Adenosine triphosphate) is released, for example in the muscles, to produce work. The product is ADP (Adenosine diphosphate) and an inorganic phosphate, orthophosphate (Pi). ADP can be further hydrolyzed to give energy, AMP (Adenosine monophosphate), and another orthophosphate (Pi).

Chapter 2. Part 2: Genes and Genomes

Energy carrier	According to ISO 13600, an energy carrier is either a substance (energy form) or a phenomenon (energy system) that can be used to produce mechanical work or heat or to operate chemical or physical processes. In the field of Energetics, however, an energy carrier corresponds only to an energy form (not an energy system) of energy input required by the various sectors of society to perform their functions. Examples of energy carriers include liquid fuel in a furnace, gasoline in a pump, electricity in a factory or a house, and hydrogen in a tank of a car.
Active transport	Active transport is the movement of a substance against its concentration gradient (from low to high concentration). In all cells, this is usually concerned with accumulating high concentrations of molecules that the cell needs, such as ions, glucose and amino acids. If the process uses chemical energy, such as from adenosine triphosphate (ATP), it is termed primary active transport.
Chemical reaction	A chemical reaction is a process that leads to the transformation of one set of chemical substances to another. Chemical reactions can be either spontaneous, requiring no input of energy, or non-spontaneous, typically following the input of some type of energy, such as heat, light or electricity. Classically, chemical reactions encompass changes that strictly involve the motion of electrons in the forming and breaking of chemical bonds, although the general concept of a chemical reaction, in particular the notion of a chemical equation, is applicable to transformations of elementary particles (such as illustrated by Feynman diagrams), as well as nuclear reactions.
Galactose	Galactose, is a type of sugar that is less sweet than glucose. It is a C-4 epimer of glucose. Galactan is a polymer of the sugar galactose found in hemicellulose.
Amino acid	Amino acids are molecules containing an amine group, a carboxylic acid group, and a side-chain that is specific to each amino acid. The key elements of an amino acid are carbon, hydrogen, oxygen, and nitrogen. They are particularly important in biochemistry, where the term usually refers to alpha-amino acids.
Diauxic growth	Diauxic growth is any cell growth characterized by cellular growth in two phases, and can be illustrated with a diauxic growth curve. Diauxic growth, meaning double growth, is caused by the presence of two sugars on a culture growth media, one of which is easier for the target bacterium to metabolize. This sugar is consumed first, which leads to rapid growth, followed by a lag phase, where the cellular machinery used to metabolize the second sugar is activated.

Stringent response	The stringent response is a stress response that occurs in bacteria and plant chloroplasts in reaction to amino-acid starvation, fatty acid limitation, iron limitation, heat shock and other stress conditions. The stringent response is signaled by the alarmone (p)ppGpp, and modulating transcription of up to 1/3 of all genes in the cell. This in turn causes the cell to divert resources away from growth and division and toward amino acid synthesis in order to promote survival until nutrient conditions improve.
Eukaryote	A eukaryote is an organism whose cells contain complex structures enclosed within membranes. Eukaryotes may more formally be referred to as the taxon Eukarya or Eukaryota. The defining membrane-bound structure that sets eukaryotic cells apart from prokaryotic cells is the nucleus, or nuclear envelope, within which the genetic material is carried.
Anti-sigma factors	In the regulation of gene expression in prokaryotes, anti-sigma factors bind to RNA polymerases and inhibit transcriptional activity. Anti-sigma factors have been found in a number of bacteria, including Escherichia coli and Salmonella, and in the T4 bacteriophage. Anti-sigma factors are antagonists to the sigma factors, which regulate numerous cell processes including flagellar production, stress response, transport and cellular growth.
Clostridium	Clostridium is a genus of Gram-positive bacteria, belonging to the Firmicutes. They are obligate anaerobes capable of producing endospores. Individual cells are rod-shaped, which gives them their name, from the Greek kloster or spindle.
Antisense RNA	Antisense RNA is a single-stranded RNA that is complementary to a messenger RNA (mRNA) strand transcribed within a cell. Antisense RNA may be introduced into a cell to inhibit translation of a complementary mRNA by base pairing to it and physically obstructing the translation machinery. This effect is therefore stoichiometric.
Direct repeat	Direct repeats are a type of genetic sequence that consists of two or more repeats of a specific sequence.

Direct repeats are nucleotide sequences present in multiple copies in the genome. There are several types of repeated sequences. |
| Hin recombinase | Hin recombinase is a 21kD protein composed of 198 amino acids that is found in the bacteria Salmonella. Hin belongs to the serine recombinase family of DNA invertases in which it relies on the active site serine to initiate DNA cleavage and recombination. The related protein, gamma-delta resolvase shares high similarity to Hin, of which much structural work has been done, including structures bound to DNA and reaction intermediates. |
| Tandem repeat | Tandem repeats occur in DNA when a pattern of two or more nucleotides is repeated and the repetitions are directly adjacent to each other. |

	An example would be:A-T-T-C-G-A-T-T-C-G-A-T-T-C-G
	in which the sequence A-T-T-C-G is repeated three times. Terminology
	When between 10 and 60 nucleotides are repeated, it is called a minisatellite.
Flagellin	Flagellin is a protein that arranges itself in a hollow cylinder to form the filament in bacterial flagellum. It has a mass of about 30,000 to 60,000 daltons. Flagellin is the principal substituent of bacterial flagellum, and is present in large amounts on nearly all flagellated bacteria.
Gene family	A gene family is a set of several similar genes, formed by duplication of a single original gene, and generally with similar biochemical functions. One such family are the genes for human haemoglobin subunits; the ten genes are in two clusters on different chromosomes, called the α-globin and β-globin loci. Genes are categorized into families based on shared nucleotide or protein sequences.
Chemotaxis	Chemotaxis is the phenomenon whereby somatic cells, bacteria, and other single-cell or multicellular organisms direct their movements according to certain chemicals in their environment. This is important for bacteria to find food (for example, glucose) by swimming towards the highest concentration of food molecules, or to flee from poisons (for example, phenol). In multicellular organisms, chemotaxis is critical to early development (e.g. movement of sperm towards the egg during fertilization) and subsequent phases of development (e.g. migration of neurons or lymphocytes) as well as in normal function.
Trypanosoma brucei	Trypanosoma brucei is a parasitic protist species that causes African trypanosomiasis (or sleeping sickness) in humans and nagana in animals in Africa. There are 3 sub-species of T. brucei: T. b. brucei, T. b. gambiense and T. b.
Methyl-accepting chemotaxis protein	MCPsignal Methyl-accepting chemotaxis protein is a transmembrane sensor protein of bacteria. Use of the MCP allows bacteria to detect concentrations of molecules in the extracellular matrix so that the bacteria may smooth swim or tumble accordingly. If the bacteria detects rising levels of attractants (nutrients) or declining levels of repellents (toxins), the bacteria will continue swimming forward, or smooth swimming.
Nitrogen cycle	The nitrogen cycle is the process by which nitrogen is converted between its various chemical forms. This transformation can be carried out to both biological and non-biological processes.

Nitrogen fixation	Nitrogen fixation is a process by which nitrogen (N_2) in the atmosphere is converted into ammonium (NH_{4+}). Atmospheric nitrogen or elemental nitrogen (N_2) is relatively inert: it does not easily react with other chemicals to form new compounds. Fixation processes free up the nitrogen atoms from their diatomic form (N_2) to be used in other ways.
Autoinducer	Autoinducers are chemical signaling molecules that are produced and used by bacteria participating in quorum sensing. Quorum sensing is a phenomenon that allows both Gram-negative and Gram-positive bacteria to sense one another and to regulate a wide variety of physiological activities. Such activities include symbiosis, virulence, motility, antibiotic production, and biofilm formation.
Vibrio harveyi	Vibrio harveyi is a species of Gram-negative, bioluminescent, marine bacteria in the genus Vibrio. V. harveyi are rod-shaped, motile (via polar flagella), facultatively anaerobic, halophilic, and competent for both fermentative and respiratory metabolism. They do not grow at 4°C or above 35° C. V. harveyi can be found free-swimming in tropical marine waters, commensally in the gut microflora of marine animals, and as both a primary and opportunistic pathogen of marine animals, including Gorgonian corals, oysters, prawns, lobsters, the common snook, barramundi, turbot, milkfish, and seahorses.
Pathogen	A pathogen (Greek: π?θος pathos, 'suffering, passion' and γεν?ς genes (-gen) 'producer of') or infectious agent -- in colloquial terms, a germ -- is a microorganism such as a virus, bacterium, prion, or fungus, that causes disease in its animal or plant host. There are several substrates including pathways wherein pathogens can invade a host; the principal pathways have different episodic time frames, but soil contamination has the longest or most persistent potential for harboring a pathogen.

Not all pathogens are negative. |
| Virulence factor | Virulence factors are molecules expressed and secreted by pathogens (bacteria, viruses, fungi and protozoa) that enable them to achieve the following:•colonization of a niche in the host (this includes adhesion to cells)•Immunoevasion, evasion of the host's immune response•Immunosuppression, inhibition of the host's immune response•entry into and exit out of cells (if the pathogen is an intracellular one)•obtain nutrition from the host.

Virulence factors are very often responsible for causing disease in the host as they inhibit certain host functions.

Pathogens possess a wide array of virulence factors. Some are intrinsic to the bacteria (e.g. capsules and endotoxin) whereas others are obtained from plasmids (e.g. some toxins). |
| Enteromorpha | Enteromorpha is an outdated sea-plant taxon. |

Chapter 2. Part 2: Genes and Genomes

Vibrio anguillarum	Vibrio anguillarum is a Gram negative, curved rod bacterium with one polar flagellum. It is an important pathogen of cultured salmonid fish, and causes the disease known as vibriosis or red pest of eels. The disease has been observed in the salmon, the bream, the eel, the mullet, the catfish and Tilapia amongst others.
Isoelectric focusing	Isoelectric focusing also known as electrofocusing, is a technique for separating different molecules by their electric charge differences. It is a type of zone electrophoresis, usually performed on proteins in a gel, that takes advantage of the fact that overall charge on the molecule of interest is a function of the pH of its surroundings. IEF involves adding an ampholyte solution into immobilized pH gradient (IPG) gels.
Isoelectric point	The isoelectric point sometimes abbreviated to IEP, is the pH at which a particular molecule or surface carries no net electrical charge. Amphoteric molecules called zwitterions contain both positive and negative charges depending on the functional groups present in the molecule. The net charge on the molecule is affected by pH of their surrounding environment and can become more positively or negatively charged due to the loss or gain of protons (H^+).
Proteome	The proteome is the entire set of proteins expressed by a genome, cell, tissue or organism. More specifically, it is the set of expressed proteins in a given type of cells or an organism at a given time under defined conditions. The term is a portmanteau of proteins and genome.
Transcriptome	The transcriptome is the set of all RNA molecules, including mRNA, rRNA, tRNA, and other non-coding RNA produced in one or a population of cells. The term can be applied to the total set of transcripts in a given organism, or to the specific subset of transcripts present in a particular cell type. Unlike the genome, which is roughly fixed for a given cell line (excluding mutations), the transcriptome can vary with external environmental conditions.
Complementary DNA	In genetics, complementary DNA is DNA synthesized from a messenger RNA (mRNA) template in a reaction catalyzed by the enzyme reverse transcriptase and the enzyme DNA polymerase. cDNA is often used to clone eukaryotic genes in prokaryotes. When scientists want to express a specific protein in a cell that does not normally express that protein (i.e., heterologous expression), they will transfer the cDNA that codes for the protein to the recipient cell.

Gel electrophoresis	Gel electrophoresis is a method used in clinical chemistry to separate proteins by charge and or size (IEF agarose, essentially size independent) and in biochemistry and molecular biology to separate a mixed population of DNA and RNA fragments by length, to estimate the size of DNA and RNA fragments or to separate proteins by charge. Nucleic acid molecules are separated by applying an electric field to move the negatively charged molecules through an agarose matrix. Shorter molecules move faster and migrate farther than longer ones because shorter molecules migrate more easily through the pores of the gel.
Polyacrylamide gel	A Polyacrylamide Gel is a separation matrix used in electrophoresis of biomolecules, such as proteins or DNA fragments. Traditional DNA sequencing techniques such as Maxam-Gilbert or Sanger methods used polyacrylamide gels to separate DNA fragments differing by a single base-pair in length so the sequence could be read. Most modern DNA separation methods now use agarose gels, except for particularly small DNA fragments.
Proteomics	Proteomics is the large-scale study of proteins, particularly their structures and functions. Proteins are vital parts of living organisms, as they are the main components of the physiological metabolic pathways of cells. The term 'proteomics' was first coined in 1997 to make an analogy with genomics, the study of the genes.
Metronidazole	Metronidazole is a nitroimidazole antibiotic medication used particularly for anaerobic bacteria and protozoa. Metronidazole is an antibiotic, amebicide, and antiprotozoal. It is the drug of choice for first episodes of mild-to-moderate Clostridium difficile infection.
RNA virus	An RNA virus is a virus that has RNA (ribonucleic acid) as its genetic material. This nucleic acid is usually single-stranded RNA (ssRNA), but may be double-stranded RNA (dsRNA). Notable human diseases caused by RNA viruses include SARS, influenza, hepatitis C and polio.
Life cycle	A life cycle is a period involving all different generations of a species succeeding each other through means of reproduction, whether through asexual reproduction or sexual reproduction (a period from one generation of organisms to the same identical). For example, a complex life cycle of Fasciola hepatica includes three different multicellular generations: 1) 'adult' hermaphroditic; 2) sporocyst; 3) redia.
Molecular biology	Molecular biology is the branch of biology that deals with the molecular basis of biological activity. This field overlaps with other areas of biology and chemistry, particularly genetics and biochemistry. Molecular biology chiefly concerns itself with understanding the interactions between the various systems of a cell, including the interactions between the different types of DNA, RNA and protein biosynthesis as well as learning how these interactions are regulated.
Chemokine receptor	Chemokine receptors are cytokine receptors found on the surface of certain cells that interact with a type of cytokine called a chemokine.

Chapter 2. Part 2: Genes and Genomes

There have been 19 distinct chemokine receptors described in mammals. Each has a 7-transmembrane (7TM) structure and couples to G-protein for signal transduction within a cell, making them members of a large protein family of G protein-coupled receptors.

Stem cell	Stem cells are biological cells found in all multicellular organisms, that can divide (through mitosis) and differentiate into diverse specialized cell types and can self-renew to produce more stem cells. In mammals, there are two broad types of stem cells: embryonic stem cells, which are isolated from the inner cell mass of blastocysts, and adult stem cells, which are found in various tissues. In adult organisms, stem cells and progenitor cells act as a repair system for the body, replenishing adult tissues.
Capsid	A capsid is the protein shell of a virus. It consists of several oligomeric structural subunits made of protein called protomers. The observable 3-dimensional morphological subunits, which may or may not correspond to individual proteins, are called capsomeres.
Lysogeny	Lysogeny, is one of two methods of viral reproduction (the lytic cycle is the other). Lysogeny is characterized by integration of the bacteriophage nucleic acid into the host bacterium's genome. The newly integrated genetic material, called a prophage can be transmitted to daughter cells at each subsequent cell division, and a later event (such as UV radiation) can release it, causing proliferation of new phages via the lytic cycle.
Nonsense suppressor	A nonsense suppressor is a tRNA mutation that suppresses the protein truncation resulting from a nonsense mutation. The nonsense suppressor is a tRNA gene (from Escherichia coli) which has an anticodon mutated to recognize the nonsense (stop) codon. As a result, when the ribosome reaches the codon in question, the nonsense suppressor will sometimes bind and incorporate its amino acid (instead of the regular stop-codon tRNA binding and causing termination), allowing translation to continue.
Polio vaccine	Two polio vaccines are used throughout the world to combat poliomyelitis (or polio). The first was developed by Jonas Salk and first tested in 1952. Announced to the world by Salk on April 12, 1955, it consists of an injected dose of inactivated (dead) poliovirus. An oral vaccine was developed by Albert Sabin using attenuated poliovirus.
Poliovirus	Poliovirus, the causative agent of poliomyelitis, is a human enterovirus and member of the family of Picornaviridae. Poliovirus is composed of an RNA genome and a protein capsid. The genome is a single-stranded positive-sense RNA genome that is about 7500 nucleotides long. The viral particle is about 30 nanometres in diameter with icosahedral symmetry.

Post-polio syndrome	Post-polio syndrome is a condition that affects approximately 25-50% of people who have previously contracted poliomyelitis--a viral infection of the nervous system--after the initial infection. Typically the symptoms appear 15-30 years after recovery from the original paralytic attack, at an age of 35 to 60. Symptoms include acute or increased muscular weakness, pain in the muscles, and fatigue. The same symptoms may also occur years after a nonparalytic polio (NPP) infection.
Gut-associated lymphoid tissue	The digestive tract's immune system is often referred to as gut-associated lymphoid tissue and works to protect the body from invasion. GALT is an example of mucosa-associated lymphoid tissue.

The digestive tract is an important component of the body's immune system. |
| Viremia | Viremia is a medical condition where viruses enter the bloodstream and hence have access to the rest of the body. It is similar to bacteremia, a condition where bacteria enter the bloodstream.

Primary viremia refers to the initial spread of virus in the blood from the first site of infection. |
| Palmaria palmata | Palmaria palmata Kuntze, also called dulse, dillisk, dilsk, red dulse, sea lettuce flakes or creathnach, is a red alga (Rhodophyta) previously referred to as Rhodymenia palmata (Linnaeus) Greville. It grows on the northern coasts of the Atlantic and Pacific oceans. It is a well-known snack food, and in Iceland, where it is known as söl, it has been an important source of fiber throughout the centuries. |
| RNA-dependent RNA polymerase | RNA-directed RNA polymerase, flaviviral

RNA-dependent RNA polymerase (RDR), or RNA replicase, is an enzyme (EC 2.7.7.48) that catalyzes the replication of RNA from an RNA template. This is in contrast to a typical DNA-dependent RNA polymerase, which catalyzes the transcription of RNA from a DNA template.

RNA-dependent RNA polymerase is an essential protein encoded in the genomes of all RNA-containing viruses with no DNA stage that have sense negative RNA. It catalyses synthesis of the RNA strand complementary to a given RNA template. |
| Hemagglutinin | Hemagglutinin

Influenza hemagglutinin (HA) or haemagglutinin is a type of hemagglutinin found on the surface of the influenza viruses. It is an antigenic glycoprotein. It is responsible for binding the virus to the cell that is being infected. |

Chapter 2. Part 2: Genes and Genomes

Pandemic	A pandemic is an epidemic of infectious disease that has spread through human populations across a large region; for instance multiple continents, or even worldwide. A widespread endemic disease that is stable in terms of how many people are getting sick from it is not a pandemic. Further, flu pandemics generally exclude recurrences of seasonal flu.
Haemophilus influenzae	Haemophilus influenzae, formerly called Pfeiffer's bacillus or Bacillus influenzae, Gram-negative, rod-shaped bacterium first described in 1892 by Richard Pfeiffer during an influenza pandemic. A member of the Pasteurellaceae family, it is generally aerobic, but can grow as a facultative anaerobe. H. influenzae was mistakenly considered to be the cause of influenza until 1933, when the viral etiology of the flu became apparent; the bacterium is colloquially known as bacterial influenza.
Influenza	Influenza, commonly referred to as the flu, is an infectious disease caused by RNA viruses of the family Orthomyxoviridae (the influenza viruses), that affects birds and mammals. The most common symptoms of the disease are chills, fever, sore throat, muscle pains, severe headache, coughing, weakness/fatigue and general discomfort. Sore throat, fever and coughs are the most frequent symptoms.
Neuraminidase	Neuraminidase enzymes are glycoside hydrolase enzymes (EC 3.2.1.18) that cleave the glycosidic linkages of neuraminic acids. Neuraminidase enzymes are a large family, found in a range of organisms. The best-known neuraminidase is the viral neuraminidase, a drug target for the prevention of the spread of influenza infection.
Vaccine	A vaccine is a biological preparation that improves immunity to a particular disease. A vaccine typically contains an agent that resembles a disease-causing microorganism, and is often made from weakened or killed forms of the microbe, its toxins or one of its surface proteins. The agent stimulates the body's immune system to recognize the agent as foreign, destroy it, and 'remember' it, so that the immune system can more easily recognize and destroy any of these microorganisms that it later encounters.
Sialic acid	Sialic acid is a generic term for the N- or O-substituted derivatives of neuraminic acid, a monosaccharide with a nine-carbon backbone. It is also the name for the most common member of this group, N-acetylneuraminic acid (Neu5Ac or NANA). Sialic acids are found widely distributed in animal tissues and to a lesser extent in other species ranging from plants and fungi to yeasts and bacteria, mostly in glycoproteins and gangliosides.
Zidovudine	Zidovudine or azidothymidine (AZT) (also called ZDV) is a nucleoside analog reverse transcriptase inhibitor (NRTI), a type of antiretroviral drug used for the treatment of HIV/AIDS. It is an analog of thymidine. AZT was the first approved treatment for HIV, sold under the names Retrovir and Retrovis.

Oncogene	An oncogene is a gene that has the potential to cause cancer. In tumor cells, they are often mutated or expressed at high levels. Most normal cells undergo a programmed form of death (apoptosis).
Rous sarcoma virus	Rous sarcoma virus is a retrovirus and is the first oncovirus to have been described: it causes sarcoma in chickens. As with all retroviruses, it reverse transcribes its RNA genome into cDNA before integration into the host DNA.
Gene therapy	Gene therapy is the use of DNA as a pharmaceutical agent to treat disease. It derives its name from the idea that DNA can be used to supplement or alter genes within an individual's cells as a therapy to treat disease. The most common form of gene therapy involves using DNA that encodes a functional, therapeutic gene in order to replace a mutated gene.
Herpes simplex	Herpes simplex is a viral disease caused by both Herpes simplex virus type 1 (HSV-1) and type 2 (HSV-2). Infection with the herpes virus is categorized into one of several distinct disorders based on the site of infection. Oral herpes, the visible symptoms of which are colloquially called cold sores or fever blisters, infects the face and mouth.
Protein A	Protein A is a 40-60 kDa MSCRAMM surface protein originally found in the cell wall of the bacterium Staphylococcus aureus. It is encoded by the spa gene and its regulation is controlled by DNA topology, cellular osmolarity, and a two-component system called ArlS-ArlR. It has found use in biochemical research because of its ability to bind immunoglobulins. It binds proteins from many of mammalian species, most notably IgGs.
Protein structure	In molecular biology protein structure describes the various levels of organization of protein molecules. Proteins are an important class of biological macromolecules present in all organisms. Proteins are polymers of amino acids.
Chemokine	Small cytokines (intecrine/chemokine), interleukin-8 like Chemokines are a family of small cytokines, or proteins secreted by cells. Their name is derived from their ability to induce directed chemotaxis in nearby responsive cells; they are chemotactic cytokines. Proteins are classified as chemokines according to shared structural characteristics such as small size (they are all approximately 8-10 kilodaltons in size), and the presence of four cysteine residues in conserved locations that are key to forming their 3-dimensional shape.
Langerhans cell	Langerhans cells are dendritic cells of the epidermis, containing large granules called Birbeck granules. They are also normally present in lymph nodes and other organs, including the stratum spinosum layer of the epidermis.

Chapter 2. Part 2: Genes and Genomes

Microglia	Microglia are a type of glial cell that are the resident macrophages of the brain and spinal cord, and thus act as the first and main form of active immune defense in the central nervous system (CNS). Microglia constitute 20% of the total glial cell population within the brain. Microglia are distributed in large non-overlapping regions throughout the brain and spinal cord.
Infectious disease	Infectious diseases, also known as transmissible diseases or communicable diseases comprise clinically evident illness (i.e., characteristic medical signs and/or symptoms of disease) resulting from the infection, presence and growth of pathogenic biological agents in an individual host organism. In certain cases, infectious diseases may be asymptomatic for much or even all of their course in a given host. In the latter case, the disease may only be defined as a 'disease' (which by definition means an illness) in hosts who secondarily become ill after contact with an asymptomatic carrier.
Opportunistic infection	An opportunistic infection is an infection caused by pathogens, particularly opportunistic pathogens--those that take advantage of certain situations--such as bacterial, viral, fungal or protozoan infections that usually do not cause disease in a healthy host, one with a healthy immune system. A compromised immune system, however, presents an 'opportunity' for the pathogen to infect.

Immunodeficiency or immunosuppression can be caused by:•Malnutrition•Recurrent infections•Immunosuppressing agents for organ transplant recipients•Advanced HIV infection•Chemotherapy for cancer•Genetic predisposition•Skin damage•Antibiotic treatment•Medical procedures•PregnancyTypes of infections

These infections include:•Acinetobacter baumanni•Aspergillus sp.•Candida albicans•Clostridium difficile•Cryptococcus neoformans•Cryptosporidium•Cytomegalovirus•Geomyces destructans•Histoplasma capsulatum•Isospora belli•Polyomavirus JC polyomavirus, the virus that causes Progressive multifocal leukoencephalopathy.•Kaposi's Sarcoma caused by Human herpesvirus 8 (HHV8), also called Kaposi's sarcoma-associated herpesvirus (KSHV)•Legionnaires' Disease (Legionella pneumophila)•Microsporidium•Mycobacterium avium complex (MAC) (Nontuberculosis Mycobacterium)•Pneumocystis jirovecii, previously known as Pneumocystis carinii f. |
| Reverse transcriptase | In the fields of molecular biology and biochemistry, a reverse transcriptase, is a DNA polymerase enzyme that transcribes single-stranded RNA into single-stranded DNA. It also is a DNA-dependent DNA polymerase which synthesizes a second strand of DNA complementary to the reverse-transcribed single-stranded cDNA after degrading the original mRNA with its RNaseH activity. Normal transcription involves the synthesis of RNA from DNA; hence, reverse transcription is the reverse of this. |

Well studied reverse transcriptases include:•HIV-1 reverse transcriptase from human immunodeficiency virus type 1 (PDB 1HMV)•M-MLV reverse transcriptase from the Moloney murine leukemia virus•AMV reverse transcriptase from the avian myeloblastosis virus•Telomerase reverse transcriptase that maintains the telomeres of eukaryotic chromosomesHistory

Reverse transcriptase was discovered by Howard Temin at the University of Wisconsin-Madison, and independently by David Baltimore in 1970 at MIT. The two shared the 1975 Nobel Prize in Physiology or Medicine with Renato Dulbecco for their discovery.

Long terminal repeat	Long terminal repeats (LTRs) are sequences of DNA that repeat hundreds or thousands of times. They are found in retroviral DNA and in retrotransposons, flanking functional genes. They are used by viruses to insert their genetic sequences into the host genomes.
Protease inhibitor	Protease inhibitors are a class of drugs used to treat or prevent infection by viruses, including HIV and Hepatitis C. Protease inhibitors prevent viral replication by inhibiting the activity of proteases, e.g.HIV-1 protease, enzymes used by the viruses to cleave nascent proteins for final assembly of new virions.

Protease inhibitors have been developed or are presently undergoing testing for treating various viruses:•HIV/AIDS: antiretroviral protease inhibitors (saquinavir, ritonavir, indinavir, nelfinavir, amprenavir etc).•Hepatitis C: Boceprevir•Hepatitis C: Telaprevir

Given the specificity of the target of these drugs there is the risk, as in antibiotics, of the development of drug-resistant mutated viruses. To reduce this risk it is common to use several different drugs together that are each aimed at different targets.

Epstein-Barr virus	The Epstein-Barr virus , also called human herpesvirus 4 (HHV-4), is a virus of the herpes family, which includes herpes simplex virus 1 and 2, and is one of the most common viruses in humans. It is best known as the cause of infectious mononucleosis. It is also associated with particular forms of cancer, particularly Hodgkin's lymphoma, Burkitt's lymphoma, nasopharyngeal carcinoma, and central nervous system lymphomas associated with HIV. Finally, there is evidence that infection with the virus is associated with a higher risk of certain autoimmune diseases, especially dermatomyositis, systemic lupus erythematosus, rheumatoid arthritis, Sjögren's syndrome, and multiple sclerosis.
Herpes simplex virus	Herpes simplex virus 1 and 2 are two members of the herpes virus family, Herpesviridae, that infect humans. Both Herpes simplex virus-1 (which produces most cold sores) and Herpes simplex virus-2 (which produces most genital herpes) are ubiquitous and contagious.

Chapter 2. Part 2: Genes and Genomes

Human papillomavirus	Human papillomavirus is a virus from the papillomavirus family that is capable of infecting humans. Like all papillomaviruses, HPVs establish productive infections only in keratinocytes of the skin or mucous membranes. While the majority of the known types of HPV cause no symptoms in most people, some types can cause warts (verrucae), while others can - in a minority of cases - lead to cancers of the cervix, vulva, vagina, penis, oropharynx and anus.
Retrotransposon	Retrotransposons (also called transposons via RNA intermediates) are genetic elements that can amplify themselves in a genome and are ubiquitous components of the DNA of many eukaryotic organisms. They are a subclass of transposon. They are particularly abundant in plants, where they are often a principal component of nuclear DNA. In maize, 49-78% of the genome is made up of retrotransposons.
Helicobacter	Helicobacter is a genus of Gram-negative bacteria possessing a characteristic helix shape. They were initially considered to be members of the Campylobacter genus, but since 1989 they have been grouped in their own genus. The Helicobacter genus belongs to class Epsilonproteobacteria, order Campylobacterales, family Helicobacteraceae and already involves >35 species.
Molecular evolution	Molecular evolution is in part a process of evolution at the scale of DNA, RNA, and proteins. Molecular evolution emerged as a scientific field in the 1960s as researchers from molecular biology, evolutionary biology and population genetics sought to understand recent discoveries on the structure and function of nucleic acids and protein. Some of the key topics that spurred development of the field have been the evolution of enzyme function, the use of nucleic acid divergence as a 'molecular clock' to study species divergence, and the origin of noncoding DNA. Recent advances in genomics, including whole-genome sequencing, high-throughput protein characterization, and bioinformatics have led to a dramatic increase in studies on the topic.
Viroid	In 1971 T.O Diener discovered a new infectious agent smaller than virus and caused potato spindle tuber disease. It was found to be a free RNA lacked the protein coat that is found in viruses, hence named viroid. Viroids are plant pathogens that consist of a short stretch (a few hundred nucleobases) of highly complementary, circular, single-stranded RNA without the protein coat that is typical for viruses.
Protein homology	Protein homology is biological homology between proteins, meaning that the proteins are derived from a common 'ancestor'. The proteins may be in different species, with the ancestral protein being the form of the protein that existed in the ancestral species (orthology). Or the proteins may be in the same species, but have evolved from a single protein whose gene was duplicated in the genome (paralogy).

Tegument	Tegument /'t?gj?m?nt/ is a terminology in helminthology for the name of the outer body covering among members of the phylum Platyhelminthes. The name is derived from a Latin word tegumentum or tegere, meaning 'to cover'. It is characteristic of all flatworms including the broad groups of tapeworms and flukes.
Bubonic plague	Bubonic plague is a zoonotic disease, circulating mainly among small rodents and their fleas, and is one of three types of infections caused by Yersinia pestis (formerly known as Pasteurella pestis), which belongs to the family Enterobacteriaceae. Without treatment, the bubonic plague kills about two out of three infected humans within 4 days.
	The term bubonic plague is derived from the Greek word βουβ?v, meaning 'groin.' Swollen lymph nodes (buboes) especially occur in the armpit and groin in persons suffering from bubonic plague.
Exocytosis	Exocytosis is the durable process by which a cell directs the contents of secretory vesicles out of the cell membrane. These membrane-bound vesicles contain soluble proteins to be secreted to the extracellular environment, as well as membrane proteins and lipids that are sent to become components of the cell membrane.
	In multicellular organisms there are two types of exocytosis: 1) Ca^{2+} triggered non-constitutive and 2) non Ca^{2+} triggered constitutive.
Listeria monocytogenes	Listeria monocytogenes, a facultative anaerobe, intracellular bacterium, is the causative agent of listeriosis. It is one of the most virulent foodborne pathogens, with 20 to 30 percent of clinical infections resulting in death. Responsible for approximately 2,500 illnesses and 500 deaths in the United States (U.S).
Yersinia pestis	Yersinia pestis is a Gram-negative rod-shaped bacterium. It is a facultative anaerobe that can infect humans and other animals.
	Human Y. pestis infection takes three main forms: pneumonic, septicemic, and the notorious bubonic plagues.
Pseudotyping	Pseudotyping is the process of producing viruses or viral vectors in combination with foreign viral envelope proteins. The result is a pseudotyped virus particle.. With this method, the foreign viral envelope proteins can be used to alter host tropism or an increased/decreased stability of the virus particles.

Chapter 2. Part 2: Genes and Genomes

Severe combined immunodeficiency	Severe Combined Immunodeficiency (SCID) is a severe immunodeficiency genetic disorder that is characterized by the complete inability of the adaptive immune system to mount, coordinate, and sustain an appropriate immune response, usually due to absent or atypical T and B lymphocytes. In humans, SCID is colloquially known as 'bubble boy' disease, as victims may require complete clinical isolation to prevent lethal infection from environmental microbes. Several forms of SCID occur in animal species.
Adenoviruses	Adenoviruses are medium-sized (90-100 nm), nonenveloped (without an outer lipid bilayer) icosahedral viruses composed of a nucleocapsid and a double-stranded linear DNA genome. There are 55 described serotypes in humans, which are responsible for 5-10% of upper respiratory infections in children, and many infections in adults as well. Viruses of the family Adenoviridae infect various species of vertebrates, including humans.
Combined immunodeficiencies	Combined immunodeficiencies are immunodeficiency disorders that involve multiple components of the immune system, including both humoral immunity and cell-mediated immunity.
Viral vector	Viral vectors are a tool commonly used by molecular biologists to deliver genetic material into cells. This process can be performed inside a living organism (in vivo) or in cell culture (in vitro). Viruses have evolved specialized molecular mechanisms to efficiently transport their genomes inside the cells they infect.
Nerve growth factor	Nerve growth factor is a small secreted protein that is important for the growth, maintenance, and survival of certain target neurons (nerve cells). It also functions as a signaling molecule. It is perhaps the prototypical growth factor, in that it is one of the first to be described.
Transfection	Transfection is the process of deliberately introducing nucleic acids into cells. The term is used notably for non-viral methods in eukaryotic cells. It may also refer to other methods and cell types, although other terms are preferred: 'transformation' is more often used to describe non-viral DNA transfer in bacteria, non-animal eukaryotic cells and plant cells - a distinctive sense of transformation refers to spontaneous genetic modifications (mutations to cancerous cells (carcinogenesis), or under stress (UV irradiation)).
Transgene	A transgene is a gene or genetic material that has been transferred naturally or by any of a number of genetic engineering techniques from one organism to another. In its most precise usage, the term transgene describes a segment of DNA containing a gene sequence that has been isolated from one organism and is introduced into a different organism.

Fusion protein	Fusion proteins or chimeric proteins are proteins created through the joining of two or more genes which originally coded for separate proteins. Translation of this fusion gene results in a single polypeptide with functional properties derived from each of the original proteins. Recombinant fusion proteins are created artificially by recombinant DNA technology for use in biological research or therapeutics.
Growth factor	A growth factor is a naturally occurring substance capable of stimulating cellular growth, proliferation and cellular differentiation. Usually it is a protein or a steroid hormone. Growth factors are important for regulating a variety of cellular processes.
Breast cancer	Breast cancer is cancer originating from breast tissue, most commonly from the inner lining of milk ducts or the lobules that supply the ducts with milk. Cancers originating from ducts are known as ductal carcinomas; those originating from lobules are known as lobular carcinomas. Prognosis and survival rate varies greatly depending on cancer type and staging.
Glioma	A glioma is a type of tumor that starts in the brain or spine. It is called a glioma because it arises from glial cells. The most common site of gliomas is the brain.
Biotechnology	Biotechnology is a field of applied biology that involves the use of living organisms and bioprocesses in engineering, technology, medicine and other fields requiring bioproducts. Biotechnology also utilizes these products for manufacturing purpose. Modern use of similar terms includes genetic engineering as well as cell and tissue culture technologies.
Genetic analysis	Genetic analysis can be used generally to describe methods both used in and resulting from the sciences of genetics and molecular biology, or to applications resulting from this research. Genetic analysis may be done to identify genetic/inherited disorders and also to make a differential diagnosis in certain somatic diseases such as cancer. Genetic analyses of cancer include detection of mutations, fusion genes, and DNA copy number changes.
Phytophthora infestans	Phytophthora infestans is an oomycete that causes the serious potato disease known as late blight or potato blight. (Early blight, caused by Alternaria solani, is also often called 'potato blight'). Late blight was a major culprit in the 1840s European, the 1845 Irish and 1846 Highland potato famines.
Western blot	The western blot is a widely used analytical technique used to detect specific proteins in the given sample of tissue homogenate or extract. It uses gel electrophoresis to separate native proteins by 3-D structure or denatured proteins by the length of the polypeptide.

Chapter 2. Part 2: Genes and Genomes

Infectious dose	Infectious dose is the amount of pathogen (measured in number of microorganisms) required to cause an infection in the host. Usually it varies according to the pathogenic agent and the consumer's age and overall health. Infectious doses for some known microorganisms •Escherichia coli : very large (10^6 - 10^8 of organisms)•Salmonella : quite large (> 10^5 of organisms)•Cholera : relatively large (10^4 - 10^6 of organisms)•Bacillus anthracis : relatively large (10^4 spores)•Campylobacter jejuni: low (500 organisms)•Shigella : very low (10s of organisms)•C.parvum : very low (10 to 30 oocysts)•Escherichia coli O157:H7 : very low (< 10 organisms)•Entamoeba coli : extremely low (from 1 cyst).
Electrophoretic mobility shift assay	An electrophoretic mobility shift assay or mobility shift electrophoresis, also referred as a gel shift assay, gel mobility shift assay, band shift assay, or gel retardation assay, is a common affinity electrophoresis technique used to study protein-DNA or protein-RNA interactions. This procedure can determine if a protein or mixture of proteins is capable of binding to a given DNA or RNA sequence, and can sometimes indicate if more than one protein molecule is involved in the binding complex. Gel shift assays are often performed in vitro concurrently with DNase footprinting, primer extension, and promoter-probe experiments when studying transcription initiation, DNA replication, DNA repair or RNA processing and maturation.
Glutamate decarboxylase	Glutamate decarboxylase is an enzyme that catalyzes the decarboxylation of glutamate to GABA and CO_2. GAD uses PLP as a cofactor. The reaction proceeds as follows:HOOC-CH_2-CH_2-CH (NH_2)-COOH → CO_2 + HOOC-CH_2-CH_2-CH_2NH_2 In mammals, GAD exists in two isoforms encoded by two different genes - GAD1 and GAD2. These isoforms are GAD_{67} and GAD_{65} with molecular weights of 67 and 65 kDa, respectively.
Affinity chromatography	Affinity chromatography is a method of separating biochemical mixtures and based on a highly specific interaction such as that between antigen and antibody, enzyme and substrate, or receptor and ligand. Affinity chromatography can be used to:•Purify and concentrate a substance from a mixture into a buffering solution•Reduce the amount of a substance in a mixture•Discern what biological compounds bind to a particular substance•Purify and concentrate an enzyme solution.Principle The immobile phase is typically a gel matrix, often of agarose; a linear sugar molecule derived from algae. Usually the starting point is an undefined heterogeneous group of molecules in solution, such as a cell lysate, growth medium or blood serum.
Primer extension	Primer extension is a technique whereby the 5' ends of RNA or DNA can be mapped.

	Primer extension can be used to determine the start site of RNA transcription for a known gene. This technique requires a radiolabelled primer (usually 20 - 50 nucleotides in length) which is complementary to a region near the 3' end of the gene.
Deoxyribonuclease I	Deoxyribonuclease I is an endonuclease coded by the human gene DNASE1. Deoxyribonuclease I is a nuclease that cleaves DNA preferentially at phosphodiester linkages adjacent to a pyrimidine nucleotide, yielding 5'-phosphate-terminated polynucleotides with a free hydroxyl group on position 3', on average producing tetranucleotides. It acts on single-stranded DNA, double-stranded DNA, and chromatin. It has been suggested to be one of the deoxyribonucleases responsible for DNA fragmentation during apoptosis.
Immunoprecipitation	Immunoprecipitation is the technique of precipitating a protein antigen out of solution using an antibody that specifically binds to that particular protein. This process can be used to isolate and concentrate a particular protein from a sample containing many thousands of different proteins. Immunoprecipitation requires that the antibody be coupled to a solid substrate at some point in the procedure.
Transcription factor	In molecular biology and genetics, a transcription factor is a protein that binds to specific DNA sequences, thereby controlling the flow of genetic information from DNA to mRNA. Transcription factors perform this function alone or with other proteins in a complex, by promoting (as an activator), or blocking (as a repressor) the recruitment of RNA polymerase (the enzyme that performs the transcription of genetic information from DNA to RNA) to specific genes.

A defining feature of transcription factors is that they contain one or more DNA-binding domains (DBDs), which attach to specific sequences of DNA adjacent to the genes that they regulate. Additional proteins such as coactivators, chromatin remodelers, histone acetylases, deacetylases, kinases, and methylases, while also playing crucial roles in gene regulation, lack DNA-binding domains, and, therefore, are not classified as transcription factors. |
| Cell physiology | Cell physiology is the biological study of the cell's mechanism and interaction in its environment. The term 'physiology' refers to all the normal functions that take place in a living organism. Absorption of water by roots, production of food in the leaves, and growth of shoots towards light are examples of plant physiology. |
| Interaction network | Interaction network is a network of nodes that are connected by features. If the feature is a physical and molecular, the interaction network is molecular interactions usually found in cells. Interaction network has become a research topic in biology in recent years due to rapid progress in high throughput data production. |

Chapter 2. Part 2: Genes and Genomes

ATP synthase	ATP synthase is an important enzyme that provides energy for the cell to use through the synthesis of adenosine triphosphate (ATP). ATP is the most commonly used 'energy currency' of cells from most organisms. It is formed from adenosine diphosphate (ADP) and inorganic phosphate (P_i), and needs energy.
Gene gun	A gene gun, originally designed for plant transformation, is a device for injecting cells with genetic information. The payload is an elemental particle of a heavy metal coated with plasmid DNA. This technique is often simply referred to as bioballistics or biolistics.
	This device is able to transform almost any type of cell, including plants, and is not limited to genetic material of the nucleus: it can also transform organelles, including plastids.
Insecticide	An insecticide is a pesticide used against insects. They include ovicides and larvicides used against the eggs and larvae of insects respectively. Insecticides are used in agriculture, medicine, industry and the household.
Parasporal body	Parasporal body is a crystalline protein that forms around a spore in some bacteria that acts as a toxin precursor when digested.
	For example, Bacillus thuringiensis and Bacillus sphaericus form a solid protein crystal, the parasporal body, next to their endospores during spore formation. The B. thuringiensis parasporal body contains protein toxins that kill over 100 species of moths by dissolving in the alkaline gut of caterpillars and destroying the epithelium.
Systemic disease	A systemic disease is one that affects a number of organs and tissues, or affects the body as a whole. Although most medical conditions will eventually involve multiple organs in advanced stage (e.g. Multiple organ dysfunction syndrome), diseases where multiple organ involvement is at presentation or in early stage are considered elsewhere.
Cystic fibrosis	Cystic fibrosis is an autosomal recessive genetic disorder affecting most critically the lungs, and also the pancreas, liver, and intestine. It is characterized by abnormal transport of chloride and sodium across an epithelium, leading to thick, viscous secretions.
	The name cystic fibrosis refers to the characteristic scarring (fibrosis) and cyst formation within the pancreas, first recognized in the 1930s.
Biopanning	Biopanning is an affinity selection technique which selects for peptides that bind to a given target. All peptide sequences obtained from biopanning using combinatorial peptide libraries have been stored in a special database with the name MimoDB, which is freely available.

Hapten	A hapten is a small molecule that can elicit an immune response only when attached to a large carrier such as a protein; the carrier may be one that also does not elicit an immune response by itself. (In general, only large molecules, infectious agents, or insoluble foreign matter can elicit an immune response in the body). Once the body has generated antibodies to a hapten-carrier adduct, the small-molecule hapten may also be able to bind to the antibody, but it will usually not initiate an immune response; usually only the hapten-carrier adduct can do this.
Phage display	Phage display is a laboratory technique for the study of protein-protein, protein-peptide, and protein-DNA interactions that uses bacteriophages to connect proteins with the genetic information that encodes them. Phage display was first described by George P. Smith in 1985, when he demonstrated the display of peptides on filamentous phage by fusing the peptide of interest on to gene III of filamentous phage. A patent by George Pieczenik claiming priority from 1985 also describes the generation of phage display libraries.
Directed evolution	Directed evolution is a method used in protein engineering to harness the power of natural selection to evolve proteins or RNA with desirable properties not found in nature. A typical directed evolution experiment involves three steps:•Diversification: The gene encoding the protein of interest is mutated and/or recombined at random to create a large library of gene variants. Techniques commonly used in this step are error-prone PCR and DNA shuffling.•Selection: The library is tested for the presence of mutants (variants) possessing the desired property using a screen or selection.
Bacillus cereus	Bacillus cereus is an endemic, soil-dwelling, Gram-positive, rod-shaped, beta hemolytic bacterium. Some strains are harmful to humans and cause foodborne illness, while other strains can be beneficial as probiotics for animals. B. cereus bacteria are facultative anaerobes, and like other members of the genus Bacillus can produce protective endospores.

Chapter 2. Part 2: Genes and Genomes

1. _____ is the phenomenon whereby somatic cells, bacteria, and other single-cell or multicellular organisms direct their movements according to certain chemicals in their environment. This is important for bacteria to find food (for example, glucose) by swimming towards the highest concentration of food molecules, or to flee from poisons (for example, phenol). In multicellular organisms, _____ is critical to early development (e.g. movement of sperm towards the egg during fertilization) and subsequent phases of development (e.g. migration of neurons or lymphocytes) as well as in normal function.

    a. Valais witch trials
    b. Chemotaxis
    c. Haplogroup
    d. Homology

2. _____ is the process by which a eukaryotic cell separates the chromosomes in its cell nucleus into two identical sets, in two separate nuclei. It is generally followed immediately by cytokinesis, which divides the nuclei, cytoplasm, organelles and cell membrane into two cells containing roughly equal shares of these cellular components. _____ and cytokinesis together define the mitotic (M) phase of the cell cycle--the division of the mother cell into two daughter cells, genetically identical to each other and to their parent cell.

    a. Binucleated cells
    b. Spindle apparatus
    c. Antisense therapy
    d. Mitosis

3. _____ is a 40-60 kDa MSCRAMM surface protein originally found in the cell wall of the bacterium Staphylococcus aureus. It is encoded by the spa gene and its regulation is controlled by DNA topology, cellular osmolarity, and a two-component system called ArlS-ArlR. It has found use in biochemical research because of its ability to bind immunoglobulins. It binds proteins from many of mammalian species, most notably IgGs.

    a. Stephen Switzer
    b. Genetic use restriction technology
    c. Protein A
    d. Genome Valley

4. _____ is a bacterial species named from Greek σταφυλ?κοκκος meaning the 'golden grape-cluster berry'. Also known as 'golden staph' and Oro staphira, it is a facultative anaerobic Gram-positive coccal bacterium. It is frequently found as part of the normal skin flora on the skin and nasal passages.

    a. Synbiotics
    b. Staphylococcus aureus
    c. Guanosine monophosphate
    d. 8-Hydroxyguanosine

5. . A _____, is a type of mutation that causes the replacement of a single base nucleotide with another nucleotide of the genetic material, DNA or RNA. The term _____ also includes insertions or deletions of a single base pair.

A point mutant is an individual that is affected by a _____.

Repeat induced _____s are recurring _____s, discussed below.

a. Polar effect
b. Postzygotic mutation
c. Resistance mutation
d. Point mutation

---

1. b
2. d
3. c
4. b
5. d

You can take the complete Chapter Practice Test

for Chapter 2. Part 2: Genes and Genomes
on all key terms, persons, places, and concepts.

Online 99 Cents

http://www.epub13.5.20451.2.cram101.com/

Use www.Cram101.com for all your study needs

including Cram101's online interactive problem solving labs in

chemistry, statistics, mathematics, and more.

	Pseudomonas aeruginosa
	Cystic fibrosis
	Amino acid
	Radioactive decay
	Catabolism
	Pyrodictium
	Delta endotoxin
	Helicobacter
	Anaerobic respiration
	Carboxylic acid
	Cell growth
	Giardia lamblia
	Glycolysis
	Passive transport
	Active transport
	Cell wall
	Sulfur metabolism
	ATP synthase
	Gibbs free energy

Chapter 3. Part 3: Metabolism and Biochemistry

CHAPTER OUTLINE: KEY TERMS, PEOPLE, PLACES, CONCEPTS

- Methanogenesis
- Methanogen
- Chemical reaction
- Adenine
- Calvin cycle
- Electron acceptor
- Energy carrier
- Glyceraldehyde 3-phosphate
- Electron donor
- Electron transfer
- Hydrolysis
- Phosphorylation
- Guanosine triphosphate
- Essential nutrient
- Lactic acid
- Nicotinamide adenine dinucleotide
- ATP hydrolysis
- Allosteric regulation
- Acne

Chapter 3. Part 3: Metabolism and Biochemistry

CHAPTER OUTLINE: KEY TERMS, PEOPLE, PLACES, CONCEPTS

- Propionibacterium acnes
- Carbohydrate catabolism
- Respiration
- Yogurt
- Cheese
- Disaccharide
- Monosaccharide
- Pectin
- Polysaccharide
- Catabolite repression
- Lactose permease
- Legionella pneumophila
- Lignin
- Putrescine
- Dairy product
- Decarboxylation
- Lac operon
- Polychlorinated biphenyl
- Deinococcus radiodurans

Chapter 3. Part 3: Metabolism and Biochemistry

CHAPTER OUTLINE: KEY TERMS, PEOPLE, PLACES, CONCEPTS

______ Methanethiol

______ Propionibacterium freudenreichii

______ Streptococcus salivarius

______ Fructose 6-phosphate

______ Glucose 6-phosphate

______ Ribulose 5-phosphate

______ Substrate-level phosphorylation

______ Erythrose 4-phosphate

______ Ribose 5-phosphate

______ Sedoheptulose 7-phosphate

______ Coenzyme A

______ Periodontal disease

______ Porphyromonas gingivalis

______ Pyruvate formate lyase

______ Fatty acid

______ Acetone

______ Agrobacterium tumefaciens

______ Alcaligenes eutrophus

______ Butyric acid

CHAPTER OUTLINE: KEY TERMS, PEOPLE, PLACES, CONCEPTS

	MacConkey agar
	Phenol red
	Shiga toxin
	Bubonic plague
	Carbon cycle
	Pyruvate dehydrogenase
	Succinyl-CoA
	Oxidative phosphorylation
	Malate
	Syphilis
	Catechol dioxygenase
	Genomic island
	Aromatic hydrocarbon
	Operon
	Polycyclic aromatic hydrocarbon
	Cytochrome
	Fuel cell
	Shewanella oneidensis
	Biofilm

_____ Electric current

_____ Frankenstein

_____ Luigi Galvani

_____ Electric power

_____ Reduction potential

_____ Rocky Mountain spotted fever

_____ Shewanella

_____ Brugia malayi

_____ Streptococcus pneumoniae

_____ Antibiotic resistance

_____ Organic acid

_____ Weak acid

_____ Liposome

_____ Valinomycin

_____ Proton pump

_____ Heme

_____ Salmonella enterica

_____ Cytochrome b

_____ Cytochrome c

CHAPTER OUTLINE: KEY TERMS, PEOPLE, PLACES, CONCEPTS

	Enterococcus faecalis
	Sodium-potassium pump
	Sodium ion
	Denitrification
	Haber process
	Haemophilus influenzae
	Neisseria gonorrhoeae
	Neisseria meningitidis
	Nitric oxide
	Biogeochemical cycle
	Nitrogen cycle
	Nitrous oxide
	Sulfate-reducing bacteria
	Marine habitats
	Extraterrestrial life
	Lithotroph
	Early Earth
	Nitrogen fixation
	Sulfur-reducing bacteria

Chapter 3. Part 3: Metabolism and Biochemistry

CHAPTER OUTLINE: KEY TERMS, PEOPLE, PLACES, CONCEPTS

_______ Sulfuric acid

_______ Molybdenum

_______ Selenium

_______ Chlorine

_______ Chlorobenzene

_______ Microbial ecology

_______ Quorum sensing

_______ Bacteriorhodopsin

_______ Halobacterium salinarum

_______ Methylotroph

_______ Proteorhodopsin

_______ Global warming

_______ Haloarchaea

_______ Photosynthesis

_______ Retinal

_______ Chloroflexus aurantiacus

_______ Algal bloom

_______ Chromophore

_______ Rhodobacter

CHAPTER OUTLINE: KEY TERMS, PEOPLE, PLACES, CONCEPTS

Bacteriochlorophyll

Carotenoid

Photosystem I

Photosystem II

Purple bacteria

Thylakoid

Chlorophyll

Lumen

Ferredoxin

Biosynthesis

Chloroplast

Phototrophic bacteria

Paper chromatography

RuBisCO

Carboxysome

Pyrobaculum

Rhodobacter sphaeroides

Thermoproteus

Glucan

_____ | Folic acid

_____ | Polyketide

_____ | Polyester

_____ | Acetyl-CoA carboxylase

_____ | Acyl carrier protein

_____ | Erythromycin

_____ | Secondary metabolite

_____ | Stringent response

_____ | Carrier protein

_____ | Filariasis

_____ | Ivermectin

_____ | Azotobacter

_____ | Nitrogen assimilation

_____ | Heterocyst

_____ | Nitrogenase

_____ | Rhizobium

_____ | Sinorhizobium meliloti

_____ | Reaction mechanism

_____ | Klebsiella pneumoniae

CHAPTER OUTLINE: KEY TERMS, PEOPLE, PLACES, CONCEPTS

Vanadium

Leghemoglobin

Nif gene

Amino acid synthesis

Glutamine synthetase

Transamination

Nonribosomal peptide

Purine

Pyrimidine

Clostridium difficile

Northern blot

Ribosomal RNA

Chocolate

Food microbiology

Soy sauce

Tempeh

Amanita phalloides

Industrial microbiology

Food preservation

Agaricus bisporus

Brown algae

Carrageenan

Kelp

Macrocystis

Pleurotus

Porphyra

Saccharomyces

Truffle

Candida albicans

Saccharomyces cerevisiae

Soylent Green

Kimchi

Lactic acid fermentation

Leuconostoc

Propionibacterium

Sauerkraut

Casein

Lactobacillus

CHAPTER OUTLINE: KEY TERMS, PEOPLE, PLACES, CONCEPTS

_____ Micelle _____

_____ Rennet _____

_____ Whey _____

_____ Chymosin _____

_____ Pasteurization _____

_____ Penicillium roqueforti _____

_____ Pepsin _____

_____ Ripening _____

_____ Lectin _____

_____ Protease inhibitor _____

_____ Rhizopus oligosporus _____

_____ Soybean _____

_____ Aspergillus oryzae _____

_____ Leuconostoc mesenteroides _____

_____ RNA polymerase _____

_____ Injera _____

_____ Drug discovery _____

_____ Putrefaction _____

_____ Rotavirus _____

_____ Psychrotrophic bacteria

_____ Crinipellis perniciosa

_____ Erwinia

_____ Serratia marcescens

_____ Theobroma cacao

_____ Seafood

_____ Listeria monocytogenes

_____ Clostridium botulinum

_____ Pathogenicity island

_____ Salmonella

_____ Botulinum toxin

_____ Endophyte

_____ Toxin

_____ Freeze-drying

_____ Mycobacterium tuberculosis

_____ Polio vaccine

_____ Vaccine

_____ Ethambutol

_____ Gene product

	Transdermal patch
	Bioprospecting
	Aspergillus nidulans
	Aspergillus niger
	Industrial fermentation
	Baculovirus
	Downstream processing
	Ti plasmid

CHAPTER HIGHLIGHTS & NOTES: KEY TERMS, PEOPLE, PLACES, CONCEPTS

Pseudomonas aeruginosa	Pseudomonas aeruginosa is a common bacterium which can cause disease in animals, including humans. It is found in soil, water, skin flora, and most man-made environments throughout the world. It thrives not only in normal atmospheres, but also in hypoxic atmospheres, and has thus colonized many natural and artificial environments.
Cystic fibrosis	Cystic fibrosis is an autosomal recessive genetic disorder affecting most critically the lungs, and also the pancreas, liver, and intestine. It is characterized by abnormal transport of chloride and sodium across an epithelium, leading to thick, viscous secretions. The name cystic fibrosis refers to the characteristic scarring (fibrosis) and cyst formation within the pancreas, first recognized in the 1930s.
Amino acid	Amino acids are molecules containing an amine group, a carboxylic acid group, and a side-chain that is specific to each amino acid.

The key elements of an amino acid are carbon, hydrogen, oxygen, and nitrogen. They are particularly important in biochemistry, where the term usually refers to alpha-amino acids.

Radioactive decay	Radioactive decay is the process by which an atomic nucleus of an unstable atom loses energy by emitting ionizing particles (ionizing radiation). The emission is spontaneous, in that the atom decays without any interaction with another particle from outside the atom (i.e., without a nuclear reaction). Usually, radioactive decay happens due to a process confined to the nucleus of the unstable atom, but, on occasion (as with the different processes of electron capture and internal conversion), an inner electron of the radioactive atom is also necessary to the process.
Catabolism	Catabolism is the set of metabolic pathways that break down molecules into smaller units and release energy. In catabolism, large molecules such as polysaccharides, lipids, nucleic acids and proteins are broken down into smaller units such as monosaccharides, fatty acids, nucleotides, and amino acids, respectively. As molecules such as polysaccharides, proteins, and nucleic acids are made from long chains of these small monomer units (mono = one + mer = part), the large molecules are called polymers (poly = many).
Pyrodictium	In taxonomy, Pyrodictium is a genus of the Pyrodictiaceae.Pyrodictium is a genera of submarine hyperthermophilic archaea whose optimal growth temperature ranges is 80 to 105 °C. They have a unique cell structure involving a network of cannulae and flat, disk-shaped cells. Pyrodictium are found in the porous walls of deep-sea vents where the temperatures inside get as high as 400 °C, while the outside marine environment is typically 3 °C. Pyrodictium is apparently able to adapt morphologically to this type of hot-cold habitat. Much research has been done on the genetics of Pyrodictium in order to understand its ability to survive and even thrive in such extreme temperatures.
Delta endotoxin	Delta endotoxin, N-terminal domain Delta endotoxins (δ-endotoxins, also called Cry and Cyt toxins) are pore-forming toxins produced by Bacillus thuringiensis species of bacteria. They are useful for their insecticidal action. During spore formation the bacteria produce crystals of this protein.
Helicobacter	Helicobacter is a genus of Gram-negative bacteria possessing a characteristic helix shape. They were initially considered to be members of the Campylobacter genus, but since 1989 they have been grouped in their own genus. The Helicobacter genus belongs to class Epsilonproteobacteria, order Campylobacterales, family Helicobacteraceae and already involves >35 species.
Anaerobic respiration	Anaerobic respiration is a form of respiration using electron acceptors other than oxygen.

Although oxygen is not used as the final electron acceptor, the process still uses a respiratory electron transport chain; it is respiration without oxygen. In order for the electron transport chain to function, an exogenous final electron acceptor must be present to allow electrons to pass through the system.

Carboxylic acid

Carboxylic acids ( ) are organic acids characterized by the presence of at least one carboxyl group. The general formula of a carboxylic acid is R-COOH, where R is some monovalent functional group. A carboxyl group is a functional group consisting of a carbonyl (RR'C=O) and a hydroxyl (R-O-H), which has the formula -C(=O)OH, usually written as -COOH or $-CO_2H$.

Carboxylic acids are Brønsted-Lowry acids because they are proton (H^+) donors.

Cell growth

The term cell growth is used in the contexts of cell development and cell division (reproduction). When used in the context of cell division, it refers to growth of cell populations, where one cell (the 'mother cell') grows and divides to produce two 'daughter cells' (M phase). When used in the context of cell development, the term refers to increase in cytoplasmic and organelle volume (G1 phase), as well as increase in genetic material before replication (G2 phase).

Giardia lamblia

Giardia lamblia is a flagellated protozoan parasite that colonizes and reproduces in the small intestine, causing giardiasis. The giardia parasite attaches to the epithelium by a ventral adhesive disc, and reproduces via binary fission. Giardiasis does not spread via the bloodstream, nor does it spread to other parts of the gastro-intestinal tract, but remains confined to the lumen of the small intestine.

Glycolysis

Glycolysis is the metabolic pathway that converts glucose $C_6H_{12}O_6$, into pyruvate, $CH_3COCOO^- + H^+$. The free energy released in this process is used to form the high-energy compounds ATP (adenosine triphosphate), $FADH_2$ and NADH (reduced nicotinamide adenine dinucleotide).

Glycolysis is a definite sequence of ten reactions involving ten intermediate compounds (one of the steps involves two intermediates).

Passive transport

Passive transport means moving biochemicals and other atomic or molecular substances across membranes. Unlike active transport, this process does not involve chemical energy, because, unlike in an active transport, the transport across membrane is always coupled with the growth of entropy of the system. So passive transport is dependent on the permeability of the cell membrane, which, in turn, is dependent on the organization and characteristics of the membrane lipids and proteins.

Active transport

Active transport is the movement of a substance against its concentration gradient (from low to high concentration). In all cells, this is usually concerned with accumulating high concentrations of molecules that the cell needs, such as ions, glucose and amino acids.

Chapter 3. Part 3: Metabolism and Biochemistry

Cell wall	The cell wall is the tough, usually flexible but sometimes fairly rigid layer that surrounds some types of cells. It is located outside the cell membrane and provides these cells with structural support and protection, in addition to acting as a filtering mechanism. A major function of the cell wall is to act as a pressure vessel, preventing over-expansion when water enters the cell.
Sulfur metabolism	Sulfur metabolism is vital for all living organisms as it is a constituent of a number of essential organic molecules like cysteine, methionine, coenzyme A, and iron-sulfur clusters. These compounds are involved in a number of essential cellular processes such as protein biosynthesis or the transfer of electrons and acyl groups. Sulfur, therefore, is an essential component of all living cells.
ATP synthase	ATP synthase is an important enzyme that provides energy for the cell to use through the synthesis of adenosine triphosphate (ATP). ATP is the most commonly used 'energy currency' of cells from most organisms. It is formed from adenosine diphosphate (ADP) and inorganic phosphate (P_i), and needs energy.
Gibbs free energy	Property database In thermodynamics, the Gibbs free energy is a thermodynamic potential that measures the 'useful' or process-initiating work obtainable from a thermodynamic system at a constant temperature and pressure (isothermal, isobaric). Just as in mechanics, where potential energy is defined as capacity to do work, similarly different potentials have different meanings. The Gibbs free energy is the maximum amount of non-expansion work that can be extracted from a closed system; this maximum can be attained only in a completely reversible process.
Methanogenesis	Methanogenesis is the formation of methane by microbes known as methanogens. Organisms capable of producing methane have been identified only from the domain Archaea, a group phylogenetically distinct from both eukaryotes and bacteria, although many live in close association with anaerobic bacteria. The production of methane is an important and widespread form of microbial metabolism.
Methanogen	Methanogens are microorganisms that produce methane as a metabolic byproduct in anoxic conditions. They are classified as archaea, a group quite distinct from bacteria. They are common in wetlands, where they are responsible for marsh gas, and in the guts of animals such as ruminants and humans, where they are responsible for the methane content of belching in ruminants and flatulence in humans.
Chemical reaction	A chemical reaction is a process that leads to the transformation of one set of chemical substances to another. Chemical reactions can be either spontaneous, requiring no input of energy, or non-spontaneous, typically following the input of some type of energy, such as heat, light or electricity.

Adenine	Adenine is a nucleobase (a purine derivative) with a variety of roles in biochemistry including cellular respiration, in the form of both the energy-rich adenosine triphosphate (ATP) and the cofactors nicotinamide adenine dinucleotide (NAD) and flavin adenine dinucleotide (FAD), and protein synthesis, as a chemical component of DNA and RNA. The shape of adenine is complementary to either thymine in DNA or uracil in RNA. Adenine forms several tautomers, compounds that can be rapidly interconverted and are often considered equivalent. However, in isolated conditions, i.e. in an inert gas matrix and in the gas phase, mainly the 9H-adenine tautomer is found. Biosynthesis Purine metabolism involves the formation of adenine and guanine.
Calvin cycle	The Calvin cycle, is a series of biochemical redox reactions that take place in the stroma of chloroplasts in photosynthetic organisms. It is also known as the dark reactions. The cycle was discovered by Melvin Calvin, James Bassham, and Andrew Benson at the University of California, Berkeley by using the radioactive isotope carbon-14. It is one of the light-independent (dark) reactions used for carbon fixation.
Electron acceptor	An electron acceptor is a chemical entity that accepts electrons transferred to it from another compound. It is an oxidizing agent that, by virtue of its accepting electrons, is itself reduced in the process. Typical oxidizing agents undergo permanent chemical alteration through covalent or ionic reaction chemistry, resulting in the complete and irreversible transfer of one or more electrons.
Energy carrier	According to ISO 13600, an energy carrier is either a substance (energy form) or a phenomenon (energy system) that can be used to produce mechanical work or heat or to operate chemical or physical processes. In the field of Energetics, however, an energy carrier corresponds only to an energy form (not an energy system) of energy input required by the various sectors of society to perform their functions. Examples of energy carriers include liquid fuel in a furnace, gasoline in a pump, electricity in a factory or a house, and hydrogen in a tank of a car.
Glyceraldehyde 3-phosphate	Glyceraldehyde 3-phosphate, GADP, GAP, TP, GALP or PGAL, is a chemical compound that occurs as an intermediate in several central metabolic pathways of all organisms. It is a phosphate ester of the 3-carbon sugar glyceraldehyde and has chemical formula $C_3H_7O_6P$.

The CAS number of glyceraldehyde 3-phosphate is 591-59-3 and that of D-glyceraldehyde 3-phosphate is 591-57-1. An intermediate in both glycolysis and gluconeogenesis Formation

D-glyceraldehyde 3-phosphate is formed from the following three compounds in reversible reactions:•Fructose-1,6-bisphosphate (F1,6BP), catalyzed by aldolase.

The numbering of the carbon atoms indicates the fate of the carbons according to their position in fructose 6-phosphate.

Electron donor	An electron donor is a chemical entity that donates electrons to another compound. It is a reducing agent that, by virtue of its donating electrons, is itself oxidized in the process.
	Typical reducing agents undergo permanent chemical alteration through covalent or ionic reaction chemistry.
Electron transfer	Electron transfer is the process by which an electron moves from an atom or a chemical species (e.g. a molecule) to another atom or chemical species. ET is a mechanistic description of the thermodynamic concept of redox, wherein the oxidation states of both reaction partners change.
	Numerous biological processes involve ET reactions.
Hydrolysis	Hydrolysis is a chemical reaction in which molecules of water (H_2O) are split into hydrogen cations (H^+, identical to protons) and hydroxide anions (OH^-). It is the type of reaction that is used to break down certain polymers, especially those made by condensation polymerization. Such polymer degradation is usually catalysed by either acid, e.g., concentrated sulfuric acid (H_2SO_4), or alkali, e.g., sodium hydroxide (NaOH).
Phosphorylation	Phosphorylation is the addition of a phosphate (PO_4^{3-}) group to a protein or other organic molecule. Phosphorylation turns many protein enzymes on and off, thereby altering their function and activity.
	Protein phosphorylation in particular plays a significant role in a wide range of cellular processes.
Guanosine triphosphate	Guanosine-5'-triphosphate (GTP) is a purine nucleoside triphosphate. It can act as a substrate for the synthesis of RNA during the transcription process. Its structure is similar to that of the guanine nucleobase, the only difference being that nucleotides like GTP have a ribose sugar and three phosphates, with the nucleobase attached to the 1' and the triphosphate moiety attached to the 5' carbons of the ribose.

It also has the role of a source of energy or an activator of substrates in metabolic reactions, like that of ATP, but more specific. It is used as a source of energy for protein synthesis and gluconeogenesis.

GTP is essential to signal transduction, particularly with G-proteins, in second-messenger mechanisms where it is converted to GDP (guanosine diphosphate) through the action of GTPases. Uses Energy Transfer

GTP is involved in energy transfer within the cell. For instance, a GTP molecule is generated by one of the enzymes in the citric acid cycle. This is tantamount to the generation of one molecule of ATP, since GTP is readily converted to ATP.Genetic Translation

During the elongation stage of translation, GTP is used as an energy source for the binding of a new amino-bound tRNA to the A site of the ribosome. GTP is also used as an energy source for the translocation of the ribosome towards the 3' end of the mRNA.Microtubule Dynamic Instability

During microtubule polymerization, each heterodimer formed by an alpha and a beta tubulin molecule carries two GTP molecules, and the GTP is hydrolyzed to GDP when the tubulin dimers are added to the plus end of the growing microtubule. Such GTP hydrolysis is not mandatory for microtubule formation, but it appears that only GDP-bound tubulin molecules are able to depolymerize. Thus, a GTP-bound tubulin serves as a cap at the tip of microtubule to protect from depolymerization; and once the GTP is hydrolyzed, the microtubule begins to depolymerize and shrink rapidly. cGTP

Cyclic guanosine triphosphate helps cyclic adenosine monophosphate (cAMP) activate cyclic nucleotide-gated ion channels in the olfactory system.

Essential nutrient	An essential nutrient is a nutrient required for normal body functioning that either cannot be synthesized by the body at all, or cannot be synthesized in amounts adequate for good health (e.g. niacin, choline), and thus must be obtained from a dietary source. Essential nutrients are also defined by the collective physiological evidence for their importance in the diet, as represented in e.g. US government approved tables for Dietary Reference Intake. Some categories of essential nutrients include vitamins, dietary minerals, essential fatty acids, and essential amino acids.
Lactic acid	Lactic acid, is a chemical compound that plays a role in various biochemical processes and was first isolated in 1780 by the Swedish chemist Carl Wilhelm Scheele. Lactic acid is a carboxylic acid with the chemical formula $C_3H_6O_3$.

Chapter 3. Part 3: Metabolism and Biochemistry

Nicotinamide adenine dinucleotide	Nicotinamide adenine dinucleotide, abbreviated NAD^+, is a coenzyme found in all living cells. The compound is a dinucleotide, since it consists of two nucleotides joined through their phosphate groups. One nucleotide contains an adenine base and the other nicotinamide.
ATP hydrolysis	ATP hydrolysis is the reaction by which chemical energy that has been stored and transported in the high-energy phosphoanhydridic bonds in ATP (Adenosine triphosphate) is released, for example in the muscles, to produce work. The product is ADP (Adenosine diphosphate) and an inorganic phosphate, orthophosphate (Pi). ADP can be further hydrolyzed to give energy, AMP (Adenosine monophosphate), and another orthophosphate (Pi).
Allosteric regulation	In biochemistry, allosteric regulation is the regulation of an enzyme or other protein by binding an effector molecule at the protein's allosteric site (that is, a site other than the protein's active site). Effectors that enhance the protein's activity are referred to as allosteric activators, whereas those that decrease the protein's activity are called allosteric inhibitors. The term allostery comes from the Greek allos , 'other', and stereos , 'solid (object)', in reference to the fact that the regulatory site of an allosteric protein is physically distinct from its active site.
Acne	Acne is a general term used for acneiform eruptions. It is usually used as a synonym for acne vulgaris, but may also refer to:•Acne aestivalis•Acne conglobata•Acne cosmetica•Acne fulminans•Acne keloidalis nuchae•Acne mechanica•Acne medicamentosa (drug-induced acne) (e.g., steroid acne)•Acne miliaris necrotica•Acne necrotica•Blackheads•Chloracne•Excoriated acne•Halogen acne•Infantile acne/Neonatal acne•Lupus miliaris disseminatus faciei•Occupational acne•Oil acne•Pomade acne•Tar acne•Tropical acne.
Propionibacterium acnes	Propionibacterium acnes is a relatively slow growing, typically aerotolerant anaerobic gram positive bacterium (rod) that is linked to the skin condition acne; it can also cause chronic blepharitis and endophthalmitis, the latter particularly following intraocular surgery. The genome of the bacterium has been sequenced and a study has shown several genes that can generate enzymes for degrading skin and proteins that may be immunogenic (activating the immune system). This bacterium is largely commensal and part of the skin flora present on most healthy adult human skin.
Carbohydrate catabolism	Carbohydrate catabolism is the breakdown of carbohydrates into smaller units. Carbohydrates literally undergo combustion to retrieve the large amounts of energy in their bonds. Energy is secured by mitochondria in the form of ATP.

Respiration	In physiology, respiration (often confused with breathing) is defined as the transport of oxygen from the outside air to the cells within tissues, and the transport of carbon dioxide in the opposite direction. This is in contrast to the biochemical definition of respiration, which refers to cellular respiration: the metabolic process by which an organism obtains energy by reacting oxygen with glucose to give water, carbon dioxide and ATP (energy). Although physiologic respiration is necessary to sustain cellular respiration and thus life in animals, the processes are distinct: cellular respiration takes place in individual cells of the organism, while physiologic respiration concerns the bulk flow and transport of metabolites between the organism and the external environment.
Yogurt	Yogurt, UK: /ˈjɒɡət/) is a dairy product produced by bacterial fermentation of milk. The bacteria used to make yogurt are known as 'yogurt cultures'. Fermentation of lactose by these bacteria produces lactic acid, which acts on milk protein to give yogurt its texture and its characteristic tang.
Cheese	Cheese is a generic term for a diverse group of milk-based food products. Cheese is produced throughout the world in wide-ranging flavors, textures, and forms. Cheese consists of proteins and fat from milk, usually the milk of cows, buffalo, goats, or sheep.
Disaccharide	A disaccharide is the carbohydrate formed when two monosaccharides undergo a condensation reaction which involves the elimination of a small molecule, such as water, from the functional groups only. Like monosaccharides, disaccharides form an aqueous solution when dissolved in water. Three common examples are sucrose, lactose, and maltose.
Monosaccharide	Monosaccharides are the most basic units of biologically important carbohydrates. They are the simplest form of sugar and are usually colorless, water-soluble, crystalline solids. Some monosaccharides have a sweet taste.
Pectin	Pectin is a structural heteropolysaccharide contained in the primary cell walls of terrestrial plants. It was first isolated and described in 1825 by Henri Braconnot. It is produced commercially as a white to light brown powder, mainly extracted from citrus fruits, and is used in food as a gelling agent particularly in jams and jellies.
Polysaccharide	Polysaccharides are long carbohydrate molecules of repeated monomer units joined together by glycosidic bonds. They range in structure from linear to highly branched. Polysaccharides are often quite heterogeneous, containing slight modifications of the repeating unit.
Catabolite repression	Carbon catabolite repression, is an important part of global control system of various bacteria and other micro-organisms. Catabolite repression allows bacteria to adapt quickly to a preferred (rapidly metabolisable) carbon and energy source first.

Chapter 3. Part 3: Metabolism and Biochemistry

Lactose permease	LacY proton/sugar symporter
	Lactose permease is a membrane protein which is a member of the major facilitator superfamily. Lactose permease can be classified as a symporter, which uses the gradient of H+ towards the cell to transport lactose in the same direction into the cell.
	The protein has twelve transmembrane helices and exhibits an internal two-fold symmetry, relating the N-terminal six helices onto the C-terminal helices.
Legionella pneumophila	Legionella pneumophila is a thin, pleomorphic, flagellated Gram-negative bacterium of the genus Legionella. L. pneumophila is the primary human pathogenic bacterium in this group and is the causative agent of legionellosis or Legionnaires' disease.
	Characterization
	L. pneumophila is non-acid-fast, non-sporulating, and morphologically a non-capsulated rod-like bacteria.
Lignin	Lignin is a complex chemical compound most commonly derived from wood, and an integral part of the secondary cell walls of plants and some algae. The term was introduced in 1819 by de Candolle and is derived from the Latin word lignum, meaning wood. It is one of the most abundant organic polymers on Earth, exceeded only by cellulose, employing 30% of non-fossil organic carbon and constituting from a quarter to a third of the dry mass of wood.
Putrescine	Putrescine is a foul-smelling organic chemical compound $NH_2(CH_2)_4NH_2$ (1,4-diaminobutane or butanediamine) that is related to cadaverine; both are produced by the breakdown of amino acids in living and dead organisms and both are toxic in large doses. The two compounds are largely responsible for the foul odor of putrefying flesh, but also contribute to the odor of such processes as bad breath and bacterial vaginosis. They are also found in semen and some microalgae, together with related molecules like spermine and spermidine.
Dairy product	Dairy products are generally defined as foods produced from cow's or domestic buffalo's milk. They are usually high-energy-yielding food products. A production plant for such processing is called a dairy or a dairy factory.
Decarboxylation	Decarboxylation is a chemical reaction that releases carbon dioxide (CO_2). Usually, decarboxylation refers to a reaction of carboxylic acids, removing a carbon atom from a carbon chain. The reverse process, which is the first chemical step in photosynthesis, is called carbonation, the addition of CO_2 to a compound.

Lac operon	The lac operon is an operon required for the transport and metabolism of lactose in Escherichia coli and some other enteric bacteria. It consists of three adjacent structural genes, lacZ, lacY and lacA. The lac operon is regulated by several factors including the availability of glucose and of lactose. Gene regulation of the lac operon was the first complex genetic regulatory mechanism to be elucidated and is one of the foremost examples of prokaryotic gene regulation.
Polychlorinated biphenyl	Polychlorinated biphenyls (PCBs; CAS number 1336-36-3) are a class of organic compounds (specifically organochlorides) with 2 to 10 chlorine atoms attached to biphenyl, which is a molecule composed of two benzene rings. The chemical formula for PCBs is $C_{12}H_{10-x}Cl_x$. Polychlorinated biphenyls were widely used as dielectric and coolant fluids, for example in transformers, capacitors, and electric motors. Due to Polychlorinated biphenyls' toxicity and classification as a persistent organic pollutant, PCB production was banned by the United States Congress in 1979 and by the Stockholm Convention on Persistent Organic Pollutants in 2001.
Deinococcus radiodurans	Deinococcus radiodurans is an extremophilic bacterium, one of the most radioresistant organisms known. It can survive cold, dehydration, vacuum, and acid, and is therefore known as a polyextremophile and has been listed as the world's toughest bacterium in The Guinness Book Of World Records. The name Deinococcus radiodurans derives from the Ancient Greek δειν?ς (deinos) and κ?κκος (kokkos) meaning 'terrible grain/berry' and the Latin radius and durare, meaning 'radiation surviving'.
Methanethiol	Methanethiol is a colorless gas with a smell like rotten cabbage. It is a natural substance found in the blood and brain of humans and other animal as well as plant tissues. It is disposed of through animal feces.
Propionibacterium freudenreichii	Propionibacterium freudenreichii is a Gram-positive, nonmotile bacterium that plays an important role in the creation of Emmental cheese, and to some extent, Leerdammer. Its concentration in Swiss-type cheeses is higher than in any other cheese. Propionibacteria are commonly found in milk and dairy products, though they have also been extracted from soil.
Streptococcus salivarius	Streptococcus salivarius is a species of spherical, Gram-positive bacteria which colonize the mouth and upper respiratory tract of humans a few hours after birth, making further exposure to the bacteria harmless. The bacteria is considered an opportunistic pathogen, rarely finding its way into the bloodstream, where it has been implicated in septicemia cases in people with neutropenia.

S. salivarius has distinct characteristics when exposed to different environmental nutrients.

Fructose 6-phosphate	Fructose 6-phosphate is fructose sugar phosphorylated on carbon 6 (i.e., is a fructosephosphate). The β-D-form of this compound is very common in cells. The vast majority of glucose and fructose entering a cell will become converted to this at some point.
Glucose 6-phosphate	Glucose 6-phosphate is glucose sugar phosphorylated on carbon 6. This compound is very common in cells as the vast majority of glucose entering a cell will become phosphorylated in this way. Because of its prominent position in cellular chemistry, glucose 6-phosphate has many possible fates within the cell. It lies at the start of two major metabolic pathways:•Glycolysis•Pentose phosphate pathway In addition to these metabolic pathways, glucose 6-phosphate may also be converted to glycogen or starch for storage.
Ribulose 5-phosphate	Ribulose 5-phosphate is one of the end-products of the pentose phosphate pathway. It is also an intermediate in the Calvin cycle. It is formed by phosphogluconate dehydrogenase, and it can be acted upon by phosphopentose isomerase and phosphopentose epimerase.
Substrate-level phosphorylation	Substrate-level phosphorylation is a type of metabolism that results in the formation and creation of adenosine triphosphate (ATP) or guanosine triphosphate (GTP) by the direct transfer and donation of a phosphoryl (PO_3) group to adenosine diphosphate (ADP) or guanosine diphosphate (GDP) from a phosphorylated reactive intermediate. Note that the phosphate group does not have to directly come from the substrate. By convention, the phosphoryl group that is transferred is referred to as a phosphate group.
Erythrose 4-phosphate	Erythrose 4-phosphate is an intermediate in the pentose phosphate pathway and the Calvin cycle. In addition, it serves as a precursor in the biosynthesis of the aromatic amino acids tyrosine, phenylalanine, and tryptophan.
Ribose 5-phosphate	Ribose 5-phosphate is both a product and an intermediate of the pentose phosphate pathway. The last step of the oxidative reactions in the pentose phosphate pathway is the production of ribulose-5-phosphate. Depending on the body's state, ribulose-5-phosphate can reversibly isomerize to ribose-5-phosphate.

Sedoheptulose 7-phosphate	Sedoheptulose 7-phosphate is an intermediate in the pentose phosphate pathway.
	It is formed by transketolase and acted upon by transaldolase.
Coenzyme A	Coenzyme A is a coenzyme, notable for its role in the synthesis and oxidation of fatty acids, and the oxidation of pyruvate in the citric acid cycle. All sequenced genomes encode enzymes that use coenzyme A as a substrate, and around 4% of cellular enzymes use it as a substrate. It is adapted from cysteamine, pantothenate, and adenosine triphosphate.
Periodontal disease	Periodontal disease is a type of disease that affects one or more of the periodontal tissues:•alveolar bone•periodontal ligament•cementum•gingiva
	While many different diseases affect the tooth-supporting structures, plaque-induced inflammatory lesions make up the vast majority of periodontal diseases and have traditionally been divided into two categories:•gingivitis or•periodontitis.
	While in some sites or individuals, gingivitis never progresses to periodontitis, data indicates that gingivitis always precedes periodontitis.
	Diagnosis
	In 1976, Page & Schroeder introduced an innovative new analysis of periodontal disease based on histopathologic and ultrastructural features of the diseased gingival tissue. Although this new classification does not correlate with clinical signs and symptoms and is admittedly 'somewhat arbitrary,' it permits a focus of attention on important pathologic aspects of the disease that were, until recently, not well understood.
Porphyromonas gingivalis	Porphyromonas gingivalis belongs to the phylum Bacteroidetes and is a non-motile, Gram-negative, rod-shaped, anaerobic pathogenic bacterium. It forms black colonies on blood agar.
	It is found in the oral cavity, where it is implicated in certain forms of periodontal disease, as well as the upper gastrointestinal tract, respiratory tract, and in the colon.
Pyruvate formate lyase	Pyruvate formate lyase (EC 2.3.1.54) is an enzyme found in Escherichia coli and other organisms. It helps regulate anaerobic glucose metabolism. Using radical non-redox chemistry, it catalyzes the reversible conversion of pyruvate and coenzyme-A into formate and acetyl-CoA. The reaction occurs as follows:
	Pyruvate formate lyase is a homodimer made of 85 kDa, 759-residue subunits.

Chapter 3. Part 3: Metabolism and Biochemistry

Fatty acid	In chemistry, especially biochemistry, a fatty acid is a carboxylic acid with a long aliphatic tail (chain), which is either saturated or unsaturated. Most naturally occurring fatty acids have a chain of an even number of carbon atoms, from 4 to 28. Fatty acids are usually derived from triglycerides or phospholipids. When they are not attached to other molecules, they are known as 'free' fatty acids.
Acetone	Acetone is the organic compound with the formula $(CH_3)_2CO$, a colorless, mobile, flammable liquid, the simplest example of the ketones. Acetone is miscible with water and serves as an important solvent in its own right, typically as the solvent of choice for cleaning purposes in the laboratory. About 6.7 million tonnes were produced worldwide in 2010, mainly for use as a solvent and production of methyl methacrylate and bisphenol A. It is a common building block in organic chemistry.
Agrobacterium tumefaciens	Agrobacterium tumefaciens is the causal agent of crown gall disease (the formation of tumours) in over 140 species of dicot. It is a rod shaped, Gram negative soil bacterium (Smith et al., 1907). Symptoms are caused by the insertion of a small segment of DNA (known as the T-DNA, for 'transfer DNA'), from a plasmid, into the plant cell, which is incorporated at a semi-random location into the plant genome.
Alcaligenes eutrophus	Alcaligenes eutrophus is a bacterial species that naturally produces polyhydroxyalkanoates (PHA). PHAs are a broad type of biodegradable polymers that can be used for biodegradable plastics. A. eutrophus specifically produces polyhydroxybutyrate (PHB), which it uses for storing carbon when in an environment with abundant carbon, but limited essential nutrients such as nitrogen or phosphorus.
Butyric acid	Butyric acid, also known under the systematic name butanoic acid, is a carboxylic acid with the structural formula $CH_3CH_2CH_2$-COOH. Salts and esters of butyric acid are known as butyrates or butanoates. Butyric acid is found in milk, especially goat, sheep and buffalo's milk, butter, Parmesan cheese, and vomit, and as a product of anaerobic fermentation (including in the colon and as body odor). It has an unpleasant smell and acrid taste, with a sweetish aftertaste (similar to ether).
MacConkey agar	MacConkey agar is a culture medium designed to grow Gram-negative bacteria and stain them for lactose fermentation. Contents It contains bile salts (to inhibit most Gram-positive bacteria, except Enterococcus and some species of Staphylococcus i.e.

Staphylococcus aureus), crystal violet dye (which also inhibits certain Gram-positive bacteria), neutral red dye (which stains microbes fermenting lactose), lactose and peptone.

Composition:

Peptic digest of animal tissues......17 g

Proteose Peptone....................3 g

Lactose............................10 g

Bile Salts.........................1.5 g

Sodium chloride....................5 g

Neutral Red........................0.03 g

Agar...............................15 g

pH.............................7.1 + or - 0.2

for 1 liter of the medium

There are many variations of McConkey agar depending on the need.

Phenol red	Phenol red is a pH indicator that is frequently used in cell biology laboratories.
	Phenol red exists as a red crystal that is stable in air. Its solubility is 0.77 grams per liter (g/L) in water and 2.9 g/L in ethanol.
Shiga toxin	Shiga toxins are a family of related toxins with two major groups, Stx1 and Stx2, whose genes are considered to be part of the genome of lambdoid prophages. The toxins are named for Kiyoshi Shiga, who first described the bacterial origin of dysentery caused by Shigella dysenteriae. The most common sources for Shiga toxin are the bacteria S. dysenteriae and the Shigatoxigenic group of Escherichia coli (STEC), which includes serotypes O157:H7, O104:H4, and other enterohemorrhagic E. coli (EHEC).
Bubonic plague	Bubonic plague is a zoonotic disease, circulating mainly among small rodents and their fleas, and is one of three types of infections caused by Yersinia pestis (formerly known as Pasteurella pestis), which belongs to the family Enterobacteriaceae. Without treatment, the bubonic plague kills about two out of three infected humans within 4 days.

Chapter 3. Part 3: Metabolism and Biochemistry

CHAPTER HIGHLIGHTS & NOTES: KEY TERMS, PEOPLE, PLACES, CONCEPTS

Carbon cycle

The carbon cycle is the biogeochemical cycle by which carbon is exchanged among the biosphere, pedosphere, geosphere, hydrosphere, and atmosphere of the Earth. It is one of the most important cycles of the Earth and allows for carbon to be recycled and reused throughout the biosphere and all of its organisms.

The global carbon budget is the balance of the exchanges (incomes and losses) of carbon between the carbon reservoirs or between one specific loop (e.g., atmosphere ↔ biosphere) of the carbon cycle.

Pyruvate dehydrogenase

Pyruvate dehydrogenase ($NADP^+$) EC 1.2.1.51 is an enzyme that should not be confused with Pyruvate dehydrogenase (acetyltransferase) EC 1.2.4.1.

It catalyzes the following reaction:Pyruvate + Coenzyme A + $NADP^+$ ⇒ acetyl-CoA + NADPH + H^+ + CO_2.

Succinyl-CoA

Succinyl-Coenzyme A, abbreviated as Succinyl-CoA, is a combination of succinic acid and coenzyme A.

Source

It is an important intermediate in the citric acid cycle, where it is synthesized from α-Ketoglutarate by α-ketoglutarate dehydrogenase through decarboxylation. During the process, coenzyme A is added.

It is also synthesized from propionyl CoA, the odd-numbered fatty acid, which cannot undergo beta-oxidation.

Oxidative phosphorylation

Oxidative phosphorylation is a metabolic pathway that uses energy released by the oxidation of nutrients to produce adenosine triphosphate (ATP). Although the many forms of life on earth use a range of different nutrients, almost all aerobic organisms carry out oxidative phosphorylation to produce ATP, the molecule that supplies energy to metabolism. This pathway is probably so pervasive because it is a highly efficient way of releasing energy, compared to alternative fermentation processes such as anaerobic glycolysis.

Malate

Malate ($^-OOC\text{-}CH_2\text{-}CH(OH)\text{-}COO^-$) is the ionized form of malic acid. It is an important chemical compound in biochemistry. In the C4 carbon fixation process, malate is a source of CO_2 in the Calvin cycle.

Syphilis

Syphilis is a sexually transmitted infection caused by the spirochete bacterium Treponema pallidum subspecies pallidum.

	The primary route of transmission is through sexual contact; however, it may also be transmitted from mother to fetus during pregnancy or at birth, resulting in congenital syphilis. Other human diseases caused by related Treponema pallidum include yaws (subspecies pertenue), pinta (subspecies carateum) and bejel (subspecies endemicum).
Catechol dioxygenase	Catechol dioxygenases are metalloprotein enzymes that carry out the oxidative cleavage of catechols. This class of enzymes incorporate dioxygen into the substrate (biochemistry). Catechol dioxygenases belong to the class of oxidoreductases and have several different substrate specificities, including catechol 1,2-dioxygenase (EC 1.13.11.1), catechol 2,3-dioxygenase (EC 1.13.11.2), and protocatechuate 3,4-dioxygenase (EC 1.13.11.3).
Genomic island	A Genomic island is part of a genome that has evidence of horizontal origins. The term is usually used in microbiology, especially with regard to bacteria. A GI can code for many functions, can be involved in symbiosis or pathogenesis, and may help an organism's adaptation.
Aromatic hydrocarbon	An aromatic hydrocarbon is a hydrocarbon with alternating double and single bonds between carbon atoms. The term 'aromatic' was assigned before the physical mechanism determining aromaticity was discovered, and was derived from the fact that many of the compounds have a sweet scent. The configuration of six carbon atoms in aromatic compounds is known as a benzene ring, after the simplest possible such hydrocarbon, benzene.
Operon	In genetics, an operon is a functioning unit of genomic DNA containing a cluster of genes under the control of a single regulatory signal or promoter. The genes are transcribed together into an mRNA strand and either translated together in the cytoplasm, or undergo trans-splicing to create monocistronic mRNAs that are translated separately, i.e. several strands of mRNA that each encode a single gene product. The result of this is that the genes contained in the operon are either expressed together or not at all.
Polycyclic aromatic hydrocarbon	Polycyclic aromatic hydrocarbons (PAHs), also known as poly-aromatic hydrocarbons or polynuclear aromatic hydrocarbons, are potent atmospheric pollutants that consist of fused aromatic rings and do not contain heteroatoms or carry substituents. Naphthalene is the simplest example of a PAH. PAHs occur in oil, coal, and tar deposits, and are produced as byproducts of fuel burning (whether fossil fuel or biomass). As a pollutant, they are of concern because some compounds have been identified as carcinogenic, mutagenic, and teratogenic.
Cytochrome	Cytochromes are, in general, membrane-bound hemoproteins that contain heme groups and carry out electron transport.

They are found either as monomeric proteins (e.g., cytochrome c) or as subunits of bigger enzymatic complexes that catalyze redox reactions. |

Chapter 3. Part 3: Metabolism and Biochemistry

Fuel cell	A fuel cell is a device that converts the chemical energy from a fuel into electricity through a chemical reaction with oxygen or another oxidizing agent. Hydrogen is the most common fuel, but hydrocarbons such as natural gas and alcohols like methanol are sometimes used. Fuel cells are different from batteries in that they require a constant source of fuel and oxygen to run, but they can produce electricity continually for as long as these inputs are supplied.
Shewanella oneidensis	Shewanella oneidensis proteobacterium was first isolated from Lake Oneida, NY in 1988, thus the derivation of the name. This species is also sometimes referred to as Shewanella oneidensis MR-1, indicating 'metal reducing', a special feature of this particular organism. Shewanella oneidesnsis is a facultative bacterium, capable of surviving and proliferating in both aerobic and anaerobic conditions.
Biofilm	A biofilm is an aggregate of microorganisms in which cells adhere to each other on a surface. These adherent cells are frequently embedded within a self-produced matrix of extracellular polymeric substance (EPS). Biofilm EPS, which is also referred to as slime (although not everything described as slime is a biofilm), is a polymeric conglomeration generally composed of extracellular DNA, proteins, and polysaccharides.
Electric current	Electric current means, depending on the context, a flow of electric charge (a phenomenon) or the rate of flow of electric charge (a quantity). This flowing electric charge is typically carried by moving electrons, in a conductor such as wire; in an electrolyte, it is instead carried by ions, and, in a plasma, by both.
	The SI unit for measuring the rate of flow of electric charge is the ampere, which is charge flowing through some surface at the rate of one coulomb per second.
Frankenstein	Frankenstein, in comics, may refer to:•Frankenstein from DC Comics' Seven Soldiers•Frankenstein the star of a short-lived series by Dell Comics•Frankenstein's Monster (Marvel Comics), from Marvel Comics' The Monster of Frankenstein •Frankenstein, a clone of the Marvel Comics character that appeared in the comic book series Nick Fury's Howling Commandos.•Frankenstein a 1940 to 1954 version by writer-artist Dick Briefer•Frankenstein Monster, a Wildstorm character who has appeared in Wetworks
	It may also refer to:•Doc Frankenstein, a series written by the Wachowski Brothers•Frankenstein, the central character in Death Race 2020 a comic book sequel to the film Death Race 2000, based on the character played by David Carradine•'Frankenstein Meets Shirley Temple', a story in A1 by Roger Langridge•Frankenstein: Monster Mayhem, a 2005 comic from Dead Dog Comics•Super Frankenstein, a parody produced by Big Bang Comics•'Universal Monsters: Frankenstein', one-shot from Dark Horse Comics•Young Frankenstein a DC character and member of the Teen Titans•Embalming -The Another Tale of Frankenstein-, a Japanese manga series written and illustrated by Nobuhiro Watsuki.

	It may also refer to the similar-sounding:•Frankenstein Mobster, a 2003 series from Image Comics
	Elseworlds titles that retell the story through well-known superheroes:•Batman: Castle of the Bat•The Superman Monster.
Luigi Galvani	Luigi Galvani was an Italian physician and physicist who lived and died in Bologna. In 1771, he discovered that the muscles of dead frogs legs twitched when struck by a spark. This was one of the first forays into the study of bioelectricity, a field that still today studies the electrical patterns and signals of the nervous system.
Electric power	Electric power is the rate at which electric energy is transferred by an electric circuit. The SI unit of power is the watt, one joule per second. Circuits
	Electric power, like mechanical power, is represented by the letter P in electrical equations.
Reduction potential	Reduction potential (also known as redox potential, oxidation / reduction potential is a measure of the tendency of a chemical species to acquire electrons and thereby be reduced. Reduction potential is measured in volts (V), or millivolts (mV). Each species has its own intrinsic reduction potential; the more positive the potential, the greater the species' affinity for electrons and tendency to be reduced.
Rocky Mountain spotted fever	Rocky Mountain spotted fever is the most lethal and most frequently reported rickettsial illness in the United States. It has been diagnosed throughout the Americas. Some synonyms for Rocky Mountain spotted fever in other countries include 'tick typhus,' 'Tobia fever' (Colombia), 'São Paulo fever' or 'febre maculosa' (Brazil), and 'fiebre manchada' (Mexico).
Shewanella	Shewanella is the sole genus included in the Shewanellaceae family of marine bacteria. Shewanella is a marine bacterium capable of modifying (or converting to an altered state) metals, by saturating them with electrons causing the metal to expand and soften, allowing them to process it, which in turn releases an electrical charge. The product is substantially less toxic and can break up in the environment.
Brugia malayi	Brugia malayi is a nematode (roundworm), one of the three causative agents of lymphatic filariasis in humans. Lymphatic filariasis, also known as elephantiasis, is a condition characterized by swelling of the lower limbs. The two other filarial causes of lymphatic filariasis are Wuchereria bancrofti and Brugia timori, which differ from B. malayi morphologically, symptomatically, and in geographical extent.

Chapter 3. Part 3: Metabolism and Biochemistry

Streptococcus pneumoniae	Streptococcus pneumoniae, is a Gram-positive, alpha-hemolytic, aerotolerant anaerobic member of the genus Streptococcus. A significant human pathogenic bacterium, S. pneumoniae was recognized as a major cause of pneumonia in the late 19th century, and is the subject of many humoral immunity studies. Despite the name, the organism causes many types of pneumococcal infections other than pneumonia.
Antibiotic resistance	Antibiotic resistance is a type of drug resistance where a microorganism is able to survive exposure to an antibiotic. While a spontaneous or induced genetic mutation in bacteria may confer resistance to antimicrobial drugs, genes that confer resistance can be transferred between bacteria in a horizontal fashion by conjugation, transduction, or transformation. Thus, a gene for antibiotic resistance that evolves via natural selection may be shared.
Organic acid	An organic acid is an organic compound with acidic properties. The most common organic acids are the carboxylic acids, whose acidity is associated with their carboxyl group -COOH. Sulfonic acids, containing the group $-SO_2OH$, are relatively stronger acids. Alcohols, with -OH, can act as acids but they are usually very weak.
Weak acid	A weak acid is an acid that dissociates incompletely. It does not release all of its hydrogens in a solution, donating only a partial amount of its protons to the solution. These acids have higher pKa than strong acids, which release all of their hydrogen atoms when dissolved in water.
Liposome	A liposome is an artificially-prepared vesicle composed of a lipid bilayer. The liposome can be used as a vehicle for administration of nutrients and pharmaceutical drugs. Liposomes can be prepared by disrupting biological membranes (such as by sonication).
Valinomycin	Valinomycin is a dodecadepsipeptide antibiotic. Valinomycin is obtained from the cells of several Streptomyces strains, among which 'S. tsusimaensis' and S. fulvissimus. It is a member of the group of natural neutral ionophores because it does not have a residual charge.
Proton pump	A proton pump is an integral membrane protein that is capable of moving protons across a cell membrane, mitochondrion, or other organelle. Mechanisms are based on conformational changes of the protein structure or on the Q cycle.

Heme	A heme is a prosthetic group that consists of an iron atom contained in the center of a large heterocyclic organic ring called a porphyrin. Not all porphyrins contain iron, but a substantial fraction of porphyrin-containing metalloproteins have heme as their prosthetic group; these are known as hemoproteins. Hemes are most commonly recognized in their presence as components of hemoglobin, the red pigment in blood, but they are also components of a number of other hemoproteins.
Salmonella enterica	Salmonella enterica is a rod-shaped flagellated, facultative anaerobic, Gram-negative bacterium, and a member of the genus Salmonella.

Most cases of salmonellosis are caused by food infected with S. enterica, which often infects cattle and poultry, though also other animals such as domestic cats and hamsters have also been shown to be sources of infection to humans. However, investigations of vacuum cleaner bags have shown that households can act as a reservoir of the bacterium; this is more likely if the household has contact with an infection source, for example members working with cattle or in a veterinary clinic. |
| Cytochrome b | Cytochrome b is the main subunit of transmembrane cytochrome bc1 and b6f complexes.

In the mitochondrion of eukaryotes and in aerobic prokaryotes, cytochrome b is a component of respiratory chain complex III (EC 1.10.2.2) - also known as the bc1 complex or ubiquinol-cytochrome c reductase. In plant chloroplasts and cyanobacteria, there is an analogous protein, cytochrome b6, a component of the plastoquinone-plastocyanin reductase (EC 1.10.99.1), also known as the b6f complex. |
Cytochrome c	The Cytochrome complex, or cyt c is a small heme protein found loosely associated with the inner membrane of the mitochondrion. It belongs to the cytochrome c family of proteins. Cytochrome c is a highly soluble protein, unlike other cytochromes, with a solubility of about 100 g/L and is an essential component of the electron transport chain, where it carries one electron.
Enterococcus faecalis	Enterococcus faecalis - formerly classified as part of the Group D Streptococcus system - is a Gram-positive commensal bacterium inhabiting the gastrointestinal tracts of humans and other mammals. It is among the main constituents of some probiotic food supplements. A commensal organism like other species in the genus Enterococcus, E. faecalis can cause life-threatening infections in humans, especially in the nosocomial (hospital) environment, where the naturally high levels of antibiotic resistance found in E. faecalis contribute to its pathogenicity.
Sodium-potassium pump	Sodium-potassium pump is an enzyme (EC 3.6.3.9) located in the plasma membrane (to be specific, an electrogenic transmembrane ATPase) in all animals.

Chapter 3. Part 3: Metabolism and Biochemistry

	Sodium-Potassium pumps are responsible for active transport for cells containing relatively high concentrations of potassium ions but low concentrations of sodium ions. The sodium-potassium pump moves these two ions in opposite directions across the plasma membrane.
Sodium ion	Sodium ion is soluble in water, and is thus present in great quantities in the Earth's oceans and other stagnant bodies of water. In these bodies it is mostly counterbalanced by the chloride ion, causing evaporated ocean water solids to consist mostly of sodium chloride, or common table salt. Sodium ion is also a component of many minerals.
Denitrification	Denitrification is a microbially facilitated process of nitrate reduction that may ultimately produce molecular nitrogen (N_2) through a series of intermediate gaseous nitrogen oxide products. This respiratory process reduces oxidized forms of nitrogen in response to the oxidation of an electron donor such as organic matter. The preferred nitrogen electron acceptors in order of most to least thermodynamically favorable include nitrate (NO_3^-), nitrite (NO_2^-), nitric oxide (NO), and nitrous oxide (N_2O) and dinitrigen [N2].
Haber process	The Haber process, is the nitrogen fixation reaction of nitrogen gas and hydrogen gas, over an enriched iron or ruthenium catalyst, which is used to industrially produce ammonia. Despite the fact that 78.1% of the air we breathe is nitrogen, the gas is relatively unavailable because it is so unreactive: nitrogen molecules are held together by strong triple bonds. It was not until the early 20th century that the Haber process was developed to harness the atmospheric abundance of nitrogen to create ammonia, which can then be oxidized to make the nitrates and nitrites essential for the production of nitrate fertilizer and explosives.
Haemophilus influenzae	Haemophilus influenzae, formerly called Pfeiffer's bacillus or Bacillus influenzae, Gram-negative, rod-shaped bacterium first described in 1892 by Richard Pfeiffer during an influenza pandemic. A member of the Pasteurellaceae family, it is generally aerobic, but can grow as a facultative anaerobe. H. influenzae was mistakenly considered to be the cause of influenza until 1933, when the viral etiology of the flu became apparent; the bacterium is colloquially known as bacterial influenza.
Neisseria gonorrhoeae	Neisseria gonorrhoeae, or gonococcus, is a species of Gram-negative coffee bean-shaped diplococci bacteria responsible for the sexually transmitted infection gonorrhea. N. gonorrhoea was first described by Albert Neisser in 1879. Microbiology

Neisseria meningitidis	Neisseria meningitidis is a heterotrophic gram-negative diplococcal bacterium best known for its role in meningitis and other forms of meningococcal disease such as meningococcemia. N. meningitidis is a major cause of morbidity and mortality during childhood in industrialized countries and is responsible for epidemics in Africa and in Asia.
	Approximately 2500 to 3500 cases of N meningitidis infection occur annually in the United States, with a case rate of about 1 in 100,000. Children younger than 5 years are at greatest risk, followed by teenagers of high school age.
Nitric oxide	Nitric oxide, is a molecule with chemical formula NO. It is a free radical and is an important intermediate in the chemical industry. Nitric oxide is a by-product of combustion of substances in the air, as in automobile engines, fossil fuel power plants, and is produced naturally during the electrical discharges of lightning in thunderstorms.
	In mammals including humans, NO is an important cellular signaling molecule involved in many physiological and pathological processes.
Biogeochemical cycle	In geography and Earth science, a biogeochemical cycle is a pathway by which a chemical element or molecule moves through both biotic (biosphere) and abiotic (lithosphere, atmosphere, and hydrosphere) compartments of Earth. A cycle is a series of change which comes back to the starting point and which can be repeated.
	The term 'biogeochemical' tells us that biological; geological and chemical factors are all involved.
Nitrogen cycle	The nitrogen cycle is the process by which nitrogen is converted between its various chemical forms. This transformation can be carried out to both biological and non-biological processes. Important processes in the nitrogen cycle include fixation, mineralization, nitrification, and denitrification.
Nitrous oxide	Nitrous oxide, commonly known as laughing gas or sweet air, is a chemical compound with the formula N_2O. It is an oxide of nitrogen. At room temperature, it is a colorless non-flammable gas, with a slightly sweet odor and taste.
Sulfate-reducing bacteria	Sulfate-reducing bacteria are those bacteria and archaea that can obtain energy by oxidizing organic compounds or molecular hydrogen (H_2) while reducing sulfate ($SO2-4$) to hydrogen sulfide (H_2S). In a sense, these organisms 'breathe' sulfate rather than oxygen, in a form of anaerobic respiration.

Chapter 3. Part 3: Metabolism and Biochemistry

Marine habitats	Marine habitats can be divided into coastal and open ocean habitats. Coastal habitats are found in the area that extends from as far as the tide comes in on the shoreline out to the edge of the continental shelf. Most marine life is found in coastal habitats, even though the shelf area occupies only seven percent of the total ocean area.
Extraterrestrial life	Extraterrestrial life is defined as life that does not originate from Earth. Referred to as alien life, or simply aliens these hypothetical forms of life range from simple bacteria-like organisms to beings far more complex than humans. The development and testing of hypotheses on extraterrestrial life is known as exobiology or astrobiology; the term astrobiology, however, includes the study of life on Earth viewed in its astronomical context.
Lithotroph	A lithotroph is an organism that uses an inorganic substrate (usually of mineral origin) to obtain reducing equivalents for use in biosynthesis (e.g., carbon dioxide fixation) or energy conservation via aerobic or anaerobic respiration. Known chemolithotrophs are exclusively microbes; No known macrofauna possesses the ability to utilize inorganic compounds as energy sources. Macrofauna and lithotrophs can form symbiotic relationships, in which case the lithotrophs are called 'prokaryotic symbionts.' An example of this is chemolithotrophic bacteria in deep sea worms or plastids, which are organelles within plant cells that may have evolved from photolithotrophic cyanobacteria-like organisms.
Early Earth	The 'Early Earth' is a term usually defined as Earth's first billion years, or gigayear. On the geologic time scale, the 'early Earth' comprises all of the Hadean eon (itself unofficially defined), as well as the Eoarchean and part of the Paleoarchean eras of the Archean eon. This period of Earth's history, being its earliest, involved the planet's condensation from a solar nebula and accretion from meteorites, as well as the formation of the earliest atmosphere and hydrosphere.
Nitrogen fixation	Nitrogen fixation is a process by which nitrogen (N_2) in the atmosphere is converted into ammonium (NH_{4+}). Atmospheric nitrogen or elemental nitrogen (N_2) is relatively inert: it does not easily react with other chemicals to form new compounds. Fixation processes free up the nitrogen atoms from their diatomic form (N_2) to be used in other ways.
Sulfur-reducing bacteria	Sulfur-reducing bacteria get their energy by reducing elemental sulfur to hydrogen sulfide. They couple this reaction with the oxidation of acetate, succinate or other organic compounds.

Sulfuric acid	Sulfuric acid is a highly corrosive strong mineral acid with the molecular formula H_2SO_4. It is a colorless to slightly yellow viscous liquid which is soluble in water at all concentrations. Sometimes, it may be dark brown as dyed during industrial production process in order to alert people's awareness to its hazards.
Molybdenum	Molybdenum is a Group 6 chemical element with the symbol Mo and atomic number 42. The name is from Neo-Latin Molybdaenum, from Ancient Greek M?λυβδος molybdos, meaning lead, since its ores were confused with lead ores. Molybdenum minerals have been known into prehistory, but the element was 'discovered' (in the sense of differentiating it as a new entity from the mineral salts of other metals) in 1778 by Carl Wilhelm Scheele. The metal was first isolated in 1781 by Peter Jacob Hjelm.
Selenium	Selenium is a chemical element with atomic number 34, chemical symbol Se, and an atomic mass of 78.96. It is a nonmetal, whose properties are intermediate between those of adjacent chalcogen elements sulfur and tellurium. It rarely occurs in its elemental state in nature, but instead is obtained as a side-product in the refining of other elements. Selenium is found in sulfide ores such as pyrite, where it partially replaces the sulfur.
Chlorine	Chlorine is the chemical element with atomic number 17 and symbol Cl. It is the second lightest halogen, with fluorine being the lightest. Chlorine is found in the periodic table in group 17. The element forms diatomic molecules under standard conditions, called dichlorine.
Chlorobenzene	Chlorobenzene is an aromatic organic compound with the chemical formula C_6H_5Cl. This colorless, flammable liquid is a common solvent and a widely used intermediate in the manufacture of other chemicals. Uses Chlorobenzene once was used in the manufacture of certain pesticides, most notably DDT by reaction with chloral (trichloroacetaldehyde), but this application has declined with the diminished use of DDT. At one time, chlorobenzene was the main precursor for the manufacture of phenol:$C_6H_5Cl + NaOH \rightarrow C_6H_5OH + NaCl$

Chapter 3. Part 3: Metabolism and Biochemistry

Microbial ecology	Microbial ecology is the ecology of microorganisms: their relationship with one another and with their environment. It concerns the three major domains of life -- Eukaryota, Archaea, and Bacteria -- as well as viruses. Microorganisms, by their omnipresence, impact the entire biosphere.
Quorum sensing	Quorum sensing is a system of stimulus and response correlated to population density. Many species of bacteria use quorum sensing to coordinate gene expression according to the density of their local population. In similar fashion, some social insects use quorum sensing to determine where to nest.
Bacteriorhodopsin	Bacteriorhodopsin is a protein used by Archaea, the most notable one being Halobacteria. It acts as a proton pump; that is, it captures light energy and uses it to move protons across the membrane out of the cell. The resulting proton gradient is subsequently converted into chemical energy.
Halobacterium salinarum	Halobacterium salinarum is an extremely halophilic marine gram-negative obligate aerobic archaeon. Despite its name, this microorganism is not a bacterium, but rather a member of the Kingdom Archaea. It is found in salted fish, hides, hypersaline lakes, and salterns.
Methylotroph	Methylotrophs are a diverse group of microorganisms that can use reduced one-carbon compounds, such as methanol or methane, as the carbon source for their growth; and multi-carbon compounds that contain no carbon bonds, such as dimethyl ether and dimethylamine. This group of microorganisms also includes those capable of assimilating reduced one-carbon compounds by way of carbon dioxide using the ribulose bisphosphate pathway. These organisms should not be confused with methanogens which on the contrary produce methane as a by-product from various one-carbon compounds such as carbon dioxide.
Proteorhodopsin	Proteorhodopsin is a photoactive retinylidene protein in marine planktonic prokaryotes and eukaryotes. Just like the homologous pigment bacteriorhodopsin found in some archaea, it consists of a transmembrane protein bound to a retinal molecule and functions as a light-driven proton pump. Some members of the family (of more than 800 types) are believed to have sensory functions.
Global warming	Global warming is the continuing rise in the average temperature of Earth's atmosphere and oceans. Global warming is caused by increased concentrations of greenhouse gases in the atmosphere, resulting from human activities such as deforestation and burning of fossil fuels. This finding is recognized by the national science academies of all the major industrialized countries and is not disputed by any scientific body of national or international standing.

Haloarchaea	Haloarchaea are microrganisms and members of the halophile community, in that they require high salt concentrations to grow. They are a distinct evolutionary branch of the Archaea, and are generally considered extremophiles, although not all members of this group can be considered as such. Haloarchaea require salt concentrations in excess of 2 M (or about 10%) to grow, and optimal growth usually occurs at much higher concentrations, typically 20-25%.
Photosynthesis	Photosynthesis is a process used by plants and other organisms to capture the sun's energy to split off water's hydrogen from oxygen. Hydrogen is combined with carbon dioxide (absorbed from air or water) to form glucose and release oxygen. All living cells in turn use fuels derived from glucose and oxidize the hydrogen and carbon to release the sun's energy and reform water and carbon dioxide in the process (cellular respiration).
Retinal	Retinal, is one of the many forms of vitamin A (the number of which varies from species to species). Retinal is a polyene chromophore, and bound to proteins called opsins, is the chemical basis of animal vision. Bound to proteins called type 1 rhodopsins, retinal allows certain microorganisms to convert light into metabolic energy.
Chloroflexus aurantiacus	Chloroflexus aurantiacus is a photosynthetic bacterium isolated from hot springs, belonging to the green non-sulfur bacteria. This organism is thermophilic and can grow at temperatures from 35 °C to 70 °C. Chloroflexus aurantiacus can survive in the dark if oxygen is available. When grown in the dark, Chloroflexus aurantiacus has a dark orange color.
Algal bloom	An algal bloom is a rapid increase or accumulation in the population of algae (typically microscopic) in an aquatic system. Algal blooms may occur in freshwater as well as marine environments. Typically, only one or a small number of phytoplankton species are involved, and some blooms may be recognized by discoloration of the water resulting from the high density of pigmented cells.
Chromophore	A chromophore is the part of a molecule responsible for its color. The color arises when a molecule absorbs certain wavelengths of visible light and transmits or reflects others. The chromophore is a region in the molecule where the energy difference between two different molecular orbitals falls within the range of the visible spectrum.
Rhodobacter	In taxonomy, Rhodobacter is a genus of the Rhodobacteraceae. The most famous species of Rhodobacter is Rhodobacter sphaeroides, which is commonly used to express proteins.
Bacteriochlorophyll	Bacteriochlorophylls are photosynthetic pigments that occur in various phototrophic bacteria.

They were discovered by Von Neil in 1932 . They are related to chlorophylls, which are the primary pigments in plants, algae, and cyanobacteria.

Carotenoid	Carotenoids are tetraterpenoid organic pigments that are naturally occurring in the chloroplasts and chromoplasts of plants and some other photosynthetic organisms like algae, some bacteria, and some types of fungus. Carotenoids can be synthesized fats and other basic organic metabolic building blocks by all these organisms. Carotenoids generally cannot be manufactured by species in the animal kingdom (although one species of aphid is known to have acquired the genes for synthesis of the carotenoid torulene from fungi by horizontal gene transfer).
Photosystem I	Photosystem I is the second photosystem in the photosynthetic light reactions of algae, plants, and some bacteria. Photosystem I is so named because it was discovered before photosystem II. Aspects of PS I were discovered in the 1950s, but the significances of these discoveries was not yet known. Louis Duysens first proposed the concepts of photosystems I and II in 1960, and, in the same year, a proposal by Fay Bendall and Robert Hill assembled earlier discoveries into a cohesive theory of serial photosynthetic reactions.
Photosystem II	Photosystem II is the first protein complex in the Light-dependent reactions. It is located in the thylakoid membrane of plants, algae, and cyanobacteria. The enzyme uses photons of light to energize electrons that are then transferred through a variety of coenzymes and cofactors to reduce plastoquinone to plastoquinol.
Purple bacteria	Purple bacteria are phototrophic, that is, capable of producing energy through photosynthesis. They are pigmented with bacteriochlorophyll a or b, together with various carotenoids, which give them colours ranging between purple, red, brown, and orange. Photosynthesis takes place at reaction centers on the cell membrane, which is folded into the cell to form sacs, tubes, or sheets, increasing the available surface area.
Thylakoid	A thylakoid is a membrane-bound compartment inside chloroplasts and cyanobacteria. They are the site of the light-dependent reactions of photosynthesis. Thylakoids consist of a thylakoid membrane surrounding a thylakoid lumen.
Chlorophyll	Chlorophyll is a green pigment found in almost all plants, algae, and cyanobacteria. Its name is derived from the Greek words χλωρος, chloros ('green') and φ?λλον, phyllon ('leaf'). Chlorophyll is an extremely important biomolecule, critical in photosynthesis, which allows plants to absorb energy from light.
Lumen	A lumen (Lat. lumen, an opening or light) (pl.

Ferredoxin	Ferredoxins are iron-sulfur proteins that mediate electron transfer in a range of metabolic reactions. The term 'ferredoxin' was coined by D.C. Wharton of the DuPont Co. and applied to the 'iron protein' first purified in 1962 by Mortenson, Valentine, and Carnahan from the anaerobic bacterium Clostridium pasteurianum.
Biosynthesis	Biosynthesis is an enzyme-catalyzed process in cells of living organisms by which substrates are converted to more complex products. The biosynthesis process often consists of several enzymatic steps in which the product of one step is used as substrate in the following step. Examples for such multi-step biosynthetic pathways are those for the production of amino acids, fatty acids, and natural products.
Chloroplast	Chloroplasts are organelles found in plant cells and other eukaryotic organisms that conduct photosynthesis. Chloroplasts capture light energy, store it in the energy storage molecules ATP and NADPH and use it in the process called photosynthesis to make organic molecules and free oxygen from carbon dioxide and water. Chloroplasts are green because they contain the chlorophyll pigment.
Phototrophic bacteria	Phototrophic bacteria are lithotrophic bacteria that use photosynthesis as their source of energy. ••.
Paper chromatography	Paper chromatography is an analytical method technique for separating and identifying mixtures that are or can be colored, especially pigments. This can also be used in secondary or primary colors in ink experiments. This method has been largely replaced by thin layer chromatography, however it is still a powerful teaching tool.
RuBisCO	Ribulose-1,5-bisphosphate carboxylase oxygenase, commonly known by the shorter name RuBisCO, is an enzyme involved in the first major step of carbon fixation, a process by which atmospheric carbon dioxide is converted by plants to energy-rich molecules such as glucose. RuBisCo is an abbreviation for Ribulose-1,5-bisphosphate carboxylase/oxygenase. In chemical terms, it catalyzes the carboxylation of ribulose-1,5-bisphosphate (also known as RuBP).
Carboxysome	Carboxysomes are bacterial microcompartments that contain enzymes involved in carbon fixation. Carboxysomes are made of polyhedral protein shells about 80 to 140 nanometres in diameter. These compartments are thought to concentrate carbon dioxide to overcome the inefficiency of RuBisCO - the predominant enzyme in carbon fixation and the rate limiting enzyme in the Calvin cycle.
Pyrobaculum	In taxonomy, Pyrobaculum is a genus of the Thermoproteaceae.

	As its latin name, Pyrobaculum suggests, the archaeon is rod-shaped and isolated from locations with high temperatures. It is Gram-negative and its cells are surrounded by an S-layer of protein subunits.
Rhodobacter sphaeroides	Rhodobacter sphaeroides is a kind of purple bacteria; a group of bacteria that can obtain energy through photosynthesis. Its best growth conditions are anaerobic phototrophy (photoheterotrophic and photoautotrophic) and aerobic chemoheterotrophy in the absence of light. R. sphaeroides is also able to fix nitrogen.
Thermoproteus	In taxonomy, Thermoproteus is a genus of the Thermoproteaceae. These prokaryotes are thermophilic sulphur-dependent organisms related to the genera Sulfolobus, Pyrodictium and Desulfurococcus. They are hydrogen-sulphur autotrophs and can grow at temperatures of up to 95 degrees Celsius. Thermoproteus is a genus of anaerobes that grow in the wild by autotrophic sulfur reduction.
Glucan	A glucan molecule is a polysaccharide of D-glucose monomers linked by glycosidic bonds. Many beta-glucans are medically important. Types The following are glucans: (The α- and β- and numbers clarify the type of O-glycosidic bond).
Folic acid	Folic acid and folate (the form naturally occurring in the body), as well as pteroyl-L-glutamic acid, pteroyl-L-glutamate, and pteroylmonoglutamic acid are forms of the water-soluble vitamin B_9. Folic acid is itself not biologically active, but its biological importance is due to tetrahydrofolate and other derivatives after its conversion from dihydrofolic acid in the liver. Vitamin B_9 (folic acid and folate inclusive) is essential to numerous bodily functions.
Polyketide	Polyketides are secondary metabolites from bacteria, fungi, plants, and animals. Polyketides are usually biosynthesized through the decarboxylative condensation of malonyl-CoA derived extender units in a similar process to fatty acid synthesis (a Claisen condensation). The polyketide chains produced by a minimal polyketide synthase are often further derivitized and modified into bioactive natural products.
Polyester	Polyester is a category of polymers which contain the ester functional group in their main chain. Although there are many polyesters, the term 'polyester' as a specific material most commonly refers to polyethylene terephthalate (PET).

Acetyl-CoA carboxylase	Acetyl-CoA carboxylase is a biotin-dependent enzyme that catalyzes the irreversible carboxylation of acetyl-CoA to produce malonyl-CoA through its two catalytic activities, biotin carboxylase (BC) and carboxyltransferase (CT). Acetyl CoA carboxylase is a multi-subunit enzyme in most prokaryotes and in the chloroplasts of most plants and algae, whereas it is a large, multi-domain enzyme in the endoplasmic reticulum of most eukaryotes. The most important function of Acetyl CoA carboxylase is to provide the malonyl-CoA substrate for the biosynthesis of fatty acids.
Acyl carrier protein	The acyl carrier protein is an important component in both fatty acid and polyketide biosynthesis with the growing chain bound during synthesis as a thiol ester at the distal thiol of a 4'-phosphopantethiene moiety. The protein is expressed in the inactive apo form and the 4'-phosphopantetheine moiety must be post-translationally attached to a conserved serine residue on the ACP by the action of holo-acyl carrier protein synthase (ACPS), a phosphopantetheinyl transferase.

4'-Phosphopantetheine is an essential prosthetic group of several acyl carrier proteins involved in pathways of primary and secondary metabolism including the acyl carrier proteins (ACP) of fatty acid synthases, ACPs of polyketide synthases, and peptidyl carrier proteins (PCP) and aryl carrier proteins (ArCP) of nonribosomal peptide synthetases (NRPS). |
Erythromycin	Erythromycin is a macrolide antibiotic that has an antimicrobial spectrum similar to or slightly wider than that of penicillin, and is often used for people who have an allergy to penicillins. For respiratory tract infections, it has better coverage of atypical organisms, including mycoplasma and Legionellosis. It was first marketed by Eli Lilly and Company, and it is today commonly known as EES (erythromycin ethylsuccinate, an ester prodrug that is commonly administered).
Secondary metabolite	Secondary metabolites are organic compounds that are not directly involved in the normal growth, development, or reproduction of an organism. Unlike primary metabolites, absence of secondary metabolites does not result in immediate death, but rather in long-term impairment of the organism's survivability, fecundity, or aesthetics, or perhaps in no significant change at all. Secondary metabolites are often restricted to a narrow set of species within a phylogenetic group.
Stringent response	The stringent response is a stress response that occurs in bacteria and plant chloroplasts in reaction to amino-acid starvation, fatty acid limitation, iron limitation, heat shock and other stress conditions. The stringent response is signaled by the alarmone (p)ppGpp, and modulating transcription of up to 1/3 of all genes in the cell. This in turn causes the cell to divert resources away from growth and division and toward amino acid synthesis in order to promote survival until nutrient conditions improve.

Chapter 3. Part 3: Metabolism and Biochemistry

Carrier protein	Carrier proteins are proteins involved in the movement of ions, small molecules, or macromolecules, such as another protein, across a biological membrane. Carrier proteins are integral membrane proteins; that is they exist within and span the membrane across which they transport substances. The proteins may assist in the movement of substances by facilitated diffusion or active transport.
Filariasis	Filariasis is a parasitic disease and is considered an infectious tropical disease, that is caused by thread-like filarial nematodes (roundworms) in the superfamily Filarioidea, also known as 'filariae'. There are 9 known filarial nematodes which use humans as their definitive host. These are divided into 3 groups according to the niche within the body that they occupy: 'lymphatic filariasis', 'subcutaneous filariasis', and 'serous cavity filariasis'.
Ivermectin	Ivermectin is a broad-spectrum antiparasitic medication. It is sold under brand names Stromectol in the United States, Mectizan in Canada by Merck and Ivexterm in Mexico by Valeant Pharmaceuticals International. Uses Ivermectin is a broad-spectrum antiparasitic agent.
Azotobacter	Azotobacter is a genus of usually motile, oval or spherical bacteria that form thick-walled cysts and may produce large quantities of capsular slime. They are aerobic, free-living soil microbes which play an important role in the nitrogen cycle in nature, binding atmospheric nitrogen, which is inaccessible to plants, and releasing it in the form of ammonium ions into the soil. Apart from being a model organism, it is used by humans for the production of biofertilizers, food additives and some biopolymers.
Nitrogen assimilation	Nitrogen assimilation is a fundamental biological process that occurs in plants and algae that are incapable of independent nitrogen fixation. The assimilation of nitrogen has marked effects on plant productivity, biomass, and crop yield, and nitrogen deficiency leads to a decrease in structural components. Initial conversion of nitrate to nitrite (by nitrate reductase) is followed by a reduction to ammonia by nitrite reductase (also called nitrite oxidoreductase).
Heterocyst	Heterocysts are specialized nitrogen-fixing cells formed by some filamentous cyanobacteria, such as Nostoc punctiforme, Cylindrospermum stagnale and Anabaena sphaerica, during nitrogen starvation. They fix nitrogen from dinitrogen (N_2) in the air using the enzyme nitrogenase, in order to provide the cells in the filament with nitrogen for biosynthesis.

Nitrogenase	Nitrogenases (EC 1.18.6.1EC 1.19.6.1) are enzymes used by some organisms to fix atmospheric nitrogen gas (N_2). It is the only known family of enzymes that accomplish this process. Dinitrogen is quite inert because of the strength of its N-N triple bond.
Rhizobium	Rhizobium is a genus of Gram-negative soil bacteria that fix nitrogen. Rhizobium forms an endosymbiotic nitrogen fixing association with roots of legumes and Parasponia.

The bacteria colonize plant cells within root nodules; here the bacteria convert atmospheric nitrogen to ammonia and then provide organic nitrogenous compounds such as glutamine or ureides to the plant. |
| Sinorhizobium meliloti | Sinorhizobium meliloti is a Gram-negative nitrogen-fixing bacterium (rhizobium). It forms a symbiotic relationship with legumes from the genera Medicago, Melilotus and Trigonella, including the model legume Medicago truncatula. This symbiosis results in a new plant organ termed a root nodule. |
| Reaction mechanism | In chemistry, a reaction mechanism is the step by step sequence of elementary reactions by which overall chemical change occurs.

Although only the net chemical change is directly observable for most chemical reactions, experiments can often be designed that suggest the possible sequence of steps in a reaction mechanism. Recently, electrospray ionization mass spectrometry has been used to corroborate the mechanism of several organic reaction proposals. |
| Klebsiella pneumoniae | Klebsiella pneumoniae is a Gram-negative, non-motile, encapsulated, lactose fermenting, facultative anaerobic, rod shaped bacterium found in the normal flora of the mouth, skin, and intestines. It is clinically the most important member of the Klebsiella genus of Enterobacteriaceae; it is closely related to K. oxytoca from which it is distinguished by being indole-negative and by its ability to grow on both melezitose and 3-hydroxybutyrate. It naturally occurs in the soil, and about 30% of strains can fix nitrogen in anaerobic condition. |
| Vanadium | Vanadium is a chemical element with the symbol V and atomic number 23. It is a hard, silvery gray, ductile and malleable transition metal. The element is found only in chemically combined form in nature, but once isolated artificially, the formation of an oxide layer stabilizes the free metal somewhat against further oxidation. Andrés Manuel del Río discovered vanadium in 1801 by analyzing a new lead-bearing mineral he called 'brown lead,' and named the new element erythronium since, upon heating, most of its salts turned from their initial color to red. |

Chapter 3. Part 3: Metabolism and Biochemistry

Leghemoglobin	Leghemoglobin is a nitrogen or oxygen carrier, because naturally occurring oxygen and nitrogen interact similarly with this protein; and a hemoprotein found in the nitrogen-fixing root nodules of leguminous plants. It is produced by legumes in response to the roots being infected by nitrogen-fixing bacteria, termed rhizobia, as part of the symbiotic interaction between plant and bacterium: roots uninfected with Rhizobium do not synthesise leghemoglobin. Leghemoglobin has close chemical and structural similarities to hemoglobin, and, like hemoglobin, is red in colour.
Nif gene	The nif gene is the gene responsible for the coding of proteins related and associated with the fixation of atmospheric nitrogen into a form of nitrogen available to plants. These genes are found in nitrogen fixing bacteria and cyanobacteria. Nif genes have both positive and negative regulators.
Amino acid synthesis	Amino acid synthesis is the set of biochemical processes (metabolic pathways) by which the various amino acids are produced from other compounds. The substrates for these processes are various compounds in the organism's diet or growth media. Not all organisms are able to synthesise all amino acids.
Glutamine synthetase	Glutamine synthetase (EC 6.3.1.2) is an enzyme that plays an essential role in the metabolism of nitrogen by catalyzing the condensation of glutamate and ammonia to form glutamine: Glutamate + ATP + NH_3 → Glutamine + ADP + phosphate + H_2O Glutamine Synthetase uses ammonia produced by nitrate reduction, amino acid degradation, and photorespiration. The amide group of glutamate is a nitrogen source for the synthesis of glutamine pathway metabolites. Other reactions may take place via Glutamine synthetase. Competition between ammonium ion and water, their binding affinities, and the concentration of ammonium ion, influences glutamine synthesis and glutamine hydrolysis.
Transamination	There are two chemical reactions known as transamination. The first is the reaction between an amino acid and an alpha-keto acid. The amino group is transferred from the former to the latter; this results in the amino acid being converted to the corresponding α-keto acid, while the reactant α-keto acid is converted to the corresponding amino acid (if the amino group is removed from an amino acid, an α-keto acid is left behind).
Nonribosomal peptide	Nonribosomal peptides (NRP) are a class of peptide secondary metabolites, usually produced by microorganisms like bacteria and fungi.

Nonribosomal peptides are also found in higher organisms, such as nudibranchs, but are thought to be made by bacteria inside these organisms. While there exist a wide range of peptides that are not synthesized by ribosomes, the term nonribosomal peptide typically refers to a very specific set of these as discussed in this article.

Purine	A purine is a heterocyclic aromatic organic compound, consisting of a pyrimidine ring fused to an imidazole ring. Purines, including substituted purines and their tautomers, are the most widely distributed kind of nitrogen-containing heterocycle in nature. Purines and pyrimidines make up the two groups of nitrogenous bases, including the two groups of nucleotide bases.
Pyrimidine	Pyrimidine is a heterocyclic aromatic organic compound similar to benzene and pyridine, containing two nitrogen atoms at positions 1 and 3 of the six-member ring. It is isomeric with two other forms of diazine: Pyridazine, with the nitrogen atoms in positions 1 and 2; and Pyrazine, with the nitrogen atoms in positions 1 and 4. A pyrimidine has many properties in common with pyridine, as the number of nitrogen atoms in the ring increases the ring pi electrons become less energetic and electrophilic aromatic substitution gets more difficult while nucleophilic aromatic substitution gets easier.
Clostridium difficile	Clostridium difficile (from the Greek kloster , spindle, and Latin difficile, difficult), also known as 'CDF/cdf', or 'C. diff', is a species of Gram-positive bacteria of the genus Clostridium that causes severe diarrhea and other intestinal disease when competing bacteria in the gut flora have been wiped out by antibiotics. Clostridia are anaerobic, spore-forming rods (bacilli). C. difficile is the most serious cause of antibiotic-associated diarrhea (AAD) and can lead to pseudomembranous colitis, a severe inflammation of the colon, often resulting from eradication of the normal gut flora by antibiotics.
Northern blot	The northern blot is a technique used in molecular biology research to study gene expression by detection of RNA in a sample. With northern blotting it is possible to observe cellular control over structure and function by determining the particular gene expression levels during differentiation, morphogenesis, as well as abnormal or diseased conditions. Northern blotting involves the use of electrophoresis to separate RNA samples by size and detection with a hybridization probe complementary to part of or the entire target sequence.
Ribosomal RNA	Ribosomal RNA is the RNA component of the ribosome, the enzyme that is the site of protein synthesis in all living cells.

	Ribosomal RNA provides a mechanism for decoding mRNA into amino acids and interacts with tRNAs during translation by providing peptidyl transferase activity. The tRNAs bring the necessary amino acids corresponding to the appropriate mRNA codon.
Chocolate	Chocolate is a raw or processed food produced from the seed of the tropical Theobroma cacao tree. Cacao has been cultivated for at least three millennia in Mexico, Central and South America, with its earliest documented use around 1100 BC. The majority of the Mesoamerican people made chocolate beverages, including the Aztecs, who made it into a beverage known as xocolatl (), a Nahuatl word meaning 'bitter water'. The seeds of the cacao tree have an intense bitter taste, and must be fermented to develop the flavor.
Food microbiology	Food microbiology is the study of the microorganisms that inhabit, create, or contaminate food. Of major importance is the study of microorganisms causing food spoilage. 'Good' bacteria, however, such as probiotics, are becoming increasingly important in food science.
Soy sauce	Soy sauce is a condiment produced by fermenting soybeans with Aspergillus oryzae or Aspergillus sojae molds, along with water and salt. After the fermentation, which yields fermented soybean paste, the paste is pressed, and two substances are obtained: a liquid, which is the soy sauce, and a cake of (wheat and) soy residue, the latter being usually reused as animal feed. Most commonly, a grain is used together with the soybeans in the fermentation process, but not always.
Tempeh	Tempeh, is a traditional soy product originally from Indonesia. It is made by a natural culturing and controlled fermentation process that binds soybeans into a cake form, similar to a very firm vegetarian burger patty. Tempeh is unique among major traditional soy foods in that it is the only one that did not originate in the Sinosphere.
Amanita phalloides	Amanita phalloides commonly known as the death cap, is a deadly poisonous basidiomycete fungus, one of many in the genus Amanita. Widely distributed across Europe, A. phalloides forms ectomycorrhizas with various broadleaved trees. In some cases, death cap has been introduced to new regions with the cultivation of non-native species of oak, chestnut, and pine.
Industrial microbiology	Industrial microbiology or microbial biotechnology encompasses the use of microorganisms in the manufacture of food or industrial products. The use of microorganisms for the production of food, either human or animal, is often considered a branch of food microbiology. The microorganisms used in industrial processes may be natural isolates, laboratory selected mutants or genetically engineered organisms.
Food preservation	Food preservation is the process of treating and handling food to stop or slow down spoilage (loss of quality, edibility or nutritional value) and thus allow for longer storage.

Preservation usually involves preventing the growth of bacteria, yeasts, fungi, and other micro-organisms (although some methods work by introducing benign bacteria, or fungi to the food), as well as retarding the oxidation of fats which cause rancidity. Food preservation can also include processes which inhibit visual deterioration that can occur during food preparation; such as the enzymatic browning reaction in apples after they are cut.

Agaricus bisporus	Agaricus bisporus--known variously as the common mushroom, button mushroom, white mushroom, table mushroom, champignon mushroom, crimini mushroom, Swiss brown mushroom, Roman brown mushroom, Italian brown, Italian mushroom, cultivated mushroom, or when mature, the Portobello mushroom--is an edible basidiomycete mushroom native to grasslands in Europe and North America. Agaricus bisporus is cultivated in more than 70 countries and is one of the most commonly and widely consumed mushrooms in the world. The common mushroom has a complicated taxonomic history. It was first described by English botanist Mordecai Cubitt Cooke in his 1871 Handbook of British Fungi, as a variety (var.
Brown algae	The Phaeophyceae or brown algae, is a large group of mostly marine multicellular algae, including many seaweeds of colder Northern Hemisphere waters. They play an important role in marine environments, both as food and for the habitats they form. For instance Macrocystis, a kelp of the order Laminariales, may reach 60 m in length, and forms prominent underwater forests.
Carrageenan	Carrageenans or carrageenins are a family of linear sulfated polysaccharides which are extracted from red seaweeds. Gelatinous extracts of the Chondrus crispus seaweed have been used as food additives for hundreds of years. Carrageenan is a vegetarian and vegan alternative to gelatin.
Kelp	Kelps are large seaweeds (algae) belonging to the brown algae (Phaeophyceae) in the order Laminariales. There are about 30 different genera. Kelps grow in underwater 'forests' (kelp forests) in shallow oceans.
Macrocystis	Macrocystis is a genus of kelp (algae). This genus contains the largest of all the phaeophyceae or brown algae. Macrocystis has pneumatocysts at the base of its blades.
Pleurotus	Pleurotus is a genus of gilled mushrooms which includes one of the most widely eaten mushrooms, P. ostreatus. Species of Pleurotus may be called oyster, abalone, or tree mushrooms, and are some of the most commonly cultivated edible mushrooms in the world.

Chapter 3. Part 3: Metabolism and Biochemistry

Porphyra	Porphyra is a foliose red algal genus of laver, comprising approximately 70 species. It grows in the intertidal zone, typically between the upper intertidal zone and the splash zone in cold waters of temperate oceans. In East Asia, it is used to produce the sea vegetable products nori and gim (in Korea), the most commonly eaten seaweed.
Saccharomyces	Saccharomyces is a genus in the kingdom of fungi that includes many species of yeast. Saccharomyces is from Greek σ?κχαρ (sugar) and μ?κης (mushroom) and means sugar fungus. Many members of this genus are considered very important in food production.
Truffle	A truffle is a fungi fruiting body that develops underground and relies on mycophagy for spore dispersal. Almost all truffles are ectomycorrhizal and are therefore usually found in close association with trees. There are hundreds of species of truffles that are big, but the fruiting body of some (mostly in the genus 'Tuber') are highly prized as a food.
Candida albicans	Candida albicans is a diploid fungus that grows both as yeast and filamentous cells and a causal agent of opportunistic oral and genital infections in humans. Systemic fungal infections (fungemias) including those by C. albicans have emerged as important causes of morbidity and mortality in immunocompromised patients (e.g., AIDS, cancer chemotherapy, organ or bone marrow transplantation). C. albicans biofilms may form on the surface of implantable medical devices.
Saccharomyces cerevisiae	Saccharomyces cerevisiae is a species of yeast. It is perhaps the most useful yeast, having been instrumental to baking and brewing since ancient times. It is believed that it was originally isolated from the skin of grapes .
Soylent Green	Soylent Green is a 1973 American science fiction film directed by Richard Fleischer and starring Charlton Heston and, in his final film, Edward G. Robinson. The film overlays the police procedural and science fiction genres as it depicts the investigation into the murder of a wealthy businessman in a dystopian future suffering from pollution, overpopulation, depleted resources, poverty, dying oceans, and a hot climate due to the greenhouse effect. Much of the population survives on processed food rations, including 'soylent green'.
Kimchi	Kimchi also spelled gimchi, kimchee, or kim chee, is a traditional fermented Korean dish made of vegetables with a variety of seasonings. There are hundreds of varieties of kimchi made with a main vegetable ingredient such as napa cabbage, radish, scallion or cucumber. Kimchi is also a main ingredient for many Korean dishes such as kimchi stew , kimchi soup , and kimchi fried rice .

Lactic acid fermentation	Lactic acid fermentation is a biological process by which sugars such as glucose, fructose, and sucrose, are converted into cellular energy and the metaboli lactate. It is an anaerobic fermentation reaction that occurs in some bacteria and animal cells, such as muscle cells. If oxygen is present in the cell, many organisms will bypass fermentation and undergo cellular respiration; however, facultative anaerobic organisms will both ferment and undergo respiration in the presence of oxygen.
Leuconostoc	Leuconostoc is a genus of Gram-positive bacteria, placed within the family of Leuconostocaceae. They are generally ovoid cocci often forming chains. Leuconostoc sp.
Propionibacterium	Propionibacterium is a genus of bacteria named for their unique metabolism: They are able to synthesize propionic acid by using unusual transcarboxylase enzymes. Its members are primarily facultative parasites and commensals of humans and other animals, living in and around the sweat glands, sebaceous glands, and other areas of the skin. They are virtually ubiquitous and do not cause problems for most people, but propionobacteria have been implicated in acne and other skin conditions.
Sauerkraut	Sauerkraut, French Choucroute, Polish Kiszona kapusta) directly translated: 'sour cabbage', is finely shredded cabbage that has been fermented by various lactic acid bacteria, including Leuconostoc, Lactobacillus, and Pediococcus. It has a long shelf-life and a distinctive sour flavor, both of which result from the lactic acid that forms when the bacteria ferment the sugars in the cabbage. It is not to be confused with coleslaw, which receives its acidic taste from vinegar.
Casein	Casein is the name for a family of related phosphoproteins (αS1, αS2, β, κ). These proteins are commonly found in mammalian milk, making up 80% of the proteins in cow milk and between 20% and 45% of the proteins in human milk. Casein has a wide variety of uses, from being a major component of cheese, to use as a food additive, to a binder for safety matches.
Lactobacillus	Lactobacillus, is a genus of Gram-positive facultative anaerobic or microaerophilic rod-shaped bacteria. They are a major part of the lactic acid bacteria group, named as such because most of its members convert lactose and other sugars to lactic acid. In humans they are present in the vagina and the gastrointestinal tract, where they are symbiotic and make up a small portion of the gut flora.
Micelle	A micelle is an aggregate of surfactant molecules dispersed in a liquid colloid. A typical micelle in aqueous solution forms an aggregate with the hydrophilic 'head' regions in contact with surrounding solvent, sequestering the hydrophobic single-tail regions in the micelle centre.

Chapter 3. Part 3: Metabolism and Biochemistry

Rennet	Rennet is usually a natural complex of enzymes produced in any mammalian stomach to digest the mother's milk, and is often used in the production of cheese. Rennet contains many enzymes, including a proteolytic enzyme (protease) that coagulates the milk, causing it to separate into solids (curds) and liquid (whey). The active enzyme in rennet is called chymosin or rennin (EC 3.4.23.4) but there are also other important enzymes in it, e.g., pepsin and lipase.
Whey	Whey is the liquid remaining after milk has been curdled and strained. It is a by-product of the manufacture of cheese or casein and has several commercial uses. Sweet whey is manufactured during the making of rennet types of hard cheese like Cheddar or Swiss.
Chymosin	Chymosin is an enzyme found in rennet. It is produced by cows in the lining of the abomasum (the fourth and final, chamber of the stomach). Chymosin is produced by gastric chief cells in infants to curdle the milk they ingest, allowing a longer residence in the bowels and better absorption.
Pasteurization	Pasteurization is a process of heating a food, usually a liquid, to a specific temperature for a definite length of time and then cooling it immediately. This process slows spoilage due to microbial growth in the food. Unlike sterilization, pasteurization is not intended to kill all micro-organisms in the food.
Penicillium roqueforti	Penicillium roqueforti is a common saprotrophic fungus from the family Trichocomaceae. Widespread in nature, it can be isolated from soil, decaying organic matter, and plants. The major industrial use of this fungus is the production of blue cheeses, flavouring agents, antifungals, polysaccharides, proteases and other enzymes.
Pepsin	Pepsin is an enzyme whose precursor form (pepsinogen) is released by the chief cells in the stomach and that degrades food proteins into peptides. It was discovered in 1836 by Theodor Schwann who also coined its name from the Greek word pepsis, meaning digestion (peptein: to digest). It was the first animal enzyme to be discovered, and, in 1929, it became one of the first enzymes to be crystallized, by John H. Northrop.
Ripening	Ripening is a process in fruits that causes them to become more palatable. In general, a fruit becomes sweeter, less green, and softer as it ripens. Even though the acidity of fruit increases as it ripens, the higher acidity level does not make the fruit seem tarter, which can lead to the misunderstanding that the riper the fruit the sweeter.
Lectin	Lectins are sugar-binding proteins (not to be confused with glycoproteins, which are proteins containing sugar chains or residues) that are highly specific for their sugar moieties. They play a role in biological recognition phenomena involving cells and proteins.

Protease inhibitor	Protease inhibitors are a class of drugs used to treat or prevent infection by viruses, including HIV and Hepatitis C. Protease inhibitors prevent viral replication by inhibiting the activity of proteases, e.g.HIV-1 protease, enzymes used by the viruses to cleave nascent proteins for final assembly of new virions. Protease inhibitors have been developed or are presently undergoing testing for treating various viruses:•HIV/AIDS: antiretroviral protease inhibitors (saquinavir, ritonavir, indinavir, nelfinavir, amprenavir etc).•Hepatitis C: Boceprevir•Hepatitis C: Telaprevir Given the specificity of the target of these drugs there is the risk, as in antibiotics, of the development of drug-resistant mutated viruses. To reduce this risk it is common to use several different drugs together that are each aimed at different targets.
Rhizopus oligosporus	Rhizopus oligosporus is a fungus of the family Mucoraceae that is a widely used starter culture for the home production of tempeh. The spores produce fluffy, white mycelia, binding the beans together to create an edible 'cake' of partly fermented soybeans. Rhizopus oligosporus produces an antibiotic that inhibits gram-positive bacteria, including the potentially harmful Staphylococcus aureus and the beneficial Bacillus subtilis (present in natto), even after the Rhizopus is consumed.
Soybean	The soybean. or soya bean (UK) (Glycine max) is a species of legume native to East Asia, widely grown for its edible bean which has numerous uses. The plant is classed as an oilseed rather than a pulse by the Food and Agricultural Organization (FAO).
Aspergillus oryzae	Aspergillus oryzae is a filamentous fungus (a mold). It is used in Chinese and Japanese cuisine to ferment soybeans. It is also used to saccharify rice, other grains, and potatoes in the making of alcoholic beverages such as huangjiu, sake, and shochu.
Leuconostoc mesenteroides	Leuconostoc mesenteroides is a species of bacteria sometimes associated with fermentation, under conditions of salinity and low temperatures.
RNA polymerase	RNA polymerase also known as DNA-dependent RNA polymerase, is an enzyme that produces RNA. In cells, RNAP is necessary for constructing RNA chains using DNA genes as templates, a process called transcription. RNA polymerase enzymes are essential to life and are found in all organisms and many viruses. In chemical terms, RNAP is a nucleotidyl transferase that polymerizes ribonucleotides at the 3' end of an RNA transcript.
Injera	Injera is a yeast-risen flatbread with a unique, slightly spongy texture. Traditionally made out of teff flour, it is a national dish in Ethiopia and Eritrea.

Chapter 3. Part 3: Metabolism and Biochemistry

Drug discovery	In the fields of medicine, biotechnology and pharmacology, drug discovery is the process by which drugs are discovered or designed. In the past most drugs have been discovered either by identifying the active ingredient from traditional remedies or by serendipitous discovery. As our understanding of disease has increased to the extent that we know how disease and infection are controlled at the molecular and physiological level, scientists are now able to try to find compounds that specifically modulate those molecules, for instance via high throughput screening.
Putrefaction	Putrefaction is one of seven stages in the decomposition of the body of a dead animal. It can be viewed, in broad terms, as the decomposition of proteins, in a process that results in the eventual breakdown of cohesion between tissues and the liquefaction of most organs. In terms of thermodynamics, all organic tissue is a stored source of chemical energy and when not maintained by the constant biochemical efforts of the living organism it will break down into simpler products.
Rotavirus	Rotavirus is the most common cause of severe diarrhoea among infants and young children. It is a genus of double-stranded RNA virus in the family Reoviridae. By the age of five, nearly every child in the world has been infected with rotavirus at least once.
Psychrotrophic bacteria	Psychrotrophic bacteria are bacteria that are capable of surviving or even thriving in a cold environment. They can be found in soils, in surface and deep sea waters, in Antarctic ecosystems, and in foods. Psychrotrophic bacteria are of particular concern to the dairy industry.
Crinipellis perniciosa	Crinipellis perniciosa is a fungus that causes 'Witches' Broom Disease' (WBD), which damages cocoa production in the Americas, and is consequently a major bane for makers of chocolate products. An infected crop may lose up to 90% of its harvest. It exists in two characteristic phases: biotrophic (expanding and infecting) and saprotrophic (dying, and producing spores).
Erwinia	Erwinia is a genus of Enterobacteriaceae bacteria containing mostly plant pathogenic species which was named for the first phytobacteriologist, Erwin Smith. It is a gram negative bacterium related to E. coli, Shigella, Salmonella and Yersinia. It is primarily a rod-shaped bacteria.
Serratia marcescens	Serratia marcescens is a species of Gram-negative, rod-shaped bacterium in the family Enterobacteriaceae. A human pathogen, S.

	marcescens is involved in nosocomial infections, particularly catheter-associated bacteremia, urinary tract infections and wound infections, and is responsible for 1.4% of nosocomial bacteremia cases in the United States. It is commonly found in the respiratory and urinary tracts of hospitalized adults and in the gastrointestinal system of children.
Theobroma cacao	Theobroma cacao also cacao tree and cocoa tree, is a small (4-8 m or 15-26 ft tall) evergreen tree in the family Sterculiaceae (alternatively Malvaceae), native to the deep tropical region of the Americas. Its seeds are used to make cocoa powder and chocolate. There are two prominent competing hypotheses about the origins of the domestication of the originally wild Theobroma cacao tree.
Seafood	Seafood refers to any sea animal or plant that is served as food and eaten by humans. Seafoods include seawater animals, such as fish and shellfish (including molluscs and crustaceans). By extension, in North America although not generally in the United Kingdom, the term seafood is also applied to similar animals from fresh water and all edible aquatic animals are collectively referred to as seafood.
Listeria monocytogenes	Listeria monocytogenes, a facultative anaerobe, intracellular bacterium, is the causative agent of listeriosis. It is one of the most virulent foodborne pathogens, with 20 to 30 percent of clinical infections resulting in death. Responsible for approximately 2,500 illnesses and 500 deaths in the United States (U.S).
Clostridium botulinum	Clostridium botulinum is a Gram-positive, rod-shaped bacterium that produces several toxins. The best known are its neurotoxins, subdivided in types A-G, that cause the flaccid muscular paralysis seen in botulism. It is also the main paralytic agent in botox.
Pathogenicity island	Pathogenicity islands (PAIs) are a distinct class of genomic islands acquired by microorganisms through horizontal gene transfer. They are incorporated in the genome of pathogenic organisms, but are usually absent from those nonpathogenic organisms of the same or closely related species. These mobile genetic elements may range from 10-200 kb and encode genes which contribute to the virulence of the respective pathogen.
Salmonella	Salmonella is a genus of rod-shaped, Gram-negative, non-spore-forming, predominantly motile enterobacteria with diameters around 0.7 to 1.5 µm, lengths from 2 to 5 µm, and flagella which grade in all directions (i.e. peritrichous). They are chemoorganotrophs, obtaining their energy from oxidation and reduction reactions using organic sources, and are facultative anaerobes. Most species produce hydrogen sulfide, which can readily be detected by growing them on media containing ferrous sulfate, such as TSI. Most isolates exist in two phases: a motile phase I and a nonmotile phase II. Cultures that are nonmotile upon primary culture may be switched to the motile phase using a Cragie tube.

Chapter 3. Part 3: Metabolism and Biochemistry

Botulinum toxin	Botulinum toxin is a protein and neurotoxin produced by the bacterium Clostridium botulinum. Botulinum toxin can cause botulism, a serious and life-threatening illness in humans and animals. When introduced intravenously in monkeys, type A (Botox Cosmetic) of the toxin exhibits an LD_{50} of 40-56 ng, type C1 around 32 ng, type D 3200 ng, and type E 88 ng; these are some of the most potent neurotoxins known.
Endophyte	An endophyte is an endosymbiont, often a bacterium or fungus, that lives within a plant for at least part of its life without causing apparent disease. Endophytes are ubiquitous and have been found in all the species of plants studied to date; however, most of these endophyte/plant relationships are not well understood. Many economically important forage and turfgrasses (e.g., Festuca spp., Lolium spp).
Toxin	A toxin is a poisonous substance produced within living cells or organisms; man-made substances created by artificial processes are thus excluded. The term was first used by organic chemist Ludwig Brieger (1849-1919). For a toxic substance not produced within living organisms, 'toxicant' and 'toxics' are also sometimes used..
Freeze-drying	Freeze-drying is a dehydration process typically used to preserve a perishable material or make the material more convenient for transport. Freeze-drying works by freezing the material and then reducing the surrounding pressure to allow the frozen water in the material to sublime directly from the solid phase to the gas phase. The origins of freeze drying Freeze-drying was first actively developed during WWII. Serum being sent to Europe for medical treatment of the wounded required refrigeration.
Mycobacterium tuberculosis	Mycobacterium tuberculosis is a pathogenic bacterial species in the genus Mycobacterium and the causative agent of most cases of tuberculosis. First discovered in 1882 by Robert Koch, M. tuberculosis has an unusual, waxy coating on the cell surface (primarily mycolic acid), which makes the cells impervious to Gram staining so acid-fast detection techniques are used instead. The physiology of M. tuberculosis is highly aerobic and requires high levels of oxygen.
Polio vaccine	Two polio vaccines are used throughout the world to combat poliomyelitis (or polio). The first was developed by Jonas Salk and first tested in 1952. Announced to the world by Salk on April 12, 1955, it consists of an injected dose of inactivated (dead) poliovirus. An oral vaccine was developed by Albert Sabin using attenuated poliovirus.
Vaccine	A vaccine is a biological preparation that improves immunity to a particular disease.

	A vaccine typically contains an agent that resembles a disease-causing microorganism, and is often made from weakened or killed forms of the microbe, its toxins or one of its surface proteins. The agent stimulates the body's immune system to recognize the agent as foreign, destroy it, and 'remember' it, so that the immune system can more easily recognize and destroy any of these microorganisms that it later encounters.
Ethambutol	Ethambutol is a bacteriostatic antimycobacterial drug prescribed to treat tuberculosis. It is usually given in combination with other tuberculosis drugs, such as isoniazid, rifampicin and pyrazinamide. It is sold under the trade names Myambutol and Servambutol.
Gene product	A gene product is the biochemical material, either RNA or protein, resulting from expression of a gene. A measurement of the amount of gene product is sometimes used to infer how active a gene is. Abnormal amounts of gene product can be correlated with disease-causing alleles, such as the overactivity of oncogenes which can cause cancer.
Transdermal patch	A transdermal patch is a medicated adhesive patch that is placed on the skin to deliver a specific dose of medication through the skin and into the bloodstream. Often, this promotes healing to an injured area of the body. An advantage of a transdermal drug delivery route over other types of medication delivery such as oral, topical, intravenous, intramuscular, etc.
Bioprospecting	Bioprospecting is an umbrella term describing the discovery of new and useful biological samples and mechanisms, typically in less-developed countries, either with or without the help of indigenous knowledge, and with or without compensation. In this way, bioprospecting includes biopiracy and also includes the search for previously unknown compounds in organisms that have never been used in traditional medicine. Biopiracy is a situation where indigenous knowledge of nature, originating with indigenous people, is exploited for commercial gain without permission from and with no compensation to the indigenous people themselves.
Aspergillus nidulans	Aspergillus nidulans is one of many species of filamentous fungi in the phylum Ascomycota. It has been an important research organism for studying eukaryotic cell biology for over 50 years, being used to study a wide range of subjects including recombination, DNA repair, mutation, cell cycle control, tubulin, chromatin, nucleokinesis, pathogenesis, and metabolism. It is one of the few species in its genus able to form sexual spores through meiosis, allowing crossing of strains in the laboratory.
Aspergillus niger	Aspergillus niger is a fungus and one of the most common species of the genus Aspergillus.

	It causes a disease called black mold on certain fruits and vegetables such as grapes, onions, and peanuts, and is a common contaminant of food. It is ubiquitous in soil and is commonly reported from indoor environments, where its black colonies can be confused with those of Stachybotrys (species of which have also been called 'black mould').
Industrial fermentation	Industrial fermentation is the intentional use of fermentation by microorganisms such as bacteria and fungi to make products useful to humans. Fermented products have applications as food as well as in general industry. Ancient fermented food processes, such as making bread, wine, cheese, curds, idli, dosa, etc., can be dated to more than 6,000 years ago.
Baculovirus	The baculoviruses are a family of large rod-shaped viruses that can be divided to two genera: nucleopolyhedroviruses (NPV) and granuloviruses (GV). While GVs contain only one nucleocapsid per envelope, NPVs contain either single (SNPV) or multiple (MNPV) nucleocapsids per envelope. The enveloped virions are further occluded in granulin matrix in GVs and polyhedrin for NPVs.
Downstream processing	Downstream processing refers to the recovery and purification of biosynthetic products, particularly pharmaceuticals, from natural sources such as animal or plant tissue or fermentation broth, including the recycling of salvageable components and the proper treatment and disposal of waste. It is an essential step in the manufacture of pharmaceuticals such as antibiotics, hormones (e.g. insulin and human growth hormone), antibodies (e.g. infliximab and abciximab) and vaccines; antibodies and enzymes used in diagnostics; industrial enzymes; and natural fragrance and flavor compounds. Downstream processing is usually considered a specialized field in biochemical engineering, itself a specialization within chemical engineering, though many of the key technologies were developed by chemists and biologists for laboratory-scale separation of biological products.
Ti plasmid	Ti plasmid is a circular plasmid that often, but not always, is a part of the genetic equipment that Agrobacterium tumefaciens and Agrobacterium rhizogenes use to transduce its genetic material to plants. Ti stands for tumor inducing. The Ti plasmid is lost when Agrobacterium is grown above 28° C. Such cured bacteria do not induce crown galls, i.e. they become avirulent.

1. _____ is an important enzyme that provides energy for the cell to use through the synthesis of adenosine triphosphate (ATP). ATP is the most commonly used 'energy currency' of cells from most organisms. It is formed from adenosine diphosphate (ADP) and inorganic phosphate (P_i), and needs energy.

   a. Emerson effect
   b. Oxygen cycle
   c. Oxygen evolution
   d. ATP synthase

2. Succinyl-Coenzyme A, abbreviated as _____, is a combination of succinic acid and coenzyme A.

   Source

   It is an important intermediate in the citric acid cycle, where it is synthesized from α-Ketoglutarate by α-ketoglutarate dehydrogenase through decarboxylation. During the process, coenzyme A is added.

   It is also synthesized from propionyl CoA, the odd-numbered fatty acid, which cannot undergo beta-oxidation.

   a. Succinyl-CoA
   b. Thomas Sydenham
   c. Johan Peter Rottler
   d. Transglutaminase

3. According to ISO 13600, an _____ is either a substance (energy form) or a phenomenon (energy system) that can be used to produce mechanical work or heat or to operate chemical or physical processes.

   In the field of Energetics, however, an _____ corresponds only to an energy form (not an energy system) of energy input required by the various sectors of society to perform their functions.

   Examples of _____s include liquid fuel in a furnace, gasoline in a pump, electricity in a factory or a house, and hydrogen in a tank of a car.

   a. Energy carrier
   b. Energy value of coal
   c. Enthalpy of mixing
   d. Enthalpy of sublimation

4. . _____ is a nucleobase (a purine derivative) with a variety of roles in biochemistry including cellular respiration, in the form of both the energy-rich adenosine triphosphate (ATP) and the cofactors nicotinamide _____ dinucleotide (NAD) and flavin _____ dinucleotide (FAD), and protein synthesis, as a chemical component of DNA and RNA. The shape of _____ is complementary to either thymine in DNA or uracil in RNA.

   _____ forms several tautomers, compounds that can be rapidly interconverted and are often considered equivalent. However, in isolated conditions, i.e. in an inert gas matrix and in the gas phase, mainly the 9H-_____ tautomer is found. Biosynthesis

Purine metabolism involves the formation of _____ and guanine.

a. Adenophostin
b. Adenosine
c. Adenosine diphosphate
d. Adenine

5. _____ is the sole genus included in the Shewanellaceae family of marine bacteria. _____ is a marine bacterium capable of modifying (or converting to an altered state) metals, by saturating them with electrons causing the metal to expand and soften, allowing them to process it, which in turn releases an electrical charge. The product is substantially less toxic and can break up in the environment.

a. Shigella
b. Shewanella
c. Shigella dysenteriae
d. Shigella flexneri

---

1. d
2. a
3. a
4. d
5. b

You can take the complete Chapter Practice Test

for Chapter 3. Part 3: Metabolism and Biochemistry
on all key terms, persons, places, and concepts.

Online 99 Cents

http://www.epub13.5.20451.3.cram101.com/

Use www.Cram101.com for all your study needs

including Cram101's online interactive problem solving labs in

chemistry, statistics, mathematics, and more.

Chapter 4. Part 4: Microbial Diversity and Ecology

CHAPTER OUTLINE: KEY TERMS, PEOPLE, PLACES, CONCEPTS

- Hyperthermophile
- Agrobacterium tumefaciens
- Ignicoccus
- Nanoarchaeota
- Nanoarchaeum equitans
- Supernova
- Amino acid
- Charles Robert Darwin
- Microbial ecology
- Quorum sensing
- Scopes Trial
- Spontaneous generation
- Stromatolite
- Sulfate-reducing bacteria
- Extraterrestrial life
- Biosphere
- Greenhouse effect
- Early Earth
- Methanogenesis

Biogeochemical cycle

Biosignature

Biotic material

Calvin cycle

Essential nutrient

Methylene blue

Bacteriorhodopsin

Haloarchaea

Proton pump

Micelle

Catabolite repression

Genetic code

Harold J. Morowitz

Sidney Altman

Ribosomal RNA

Ames test

Natural selection

Panspermia

Molecular clock

Phylogenetic tree

Bubonic plague

Buchnera

Escherichia

Photobacterium profundum

Photorhabdus luminescens

Shewanella

Yersinia pestis

Legionella pneumophila

Sphingomonas

Sphingomonas paucimobilis

Biofilm

Thermoplasma

Epulopiscium fishelsoni

Genomic island

Horizontal gene transfer

Pyrococcus

Membrane lipids

Aeropyrum

_____ | Retrotransposon _____

_____ | Transposable element _____

_____ | Transposon _____

_____ | Bacillus anthracis _____

_____ | Bacillus thuringiensis _____

_____ | Helicobacter pylori _____

_____ | Taxonomy _____

_____ | Genetic variation _____

_____ | Genome _____

_____ | Bacillus cereus _____

_____ | Food web _____

_____ | Haemophilus influenzae _____

_____ | Neisseria meningitidis _____

_____ | Ecological niche _____

_____ | Food chain _____

_____ | Pan-genome _____

_____ | Species name _____

_____ | Streptococcus pneumoniae _____

_____ | Taxon _____

Metagenomics

Methanobrevibacter smithii

Pelagibacter ubique

Rocky Mountain spotted fever

Trophic level

Gut flora

DNA Research

Pathogenicity island

Shiga toxin

Virulence factor

Parasitism

Nematode

Endosymbiont

Brugia malayi

Filariasis

Nucleomorph

Bifidobacterium

Gel electrophoresis

Polyacrylamide gel

Aquifex

Peptidoglycan

Phylum

Pseudopeptidoglycan

Cell wall

Chitin

Cyanobacteria

Deinococcus radiodurans

Deinococcus-Thermus

Thermus aquaticus

Actinobacteria

Bacteroidetes

Endospore

Firmicutes

Gram-positive bacteria

Nitrospira

Nitrospirae

Planctomycetes

Verrucomicrobia

Aquifex pyrophilus

Aquifex aeolicus

Chlorosome

Thermotoga maritima

Anabaena

Phycoerythrin

Prochlorococcus

Thylakoid

Akinete

Heterocyst

Microbial mat

Algal bloom

Primary producers

Trichodesmium

Clostridium

Bacillus subtilis

DNA polymerase

Reporter gene

Starvation response

Chapter 4. Part 4: Microbial Diversity and Ecology

CHAPTER OUTLINE: KEY TERMS, PEOPLE, PLACES, CONCEPTS

- _______________ Acetone
- _______________ Clostridium acetobutylicum
- _______________ Clostridium botulinum
- _______________ Delta endotoxin
- _______________ Insecticide
- _______________ Clostridium difficile
- _______________ Botulinum toxin
- _______________ Photosystem I
- _______________ Lactic acid
- _______________ Lactobacillales
- _______________ Lactobacillus
- _______________ Lactococcus
- _______________ Leuconostoc
- _______________ Listeria
- _______________ Listeria monocytogenes
- _______________ Nosocomial infection
- _______________ Probiotic
- _______________ Sauerkraut
- _______________ Staphylococcus aureus

_____ Staphylococcus epidermidis

_____ Substrate-level phosphorylation

_____ Cheese

_____ Toxic shock syndrome

_____ Yogurt

_____ Anaerobic respiration

_____ Enterococcus faecalis

_____ Streptococcus

_____ Mycoplasma

_____ Corynebacterium

_____ Geosmin

_____ Mycobacterium

_____ Ecology

_____ Actinomyces

_____ Frankia

_____ Mycobacterium tuberculosis

_____ Mycolic acid

_____ Actinomycosis

_____ Cell envelope

Chromosome

Endophyte

Arthrobacter

Chromium

Corynebacterium diphtheriae

Diphtheria

Metabolism

Alphaproteobacteria

Proteorhodopsin

Purple sulfur bacteria

Retinal

Rhodobacter sphaeroides

Photoheterotroph

Arsenic

Caulobacter crescentus

Selenium

Cell division

Bradyrhizobium

Carbon monoxide

Chapter 4. Part 4: Microbial Diversity and Ecology

CHAPTER OUTLINE: KEY TERMS, PEOPLE, PLACES, CONCEPTS

- Roseobacter
- Rhizobium
- Sinorhizobium meliloti
- Aphotic zone
- Nitrogen fixation
- Photic zone
- Filopodia
- Leghemoglobin
- Marine habitats
- Q fever
- Rickettsia
- Rickettsia rickettsii
- Biochemical oxygen demand
- Betaproteobacteria
- Diplococcus
- Neisseria gonorrhoeae
- Nitrogen cycle
- Nitrosomonas
- Antibiotic resistance

Asymptomatic carrier

Enterobacteriaceae

Gammaproteobacteria

Enterobacter

Erwinia

Proteus mirabilis

Proteus vulgaris

Pseudomonadaceae

Pseudomonas aeruginosa

Pseudomonas fluorescens

Shewanella oneidensis

Deltaproteobacteria

Myxobacteria

Myxococcus xanthus

Sorangium cellulosum

Slime mold

Agrobacterium

Bdellovibrio

Epsilonproteobacteria

CHAPTER OUTLINE: KEY TERMS, PEOPLE, PLACES, CONCEPTS

Helicobacter

Bacteroides

Bacteroides fragilis

Chlorobium

Flavobacterium

Obligate anaerobe

Borrelia burgdorferi

Chlamydia

Chlamydia trachomatis

Chlamydomonas reinhardtii

Leptospirosis

Confocal microscopy

Methanomicrobiales

Extreme environment

Periodontal disease

DNA gyrase

Halobacterium

Halorhodopsin

Pyrococcus furiosus

Chapter 4. Part 4: Microbial Diversity and Ecology

CHAPTER OUTLINE: KEY TERMS, PEOPLE, PLACES, CONCEPTS

______ S-layer

______ Cell membrane

______ Phase variation

______ RNA polymerase

______ Sigma factor

______ TATA-binding protein

______ Transcription factor

______ Transfer RNA

______ Archaea

______ Crenarchaeota

______ Euryarchaeota

______ Histone

______ Archaeoglobus

______ Ferroplasma

______ Thermococcales

______ High pressure

______ Thermophile

______ Korarchaeota

______ Thermoprotei

CHAPTER OUTLINE: KEY TERMS, PEOPLE, PLACES, CONCEPTS

_____ Habitat

_____ Desulfurococcales

_____ Pyrodictium

_____ Sulfolobales

_____ Sulfolobus

_____ Sulfur metabolism

_____ Fusellovirus

_____ Sulfolobus turreted icosahedral virus

_____ Caldisphaera

_____ Cenarchaeum symbiosum

_____ Thermoproteales

_____ Thermoproteus

_____ Mesophile

_____ Benthos

_____ Methanobacteriales

_____ Methanogen

_____ Methanobacterium

_____ Methanosarcina

_____ Methanococcus

Methanomicrobium

Methanosaeta

Wetland

Cattle

Landfill

Termite

Sulfur-reducing bacteria

Global warming

Molybdenum

Halobacteriales

Halophile

Carotenoid

Halocin

Taq polymerase

Thermococcus

Chain reaction

Polymerase chain reaction

Archaeoglobales

Okazaki fragment

Chapter 4. Part 4: Microbial Diversity and Ecology

CHAPTER OUTLINE: KEY TERMS, PEOPLE, PLACES, CONCEPTS

- Thermococcus litoralis
- Thermoplasmatales
- Dinoflagellate
- Eukaryote
- Vorticella
- Red tide
- Convergent evolution
- Mycology
- Fairy ring
- Giardia
- Immune system
- Blood cell
- Amoebozoa
- Elongation factor
- Heterokont
- Oomycete
- Opisthokont
- Rhizaria
- Trypanosoma brucei

_____ Zoospore

_____ Chloroflexus aurantiacus

_____ Chytridiomycota

_____ Alveolate

_____ Green algae

_____ Mixotroph

_____ Phytoplankton

_____ Red algae

_____ Zooplankton

_____ Kleptoplasty

_____ Ciliate

_____ Euglenozoa

_____ Kelp

_____ Ostreococcus tauri

_____ Picoeukaryote

_____ Mycelium

_____ Nystatin

_____ Blastomycosis

_____ Budding

_____ | Candida albicans |

_____ | Meiosis |

_____ | Pneumocystis jirovecii |

_____ | Saccharomyces cerevisiae |

_____ | Dimorphic fungi |

_____ | Life cycle |

_____ | Vaginal flora |

_____ | Penicillium |

_____ | Reproduction |

_____ | Batrachochytrium dendrobatidis |

_____ | Hydrogenosome |

_____ | Mucor |

_____ | Rhizopus |

_____ | Sporangium |

_____ | Ascospore |

_____ | Aspergillus |

_____ | Zygospore |

_____ | Mycotoxin |

_____ | Truffle |

Microsporum

Trichophyton

Amanita phalloides

Basidiospore

Chlamydomonas

Encephalitozoon intestinalis

Phytophthora infestans

Conidia

Pyrenoid

Spirogyra

Caulerpa taxifolia

Porphyra

Volvox

Coral bleaching

Coral reef

Frustule

Brown algae

Kelp forest

Acanthamoeba

	Dictyostelium discoideum
	Radiolarian
	Contractile vacuole
	Telomere
	Sea anemone
	Toxoplasma gondii
	Zooxanthella
	Plasmodium falciparum
	Malaria
	Kinetoplast
	Giardia lamblia
	Leishmania major
	Leishmaniasis
	Trypanosoma cruzi
	Microbial metabolism
	Protein
	Desulfovibrio
	Copepod
	Vibrio cholerae

Chapter 4. Part 4: Microbial Diversity and Ecology

CHAPTER OUTLINE: KEY TERMS, PEOPLE, PLACES, CONCEPTS

______ Assimilation

______ Carbon cycle

______ Dissimilation

______ Ecosystem

______ Consumers

______ Decomposer

______ Extremophile

______ Detritus

______ Electron acceptor

______ Lyme disease

______ Commensalism

______ Neuston

______ Pelagic zone

______ Gas gangrene

______ Oil spill

______ Thermocline

______ Picoplankton

______ Marine snow

______ MHC restriction

_____ | Psychrophile

_____ | Drug resistance

_____ | Littoral zone

_____ | Epilimnion

_____ | Eutrophication

_____ | Soil microbiology

_____ | Rhizosphere

_____ | Radioactive decay

_____ | Ribosomal protein

_____ | Soil food web

_____ | Lignin

_____ | Hydric soil

_____ | Atmospheric methane

_____ | Flavonoid

_____ | Haber process

_____ | Neotyphodium coenophialum

_____ | Rhizobia

_____ | Salmonella enterica

_____ | Soybean

Stenotrophomonas

B cell

Fistula

Prevotella

Chlorophyll

Fossil fuel

Iron-sulfur cluster

Biogeochemistry

Micronutrient

Sulfur cycle

Oxidation state

Mesocosm

Stable isotope

Reservoir

Deforestation

Nitrosopumilus

Carbon sink

Water cycle

Dead zone

Chapter 4. Part 4: Microbial Diversity and Ecology

CHAPTER OUTLINE: KEY TERMS, PEOPLE, PLACES, CONCEPTS

- ______ Hypoxia
- ______ Sludge
- ______ Treatment wetland
- ______ Nitrification
- ______ Nitrogenase
- ______ Methemoglobinemia
- ______ Nitrobacter
- ______ Denitrification
- ______ Nitrous oxide
- ______ Iron cycle
- ______ Siderophore
- ______ Iron fertilization
- ______ Astrobiology
- ______ Extrasolar planet
- ______ Terraforming

CHAPTER HIGHLIGHTS & NOTES: KEY TERMS, PEOPLE, PLACES, CONCEPTS

Chapter 4. Part 4: Microbial Diversity and Ecology

Hyperthermophile	A hyperthermophile is an organism that thrives in extremely hot environments-- from 60 degrees C (140 degrees F) upwards. An optimal temperature for the existence of hyperthermophiles is above 80°C (176°F). Hyperthermophiles are a subset of extremophiles, micro-organisms within the domain Archaea, although some bacteria are able to tolerate temperatures of around 100°C (212° F), as well.
Agrobacterium tumefaciens	Agrobacterium tumefaciens is the causal agent of crown gall disease (the formation of tumours) in over 140 species of dicot. It is a rod shaped, Gram negative soil bacterium (Smith et al., 1907). Symptoms are caused by the insertion of a small segment of DNA (known as the T-DNA, for 'transfer DNA'), from a plasmid, into the plant cell, which is incorporated at a semi-random location into the plant genome.
Ignicoccus	Ignicoccus is a genus of Archaea living in marine hydrothermal vents. They were discovered in Kolbeinsey Ridge north of Iceland and in the Pacific Ocean (at 9 degrees N, 104 degrees W) in 2000 (Huber et al., 2000). According to the comparisons of 16S rRNA genes, Ignicoccus represents a new, deeply branching lineage within the family of the Desulfurococcaceae (Huber et al., 2002).
Nanoarchaeota	In taxonomy, the Nanoarchaeota are a phylum of the Archaea. This phylum currently has only one representative, Nanoarchaeum equitans.
Nanoarchaeum equitans	Nanoarchaeum equitans is a marine Archaea that was discovered in 2002 in a hydrothermal vent off the coast of Iceland on the Kolbeinsky Ridge by Karl Stetter. Strains of this microbe were also found on the Sub-polar Mid Oceanic Ridge, and in the Obsidian Pool in Yellowstone National Park. Since it grows in temperatures approaching boiling, at about 80 degrees Celsius, it is considered to be a thermophile.
Supernova	A supernova is a stellar explosion that is more energetic than a nova. Supernovae are extremely luminous and cause a burst of radiation that often briefly outshines an entire galaxy, before fading from view over several weeks or months. During this short interval a supernova can radiate as much energy as the Sun is expected to emit over its entire life span.
Amino acid	Amino acids are molecules containing an amine group, a carboxylic acid group, and a side-chain that is specific to each amino acid. The key elements of an amino acid are carbon, hydrogen, oxygen, and nitrogen. They are particularly important in biochemistry, where the term usually refers to alpha-amino acids.
Charles Robert Darwin	Charles Robert Darwin FRS (12 February 1809 - 19 April 1882) was an English naturalist.

He established that all species of life have descended over time from common ancestry, and proposed the scientific theory that this branching pattern of evolution resulted from a process that he called natural selection.

He published his theory with compelling evidence for evolution in his 1859 book On the Origin of Species. The scientific community and much of the general public came to accept evolution as a fact in his lifetime.

Microbial ecology	Microbial ecology is the ecology of microorganisms: their relationship with one another and with their environment. It concerns the three major domains of life -- Eukaryota, Archaea, and Bacteria -- as well as viruses. Microorganisms, by their omnipresence, impact the entire biosphere.
Quorum sensing	Quorum sensing is a system of stimulus and response correlated to population density. Many species of bacteria use quorum sensing to coordinate gene expression according to the density of their local population. In similar fashion, some social insects use quorum sensing to determine where to nest.
Scopes Trial	The Scopes Trial, formally known as The State of Tennessee v. John Thomas Scopes and informally known as the Scopes Monkey Trial, was a landmark American legal case in 1925 in which high school science teacher, John Scopes, was accused of violating Tennessee's Butler Act, which made it unlawful to teach evolution in any state-funded school. Scopes was found guilty, but the verdict was overturned on a technicality and he went free. The trial drew intense national publicity, as national reporters flocked to the small town of Dayton, Tennessee, to cover the big-name lawyers representing each side.
Spontaneous generation	Spontaneous generation is an obsolete principle regarding the origin of life from inanimate matter, which held that this process was a commonplace and everyday occurrence, as distinguished from univocal generation, or reproduction from parent(s). The hypothesis was synthesized by Aristotle, who compiled and expanded the work of prior natural philosophers and the various ancient explanations of the appearance of organisms; it held sway for two millennia. It is generally accepted to have been ultimately disproven in the 19th century by the experiments of Louis Pasteur, expanding upon the experiments of other scientists before him (such as Francesco Redi who had performed similar experiments in the 17th century).

Chapter 4. Part 4: Microbial Diversity and Ecology

Stromatolite	Stromatolites or stromatoliths are layered accretionary structures formed in shallow water by the trapping, binding and cementation of sedimentary grains by biofilms of microorganisms, especially cyanobacteria (commonly known as blue-green algae). Stromatolites provide some of the most ancient records of life on Earth. A variety of stromatolite morphologies exist including conical, stratiform, branching, domal, and columnar types.
Sulfate-reducing bacteria	Sulfate-reducing bacteria are those bacteria and archaea that can obtain energy by oxidizing organic compounds or molecular hydrogen (H_2) while reducing sulfate ($SO2-4$) to hydrogen sulfide (H_2S). In a sense, these organisms 'breathe' sulfate rather than oxygen, in a form of anaerobic respiration. Sulfate-reducing bacteria can be traced back to 3.5 billion years ago and are considered to be among the oldest forms of microorganisms, having contributed to the sulfur cycle soon after life emerged on Earth.
Extraterrestrial life	Extraterrestrial life is defined as life that does not originate from Earth. Referred to as alien life, or simply aliens these hypothetical forms of life range from simple bacteria-like organisms to beings far more complex than humans. The development and testing of hypotheses on extraterrestrial life is known as exobiology or astrobiology; the term astrobiology, however, includes the study of life on Earth viewed in its astronomical context.
Biosphere	The biosphere is the global sum of all ecosystems. It can also be called the zone of life on Earth, a closed (apart from solar and cosmic radiation) and self-regulating system. From the broadest biophysiological point of view, the biosphere is the global ecological system integrating all living beings and their relationships, including their interaction with the elements of the lithosphere, hydrosphere and atmosphere.
Greenhouse effect	The greenhouse effect is a process by which thermal radiation from a planetary surface is absorbed by atmospheric greenhouse gases, and is re-radiated in all directions. Since part of this re-radiation is back towards the surface, energy is transferred to the surface and the lower atmosphere. As a result, the temperature there is higher than it would be if direct heating by solar radiation were the only warming mechanism.
Early Earth	The 'Early Earth' is a term usually defined as Earth's first billion years, or gigayear. On the geologic time scale, the 'early Earth' comprises all of the Hadean eon (itself unofficially defined), as well as the Eoarchean and part of the Paleoarchean eras of the Archean eon.

Methanogenesis	Methanogenesis is the formation of methane by microbes known as methanogens. Organisms capable of producing methane have been identified only from the domain Archaea, a group phylogenetically distinct from both eukaryotes and bacteria, although many live in close association with anaerobic bacteria. The production of methane is an important and widespread form of microbial metabolism.
Biogeochemical cycle	In geography and Earth science, a biogeochemical cycle is a pathway by which a chemical element or molecule moves through both biotic (biosphere) and abiotic (lithosphere, atmosphere, and hydrosphere) compartments of Earth. A cycle is a series of change which comes back to the starting point and which can be repeated. The term 'biogeochemical' tells us that biological; geological and chemical factors are all involved.
Biosignature	A biosignature is any substance - such as an element, isotope, or molecule, or phenomenon - that provides scientific evidence of past or present life. Measurable attributes of life include its complex physical and chemical structures and also its utilization of free energy and the production of biomass and wastes. Due to its unique characteristics, a biosignature can be interpreted as having been produced by living organisms, however, it is important that they not be considered definitive because there is no way of knowing in advance which ones are universal to life and which ones are unique to the peculiar circumstances of life on Earth.
Biotic material	Biotic material is originated from living organisms. Most such materials contain carbon and are capable of decay. Examples of biotic materials are wood, linoleum, straw, humus, manure, bark, crude oil, cotton, spider silk, chitin, fibrin, and bone.
Calvin cycle	The Calvin cycle, is a series of biochemical redox reactions that take place in the stroma of chloroplasts in photosynthetic organisms. It is also known as the dark reactions. The cycle was discovered by Melvin Calvin, James Bassham, and Andrew Benson at the University of California, Berkeley by using the radioactive isotope carbon-14. It is one of the light-independent (dark) reactions used for carbon fixation.
Essential nutrient	An essential nutrient is a nutrient required for normal body functioning that either cannot be synthesized by the body at all, or cannot be synthesized in amounts adequate for good health (e.g. niacin, choline), and thus must be obtained from a dietary source. Essential nutrients are also defined by the collective physiological evidence for their importance in the diet, as represented in e.g. US government approved tables for Dietary Reference Intake.

Chapter 4. Part 4: Microbial Diversity and Ecology

Methylene blue	Methylene blue is a heterocyclic aromatic chemical compound with the molecular formula $C_{16}H_{18}N_3SCl$. It has many uses in a range of different fields, such as biology and chemistry. At room temperature it appears as a solid, odorless, dark green powder, that yields a blue solution when dissolved in water.
Bacteriorhodopsin	Bacteriorhodopsin is a protein used by Archaea, the most notable one being Halobacteria. It acts as a proton pump; that is, it captures light energy and uses it to move protons across the membrane out of the cell. The resulting proton gradient is subsequently converted into chemical energy.
Haloarchaea	Haloarchaea are microrganisms and members of the halophile community, in that they require high salt concentrations to grow. They are a distinct evolutionary branch of the Archaea, and are generally considered extremophiles, although not all members of this group can be considered as such. Haloarchaea require salt concentrations in excess of 2 M (or about 10%) to grow, and optimal growth usually occurs at much higher concentrations, typically 20-25%.
Proton pump	A proton pump is an integral membrane protein that is capable of moving protons across a cell membrane, mitochondrion, or other organelle. Mechanisms are based on conformational changes of the protein structure or on the Q cycle. In cell respiration, the proton pump uses energy to transport protons from the matrix of the mitochondrion to the inner and outer mitochondrial membranes.
Micelle	A micelle is an aggregate of surfactant molecules dispersed in a liquid colloid. A typical micelle in aqueous solution forms an aggregate with the hydrophilic 'head' regions in contact with surrounding solvent, sequestering the hydrophobic single-tail regions in the micelle centre. This phase is caused by the insufficient packing issues of single-tailed lipids in a bilayer.
Catabolite repression	Carbon catabolite repression, is an important part of global control system of various bacteria and other micro-organisms. Catabolite repression allows bacteria to adapt quickly to a preferred (rapidly metabolisable) carbon and energy source first. This is usually achieved through inhibition of synthesis of enzymes involved in catabolism of carbon sources other than the preferred one.
Genetic code	The genetic code is the set of rules by which information encoded in genetic material (DNA or mRNA sequences) is translated into proteins (amino acid sequences) by living cells. The code defines how sequences of three nucleotides, called codons, specify which amino acid will be added next during protein synthesis.

Harold J. Morowitz	Harold J. Morowitz is an American biophysicist who studies the application of thermodynamics to living systems. Author of numerous books and articles, his work includes technical monographs as well as essays. He is a leading authority on the origin of life, his primary research interest for more than fifty years.
Sidney Altman	Sidney Altman is a Canadian American molecular biologist, who is currently the Sterling Professor of Molecular, Cellular, and Developmental Biology and Chemistry at Yale University. In 1989 he shared the Nobel Prize in Chemistry with Thomas R. Cech for their work on the catalytic properties of RNA. Altman was born on May 7, 1939 in Montreal, Quebec, Canada. His parents were immigrants to Canada, each coming from Eastern Europe as a young adult, in the 1920s.
Ribosomal RNA	Ribosomal RNA is the RNA component of the ribosome, the enzyme that is the site of protein synthesis in all living cells. Ribosomal RNA provides a mechanism for decoding mRNA into amino acids and interacts with tRNAs during translation by providing peptidyl transferase activity. The tRNAs bring the necessary amino acids corresponding to the appropriate mRNA codon.
Ames test	The Ames test is a biological assay to assess the mutagenic potential of chemical compounds. A positive test indicates that the chemical is mutagenic and therefore may act as a carcinogen, since cancer is often linked to mutation. However, a number of false-positives and false-negatives are known.
Natural selection	Natural selection is the gradual, non-random, process by which biological traits become either more or less common in a population as a function of differential reproduction of their bearers. It is a key mechanism of evolution. Variation exists within all populations of organisms.
Panspermia	Panspermia (Greek: πανσπερμ?α from π?ς/π?ν (pas/pan) 'all' and σπ?ρμα (sperma) 'seed') is the hypothesis that life exists throughout the Universe, distributed by meteoroids, asteroids and planetoids. Panspermia proposes that life that can survive the effects of space, such as extremophile archaea, become trapped in debris that is ejected into space after collisions between planets that harbor life and Small Solar System Bodies (SSSB). Bacteria may travel dormant for an extended amount of time before colliding randomly with other planets or intermingling with protoplanetary disks.

Chapter 4. Part 4: Microbial Diversity and Ecology

Molecular clock	The molecular clock (based on the molecular clock hypothesis (MCH)) is a technique in molecular evolution that uses fossil constraints and rates of molecular change to deduce the time in geologic history when two species or other taxa diverged. It is used to estimate the time of occurrence of events called speciation or radiation. The molecular data used for such calculations is usually nucleotide sequences for DNA or amino acid sequences for proteins.
Phylogenetic tree	A phylogenetic tree is a branching diagram or 'tree' showing the inferred evolutionary relationships among various biological species or other entities based upon similarities and differences in their physical and/or genetic characteristics. The taxa joined together in the tree are implied to have descended from a common ancestor. In a rooted phylogenetic tree, each node with descendants represents the inferred most recent common ancestor of the descendants, and the edge lengths in some trees may be interpreted as time estimates.
Bubonic plague	Bubonic plague is a zoonotic disease, circulating mainly among small rodents and their fleas, and is one of three types of infections caused by Yersinia pestis (formerly known as Pasteurella pestis), which belongs to the family Enterobacteriaceae. Without treatment, the bubonic plague kills about two out of three infected humans within 4 days. The term bubonic plague is derived from the Greek word βουβ?ν, meaning 'groin.' Swollen lymph nodes (buboes) especially occur in the armpit and groin in persons suffering from bubonic plague.
Buchnera	Buchnera aphidicola a member of the Proteobacteria, is the primary endosymbiont of aphids (A. pisum). It is believed that Buchnera was once a free living gram negative ancestor similar to a modern Enterobacteriaceae such as Escherichia coli. Buchnera are 3 µm in diameter and have some of the key characteristics of their Enterobacteriaceae relatives such as a gram-negative cell wall.
Escherichia	Escherichia is a genus of Gram-negative, non-spore forming, facultatively anaerobic, rod-shaped bacteria from the family Enterobacteriaceae. In those species which are inhabitants of the gastrointestinal tracts of warm-blooded animals, Escherichia species provide a portion of the microbially-derived vitamin K for their host. A number of the species of Escherichia are pathogenic.
Photobacterium profundum	Photobacterium profundum is a deep sea Gammaproteobacterium, belonging to the family Vibrionaceae and genus Photobacterium. Like other members of this genus, P. profundum is a marine organism and has two circular chromosomes. P.

Photorhabdus luminescens	Photorhabdus luminescens is a Gammaproteobacteria, belonging to the family Enterobacteriaceae, and is a symbiotic pathogen of insects. It lives in the gut of an entomopathogenic nematode of the family Heterorhabditidae. When the nematode infects an insect, P. luminescens is released into the blood stream and rapidly kills the insect host (within 48 hours) by producing toxins such as TcA. It also secretes enzymes which break down the body of the infected insect and bioconvert it into nutrients which can be utilised by both nematode and bacteria.
Shewanella	Shewanella is the sole genus included in the Shewanellaceae family of marine bacteria. Shewanella is a marine bacterium capable of modifying (or converting to an altered state) metals, by saturating them with electrons causing the metal to expand and soften, allowing them to process it, which in turn releases an electrical charge. The product is substantially less toxic and can break up in the environment.
Yersinia pestis	Yersinia pestis is a Gram-negative rod-shaped bacterium. It is a facultative anaerobe that can infect humans and other animals. Human Y. pestis infection takes three main forms: pneumonic, septicemic, and the notorious bubonic plagues.
Legionella pneumophila	Legionella pneumophila is a thin, pleomorphic, flagellated Gram-negative bacterium of the genus Legionella. L. pneumophila is the primary human pathogenic bacterium in this group and is the causative agent of legionellosis or Legionnaires' disease. Characterization L. pneumophila is non-acid-fast, non-sporulating, and morphologically a non-capsulated rod-like bacteria.
Sphingomonas	Sphingomonas was defined in 1990 as a group of Gram-negative, rod-shaped, chemoheterotrophic, strictly aerobic bacteria. They possess ubiquinone 10 as their major respiratory quinone, contain glycosphingolipids (GSLs) instead of lipopolysaccharide in their cell envelopes, and typically produce yellow-pigmented colonies. By 2001, the genus included more than 20 species that were quite diverse in terms of their phylogenetic, ecological, and physiological properties.
Sphingomonas paucimobilis	Sphingomonas paucimobilis is an aerobic Gram-negative soil bacillus that has a single polar flagellum with slow motility.

S. paucimobilis is able to degrade lignin-related biphenyl chemical compounds.

Biofilm	A biofilm is an aggregate of microorganisms in which cells adhere to each other on a surface. These adherent cells are frequently embedded within a self-produced matrix of extracellular polymeric substance (EPS). Biofilm EPS, which is also referred to as slime (although not everything described as slime is a biofilm), is a polymeric conglomeration generally composed of extracellular DNA, proteins, and polysaccharides.
Thermoplasma	In taxonomy, Thermoplasma is a genus of the Thermoplasmataceae. Thermoplasma is a genus of archaea. It belongs to the Thermoplasmata, which thrive in acidic and high-temperature environments.
Epulopiscium fishelsoni	Epulopiscium fishelsoni is a Gram-positive bacterium that has a symbiotic relationship with the surgeonfish. It is most well-known for its large size, ranging from 200-700 µm in length, and about 80 µm in diameter. Until the discovery of Thiomargarita namibiensis in 1999, it was the largest bacterium known.
Genomic island	A Genomic island is part of a genome that has evidence of horizontal origins. The term is usually used in microbiology, especially with regard to bacteria. A GI can code for many functions, can be involved in symbiosis or pathogenesis, and may help an organism's adaptation.
Horizontal gene transfer	Horizontal gene transfer also lateral gene transfer (LGT) or transposition refers to the transfer of genetic material between organisms other than vertical gene transfer. Vertical transfer occurs when there is gene exchange from the parental generation to the offspring. LGT is then a mechanism of gene exchange that happens independently of reproduction.
Pyrococcus	Pyrococcus is a genus of Thermococcaceaen archaean. Pyrococcus has similar characteristics of other thermoautotrophican archaea such as Archaeoglobus, and Methanococcus in the respect that they are all thermophilic and anaerobic. Pyrococcus differs, however, because its optimal growth temperature is nearly 100 °C and dwells at a greater sea depth than the other archaea.
Membrane lipids	The three major classes of membrane lipids are phospholipids, glycolipids, and cholesterol. Phospholipids and glycolipids consist of two long, nonpolar (hydrophobic) hydrocarbon chains linked to a hydrophilic head group. The heads of phospholipids are phosphorylated and they consist of either:•

	Glycerol (and hence the name phosphoglycerides given to this group of lipids).•Sphingosine (with only one member - sphingomyelin).Glycolipids
	The heads of glycolipids contain a sphingosine with one or several sugar units attached to it.
Aeropyrum	In taxonomy, Aeropyrum is a genus of the Desulfurococcaceae.
	The name Aeropyrum derives from:Greek noun aer, aeros , air; Greek neuter gender noun pur, fire; New Latin neuter gender noun Aeropyrum, air fire, referring to the hyperthermophilic respirative character of the organism. Species
	The genus contains 2 species , namely•A. camini •A. pernix (Sako et al. 1996, (Type species of the genus).; Latin neuter gender adjective pernix, nimble, active, agile, indicating high motility in microscopic inspection)..
Retrotransposon	Retrotransposons (also called transposons via RNA intermediates) are genetic elements that can amplify themselves in a genome and are ubiquitous components of the DNA of many eukaryotic organisms. They are a subclass of transposon. They are particularly abundant in plants, where they are often a principal component of nuclear DNA. In maize, 49-78% of the genome is made up of retrotransposons.
Transposable element	A transposable element is a DNA sequence that can change its relative position (self-transpose) within the genome of a single cell. The mechanism of transposition can be either 'copy and paste' or 'cut and paste'. Transposition can create phenotypically significant mutations and alter the cell's genome size.
Transposon	Transposons are sequences of DNA that can move or transpose themselves to new positions within the genome of a single cell. The mechanism of transposition can be either 'copy and paste' or 'cut and paste'. Transposition can create phenotypically significant mutations and alter the cell's genome size.
Bacillus anthracis	Bacillus anthracis is a Gram-positive spore-forming, rod-shaped bacterium, with a width of 1-1.2µm and a length of 3-5µm. It can be grown in an ordinary nutrient medium under aerobic or anaerobic conditions.
	It is the only bacterium known to synthesize a protein capsule (D-glutamate), and the only pathogenic bacterium to carry its own adenylyl cyclase virulence factor (edema factor).
Bacillus thuringiensis	Bacillus thuringiensis is a Gram-positive, soil-dwelling bacterium, commonly used as a biological pesticide; alternatively, the Cry toxin may be extracted and used as a pesticide. B.

thuringiensis also occurs naturally in the gut of caterpillars of various types of moths and butterflies, as well as on the dark surface of plants.

During sporulation many Bt strains produce crystal proteins (proteinaceous inclusions), called δ-endotoxins, that have insecticidal action.

Helicobacter pylori	Helicobacter pylori previously named Campylobacter pyloridis, is a Gram-negative, microaerophilic bacterium found in the stomach. It was identified in 1982 by Barry Marshall and Robin Warren, who found that it was present in patients with chronic gastritis and gastric ulcers, conditions that were not previously believed to have a microbial cause. It is also linked to the development of duodenal ulcers and stomach cancer.
Taxonomy	Taxonomy is the science of identifying and naming species, and arranging them into a classification. The field of taxonomy, sometimes referred to as 'biological taxonomy', revolves around the description and use of taxonomic units, known as taxa . A resulting taxonomy is a particular classification ('the taxonomy of ...'), arranged in a hierarchical structure or classification scheme.
Genetic variation	Genetic variation, variation in alleles of genes, occurs both within and among populations. Genetic variation is important because it provides the 'raw material' for natural selection. Genetic variation is brought about by mutation, which is a change in the chemical structure of a gene.
Genome	In modern molecular biology and genetics, the genome is the entirety of an organism's hereditary information. It is encoded either in DNA or, for many types of virus, in RNA. The genome includes both the genes and the non-coding sequences of the DNA/RNA.

The term was adapted in 1920 by Hans Winkler, Professor of Botany at the University of Hamburg, Germany. The Oxford English Dictionary suggests the name to be a blend of the words gene and chromosome. |
| Bacillus cereus | Bacillus cereus is an endemic, soil-dwelling, Gram-positive, rod-shaped, beta hemolytic bacterium. Some strains are harmful to humans and cause foodborne illness, while other strains can be beneficial as probiotics for animals. B. cereus bacteria are facultative anaerobes, and like other members of the genus Bacillus can produce protective endospores. |
| Food web | A food web depicts feeding connections (what eats what) in an ecological community. Ecologists can broadly lump all life forms into one of two categories called trophic levels: 1) the autotrophs, and 2) the heterotrophs. To maintain their bodies, grow, develop, and to reproduce, autotrophs produce organic matter from inorganic substances, including both minerals and gases such as carbon dioxide. |

Haemophilus influenzae	Haemophilus influenzae, formerly called Pfeiffer's bacillus or Bacillus influenzae, Gram-negative, rod-shaped bacterium first described in 1892 by Richard Pfeiffer during an influenza pandemic. A member of the Pasteurellaceae family, it is generally aerobic, but can grow as a facultative anaerobe. H. influenzae was mistakenly considered to be the cause of influenza until 1933, when the viral etiology of the flu became apparent; the bacterium is colloquially known as bacterial influenza.
Neisseria meningitidis	Neisseria meningitidis is a heterotrophic gram-negative diplococcal bacterium best known for its role in meningitis and other forms of meningococcal disease such as meningococcemia. N. meningitidis is a major cause of morbidity and mortality during childhood in industrialized countries and is responsible for epidemics in Africa and in Asia. Approximately 2500 to 3500 cases of N meningitidis infection occur annually in the United States, with a case rate of about 1 in 100,000. Children younger than 5 years are at greatest risk, followed by teenagers of high school age.
Ecological niche	An ecological niche is a term describing the relational position of a species or population in its ecosystem to each other; e.g. a dolphin could potentially be in another ecological niche from one that travels in a different pod if the members of these pods utilize significantly different food resources and foraging methods. A shorthand definition of niche is how an organism makes a living. The ecological niche describes how an organism or population responds to the distribution of resources and competitors (e.g., by growing when resources are abundant, and when predators, parasites and pathogens are scarce) and how it in turn alters those same factors (e.g., limiting access to resources by other organisms, acting as a food source for predators and a consumer of prey).
Food chain	A food chain is somewhat a linear sequence of links in a food web starting from a trophic species that eats no other species in the web and ends at a trophic species that is eaten by no other species in the web. A food chain differs from a food web, because the complex polyphagous network of feeding relations are aggregated into trophic species and the chain only follows linear monophagous pathways. A common metric used to quantify food web trophic structure is food chain length.
Pan-genome	In molecular biology a pan-genome describes the full complement of genes in a species (typically applied to bacteria and archaea, which can have large variation in gene content among closely related strains). It is a superset of all the genes in all the strains of a species. The significance of the pangenome arises in an evolutionary context, especially with relevance to metagenomics, but is also used in a broader genomics context.
Species name	Species name is a technical term in biological taxonomy.

Chapter 4. Part 4: Microbial Diversity and Ecology

	Unlike the closely related term 'specific name', it has the same meaning in the different nomenclature codes, and refers to the two-part name that is used for a species in binomial nomenclature. For example, Silene latifolia is the species name for a plant commonly called white campion, and Bos primigenius is the species name for a type of cattle commonly called zebu.
Streptococcus pneumoniae	Streptococcus pneumoniae, is a Gram-positive, alpha-hemolytic, aerotolerant anaerobic member of the genus Streptococcus. A significant human pathogenic bacterium, S. pneumoniae was recognized as a major cause of pneumonia in the late 19th century, and is the subject of many humoral immunity studies. Despite the name, the organism causes many types of pneumococcal infections other than pneumonia.
Taxon	A taxon is a group of (one or more) organisms, which a taxonomist adjudges to be a unit. Usually a taxon is given a name and a rank, although neither is a requirement. Defining what belongs or does not belong to such a taxonomic group is done by a taxonomist with the science of taxonomy.
Metagenomics	Metagenomics is the study of metagenomes, genetic material recovered directly from environmental samples. The broad field may also be referred to as environmental genomics, ecogenomics or community genomics. While traditional microbiology and microbial genome sequencing and genomics rely upon cultivated clonal cultures, early environmental gene sequencing cloned specific genes (often the 16S rRNA gene) to produce a profile of diversity in a natural sample.
Methanobrevibacter smithii	Methanobrevibacter smithii is the dominant archaeon in the human gut. It is important for the efficient digestion of polysaccharides (complex sugars) because it consumes end products of bacterial fermentation. Methanobrevibacter smithii is a single-celled micro-organism from the Archaea domain.
Pelagibacter ubique	Pelagibacter, with the single species P. ubique, was isolated in 2002 and given a specific name, although it has not yet been validly published according to the bacteriological code. It is an abundant member of the SAR11 clade in the phylum Alphaproteobacteria. SAR11 members are highly dominant organisms found in both salt and fresh water worldwide -- possibly the most numerous bacteria in the world and were originally known only from their rRNA genes, which were first identified in environmental samples from the Sargasso Sea in 1990 by Stephen Giovannoni's laboratory in the Department of Microbiology at Oregon State University and later found in oceans worldwide. P. ubique and its relatives may be the most abundant organisms in the ocean, and quite possibly the most abundant bacteria in the entire world.

It can make up about 25% of all microbial plankton cells, and in the summer they may account for approximately half the cells present in temperate ocean surface water. The total abundance of P. ubique and relatives is estimated to be about 2×10^{28} microbes.

It is rod or crescent shaped and one of the smallest self-replicating cells known, with a length of 0.37-0.89 μm and a diameter of only 0.12-0.20 μm. 30% of the cell's volume is taken up by its genome. It is gram negative. It recycles dissolved organic carbon. It undergoes regular seasonal cycles in abundance - in summer reaching ~50% of the cells in the temperate ocean surface waters. Thus it plays a major role in the Earth's carbon cycle.

Its discovery was the subject of 'Oceans of Microbes', Episode 5 of 'Intimate Strangers: Unseen Life on Earth' by PBS Cultivation

Several strains of Pelagibacter ubique have been cultured thanks to improved isolation techniques.

Rocky Mountain spotted fever	Rocky Mountain spotted fever is the most lethal and most frequently reported rickettsial illness in the United States. It has been diagnosed throughout the Americas. Some synonyms for Rocky Mountain spotted fever in other countries include 'tick typhus,' 'Tobia fever' (Colombia), 'São Paulo fever' or 'febre maculosa' (Brazil), and 'fiebre manchada' (Mexico).
Trophic level	The trophic level of an organism is the position it occupies in a food chain. The word trophic derives from the Greek τροφ? (trophe) referring to food or feeding. A food chain represents a succession of organisms that eat another organism and are, in turn, eaten themselves.
Gut flora	Gut flora consists of microorganisms that live in the digestive tracts of animals and is the largest reservoir of human flora. In this context, gut is synonymous with intestinal, and flora with microbiota and microflora. The human body, consisting of about 10 trillion cells, carries about ten times as many microorganisms in the intestines.
DNA Research	DNA Research is an international, peer reviewed journal of genomics and DNA research.
Pathogenicity island	Pathogenicity islands (PAIs) are a distinct class of genomic islands acquired by microorganisms through horizontal gene transfer. They are incorporated in the genome of pathogenic organisms, but are usually absent from those nonpathogenic organisms of the same or closely related species. These mobile genetic elements may range from 10-200 kb and encode genes which contribute to the virulence of the respective pathogen.

Chapter 4. Part 4: Microbial Diversity and Ecology

Shiga toxin	Shiga toxins are a family of related toxins with two major groups, Stx1 and Stx2, whose genes are considered to be part of the genome of lambdoid prophages. The toxins are named for Kiyoshi Shiga, who first described the bacterial origin of dysentery caused by Shigella dysenteriae. The most common sources for Shiga toxin are the bacteria S. dysenteriae and the Shigatoxigenic group of Escherichia coli (STEC), which includes serotypes O157:H7, O104:H4, and other enterohemorrhagic E. coli (EHEC).
Virulence factor	Virulence factors are molecules expressed and secreted by pathogens (bacteria, viruses, fungi and protozoa) that enable them to achieve the following:•colonization of a niche in the host (this includes adhesion to cells)•Immunoevasion, evasion of the host's immune response•Immunosuppression, inhibition of the host's immune response•entry into and exit out of cells (if the pathogen is an intracellular one)•obtain nutrition from the host.

Virulence factors are very often responsible for causing disease in the host as they inhibit certain host functions.

Pathogens possess a wide array of virulence factors. Some are intrinsic to the bacteria (e.g. capsules and endotoxin) whereas others are obtained from plasmids (e.g. some toxins). |
| Parasitism | Parasitism is a type of non mutual relationship between organisms of different species where one organism, the parasite, benefits at the expense of the other, the host. Traditionally parasite referred to organisms with lifestages that needed more than one host (e.g. Taenia solium). These are now called macroparasites (typically protozoa and helminths). |
| Nematode | The nematodes ( ) or roundworms (phylum Nematoda) are the most diverse phylum of pseudocoelomates, and one of the most diverse of all animals. Nematode species are very difficult to distinguish; over 28,000 have been described, of which over 16,000 are parasitic. The total number of nematode species has been estimated to be about 1,000,000. Unlike cnidarians or flatworms, roundworms have tubular digestive systems with openings at both ends. |
| Endosymbiont | An endosymbiont is any organism that lives within the body or cells of another organism, i.e. forming an endosymbiosis . Examples are nitrogen-fixing bacteria (called rhizobia) which live in root nodules on legume roots, single-celled algae inside reef-building corals, and bacterial endosymbionts that provide essential nutrients to about 10-15% of insects.

Many instances of endosymbiosis are obligate; that is, either the endosymbiont or the host cannot survive without the other, such as the gutless marine worms of the genus Riftia, which get nutrition from their endosymbiotic bacteria. |
| Brugia malayi | Brugia malayi is a nematode (roundworm), one of the three causative agents of lymphatic filariasis in humans. |

Lymphatic filariasis, also known as elephantiasis, is a condition characterized by swelling of the lower limbs. The two other filarial causes of lymphatic filariasis are Wuchereria bancrofti and Brugia timori, which differ from B. malayi morphologically, symptomatically, and in geographical extent.

Filariasis	Filariasis is a parasitic disease and is considered an infectious tropical disease, that is caused by thread-like filarial nematodes (roundworms) in the superfamily Filarioidea, also known as 'filariae'. There are 9 known filarial nematodes which use humans as their definitive host. These are divided into 3 groups according to the niche within the body that they occupy: 'lymphatic filariasis', 'subcutaneous filariasis', and 'serous cavity filariasis'.
Nucleomorph	Nucleomorphs are small, reduced eukaryotic nuclei found in certain plastids. So far, only two groups of organisms are known to contain a nucleomorph: the cryptomonads of the supergroup Chromista and the chlorarachniophytes of the supergroup Rhizaria, both of which have examples of sequenced nucleomorph genomes. The nucleomorphs support the endosymbiotic theory, and are an evidence that the plastids of these organisms are so-called complex plastids.
Bifidobacterium	Bifidobacterium is a genus of Gram-positive, non-motile, often branched anaerobic bacteria. They are ubiquitous, endosymbiotic inhabitants of the gastrointestinal tract, vagina and mouth (B. dentium) of mammals and other animals. Bifidobacteria are one of the major genera of bacteria that make up the colon flora in mammals.
Gel electrophoresis	Gel electrophoresis is a method used in clinical chemistry to separate proteins by charge and or size (IEF agarose, essentially size independent) and in biochemistry and molecular biology to separate a mixed population of DNA and RNA fragments by length, to estimate the size of DNA and RNA fragments or to separate proteins by charge. Nucleic acid molecules are separated by applying an electric field to move the negatively charged molecules through an agarose matrix. Shorter molecules move faster and migrate farther than longer ones because shorter molecules migrate more easily through the pores of the gel.
Polyacrylamide gel	A Polyacrylamide Gel is a separation matrix used in electrophoresis of biomolecules, such as proteins or DNA fragments. Traditional DNA sequencing techniques such as Maxam-Gilbert or Sanger methods used polyacrylamide gels to separate DNA fragments differing by a single base-pair in length so the sequence could be read. Most modern DNA separation methods now use agarose gels, except for particularly small DNA fragments.
Aquifex	Aquifex is a genus of bacteria, one of the few in the phylum Aquificae. The two species generally classified in Aquifex are A. pyrophilus and A. aeolicus. Both are highly thermophilic, growing best in water temperature of 85 °C to 95 °C.

Chapter 4. Part 4: Microbial Diversity and Ecology

Peptidoglycan	Peptidoglycan, is a polymer consisting of sugars and amino acids that forms a mesh-like layer outside the plasma membrane of bacteria (but not Archaea), forming the cell wall. The sugar component consists of alternating residues of β-(1,4) linked N-acetylglucosamine and N-acetylmuramic acid. Attached to the N-acetylmuramic acid is a peptide chain of three to five amino acids.
Phylum	In biology, a phylum is a taxonomic rank below kingdom and above class. 'Phylum' is equivalent to the botanical term division. The kingdom Animalia contains approximately 35 phyla; the kingdom Plantae contains 12 divisions.
Pseudopeptidoglycan	Pseudopeptidoglycan is a major cell wall component of some archaea that differs from bacterial peptidoglycan in chemical structure, but resembles eubacterial peptidoglycan in morphology, function, and physical structure. The basic components are N-acetylglucosamine and N-acetyltalosaminuronic acid (peptidoglycan has N-acetylmuramic acid instead), which are linked by β-1,3-glycosidic bonds. Lysozyme, a host defense mechanism, is ineffective against organisms with pseudopeptidoglycan cell walls.
Cell wall	The cell wall is the tough, usually flexible but sometimes fairly rigid layer that surrounds some types of cells. It is located outside the cell membrane and provides these cells with structural support and protection, in addition to acting as a filtering mechanism. A major function of the cell wall is to act as a pressure vessel, preventing over-expansion when water enters the cell.
Chitin	Chitin is a long-chain polymer of a N-acetylglucosamine, a derivative of glucose, and is found in many places throughout the natural world. It is the main component of the cell walls of fungi, the exoskeletons of arthropods such as crustaceans (e.g., crabs, lobsters and shrimps) and insects, the radulas of mollusks, and the beaks of cephalopods, including squid and octopuses. In terms of structure, chitin may be compared to the polysaccharide cellulose and, in terms of function, to the protein keratin.
Cyanobacteria	Cyanobacteria is a phylum of bacteria that obtain their energy through photosynthesis. The name 'cyanobacteria' comes from the color of the bacteria (Greek: κυαν?ς (kyanós) = blue). The ability of cyanobacteria to perform oxygenic photosynthesis is thought to have converted the early reducing atmosphere into an oxidizing one, which dramatically changed the composition of life forms on Earth by stimulating biodiversity and leading to the near-extinction of oxygen-intolerant organisms.
Deinococcus radiodurans	Deinococcus radiodurans is an extremophilic bacterium, one of the most radioresistant organisms known. It can survive cold, dehydration, vacuum, and acid, and is therefore known as a polyextremophile and has been listed as the world's toughest bacterium in The Guinness Book Of World Records.

Deinococcus-Thermus	The Deinococcus-Thermus are a small group of bacteria composed of cocci highly resistant to environmental hazards. There are two main groups. •The Deinococcales include two families, each with a single genus, Deinococcus and Truepera, the former with several species that are resistant to radiation; they have become famous for their ability to eat nuclear waste and other toxic materials, survive in the vacuum of space and survive extremes of heat and cold.•The Thermales include several genera resistant to heat (Marinithermus, Meiothermus, Oceanithermus, Thermus, Vulcanithermus).
Thermus aquaticus	Thermus aquaticus is a species of bacterium that can tolerate high temperatures, one of several thermophilic bacteria that belong to the Deinococcus-Thermus group. It is the source of the heat-resistant enzyme Taq DNA polymerase, one of the most important enzymes in molecular biology because of its use in the polymerase chain reaction (PCR) DNA amplification technique. When studies of biological organisms in hot springs began in the 1960s, scientists thought that the life of thermophilic bacteria could not be sustained in temperatures above about 55° Celsius (131° Fahrenheit).
Actinobacteria	Actinobacteria are a group of Gram-positive bacteria with high guanine and cytosine content. They can be terrestrial or aquatic. Actinobacteria is one of the dominant phyla of the bacteria.
Bacteroidetes	The phylum Bacteroidetes is composed of three large classes of gram-negative, nonsporeforming, anaerobic, and rod-shaped bacteria that are widely distributed in the environment, including in soil, in sediments, sea water and in the guts and on the skin of animals. By far, the ones in the Bacteroidia class are the most well-studied, including the genus Bacteroides (an abundant organism in the feces of warm-blooded animals including humans), and Porphyromonas, a group of organisms inhabiting the human oral cavity. The class Bacteroidia was formally called Bacteroidetes as it was until recently the only class in the phylum, the name was changed in the fourth volume of Bergey's Manual of Systematic Bacteriology.
Endospore	An endospore is a dormant, tough, and non-reproductive structure produced by certain bacteria from the Firmicute phylum. The name 'endospore' is suggestive of a spore or seed-like form (endo means within), but it is not a true spore (i.e. not an offspring). It is a stripped-down, dormant form to which the bacterium can reduce itself.
Firmicutes	The Firmicutes are a phylum of bacteria, most of which have Gram-positive cell wall structure. A few, however, such as Megasphaera, Pectinatus, Selenomonas and Zymophilus, have a porous pseudo-outer-membrane that causes them to stain Gram-negative.

Chapter 4. Part 4: Microbial Diversity and Ecology

Gram-positive bacteria	Gram-positive bacteria are those that are stained dark blue or violet by Gram staining. This is in contrast to Gram-negative bacteria, which cannot retain the crystal violet stain, instead taking up the counterstain (safranin or fuchsine) and appearing red or pink. Gram-positive organisms are able to retain the crystal violet stain because of the high amount of peptidoglycan in the cell wall.
Nitrospira	Nitrospira is a genus of bacteria in the phylum Nitrospirae. The first member of this genus was described 1986 by Watson et al. isolated from the Gulf of Maine. The bacterium was named Nitrospira marina.
Nitrospirae	Nitrospirae is a phylum of bacteria. It contains only one class, Nitrospira, which itself contains one order (Nitrospirales) and one family (Nitrospiraceae). It includes multiple genera, such as Nitrospira, the largest.
Planctomycetes	Planctomycetes are a phylum of aquatic bacteria and are found in samples of brackish, and marine and fresh water. They reproduce by budding. In structure, the organisms of this group are ovoid and have a holdfast, called the stalk, at the nonreproductive end that helps them to attach to each other during budding.
Verrucomicrobia	Verrucomicrobia is a recently described phylum of bacteria. This phylum contains only a few described species . The species identified have been isolated from fresh water and soil environments and human feces.
Aquifex pyrophilus	Aquifex pyrophilus is a rod-shaped bacterium with a length of 2 to 6 micrometers and a diameter of around half a micrometer. It is one of a handful of species in the Aquificae phylum, an unusual group of thermophilic bacteria that are thought to be some of the oldest species in the bacteria domain. A. pyrophilus grows best in water between 85 to 95 °C, and can be found near underwater volcanoes or hot springs.
Aquifex aeolicus	Aquifex aeolicus is a rod-shaped prokaryote with a length of 2 to 6 micrometers and a diameter of around half a micrometer. It is one of a handful of species in the Aquificae phylum, an unusual group of thermophilic bacteria that are thought to be some of the oldest species of bacteria. A. aeolicus grows best in water between 85 to 95 °C, and can be found near underwater volcanoes or hot springs.
Chlorosome	A Chlorosome is a photosynthetic antenna complex found in green sulfur bacteria (GSB) and some green filamentous anoxygenic phototrophs (FAP) (Chloroflexaceae, Oscillochloridaceae). They differ from other antenna complexes by their large size and lack of protein matrix supporting the photosynthetic pigments.

Thermotoga maritima	Thermotoga maritima is a hyperthermophilic organism that is a member of the order Thermotogales. First discovered in the sediment of a marine geothermal area near Vulcano, Italy, Thermotoga maritima resides in hot springs as well as hydrothermal vents. The ideal environment for the organism is a water temperature of 80 °C (176 °F), though it is capable of growing in waters of 55-90 °C (131-194 °F).
Anabaena	Anabaena is a genus of filamentous cyanobacteria that exists as plankton. It is known for its nitrogen fixing abilities, and they form symbiotic relationships with certain plants, such as the mosquito fern. They are one of four genera of cyanobacteria that produce neurotoxins, which are harmful to local wildlife, as well as farm animals and pets.
Phycoerythrin	Phycoerythrin, alpha/beta chain Phycoerythrin is a red protein from the light-harvesting phycobiliprotein family, present in cyanobacteria, red algae and cryptomonads. Like all phycobiliproteins, phycoerythrin is composed of a protein part, organised in a hexameric structure of alpha and beta chains, covalently binding chromophores called phycobilins. In the phycoerythrin family, the phycobilins are: phycoerythrobilin, the typical phycoerythrin acceptor chromophore, and sometimes phycourobilin (marine organisms).
Prochlorococcus	Prochlorococcus is a genus of very small (0.6 μm) marine cyanobacteria with an unusual pigmentation (chlorophyll b). These bacteria belong to the photosynthetic picoplankton and are probably the most abundant photosynthetic organism on Earth. Microbes of the genus Prochlorococcus are among the major primary producers in the ocean, responsible for at least 20% of atmospheric oxygen.
Thylakoid	A thylakoid is a membrane-bound compartment inside chloroplasts and cyanobacteria. They are the site of the light-dependent reactions of photosynthesis. Thylakoids consist of a thylakoid membrane surrounding a thylakoid lumen.
Akinete	An akinete is a thick-walled dormant cell derived from the enlargement of a vegetative cell. It serves as a survival structure. It is a resting cell of cyanobacteria and unicellular and filamentous green algae.
Heterocyst	Heterocysts are specialized nitrogen-fixing cells formed by some filamentous cyanobacteria, such as Nostoc punctiforme, Cylindrospermum stagnale and Anabaena sphaerica, during nitrogen starvation. They fix nitrogen from dinitrogen (N_2) in the air using the enzyme nitrogenase, in order to provide the cells in the filament with nitrogen for biosynthesis.

Chapter 4. Part 4: Microbial Diversity and Ecology

Microbial mat	A microbial mat is a multi-layered sheet of micro-organisms, mainly bacteria and archaea. Microbial mats grow at interfaces between different types of material, mostly on submerged or moist surfaces but a few survive in deserts. They colonize environments ranging in temperature from -40°C to +120°C. A few are found as endosymbionts of animals.
Algal bloom	An algal bloom is a rapid increase or accumulation in the population of algae (typically microscopic) in an aquatic system. Algal blooms may occur in freshwater as well as marine environments. Typically, only one or a small number of phytoplankton species are involved, and some blooms may be recognized by discoloration of the water resulting from the high density of pigmented cells.
Primary producers	Primary producers are those organisms in an ecosystem that produce biomass from inorganic compounds (autotrophs) However, there are examples of archea (unicellular organisms) that produce biomass from the oxidation of inorganic chemical compounds (chemoautotrophs) in hydrothermal vents in the deep ocean.
Trichodesmium	Trichodesmium, is a genus of filamentous cyanobacteria. They are found in nutrient poor tropical and subtropical ocean waters (particularly around Australia, where they were first described by Captain Cook). Trichodesmium fixes atmospheric nitrogen into ammonium, usable also for other organisms.
Clostridium	Clostridium is a genus of Gram-positive bacteria, belonging to the Firmicutes. They are obligate anaerobes capable of producing endospores. Individual cells are rod-shaped, which gives them their name, from the Greek kloster or spindle.
Bacillus subtilis	Bacillus subtilis, known also as the hay bacillus or grass bacillus, is a Gram-positive, catalase-positive bacterium commonly found in soil. A member of the genus Bacillus, B. subtilis is rod-shaped, and has the ability to form a tough, protective endospore, allowing the organism to tolerate extreme environmental conditions. Unlike several other well-known species, B. subtilis has historically been classified as an obligate aerobe, though recent research has demonstrated that this is not strictly correct.
DNA polymerase	A DNA polymerase is an enzyme (the suffix -ase is used to identify enzymes) that helps catalyze the polymerization of deoxyribonucleotides into a DNA strand. DNA polymerases are best known for their feedback role in DNA replication, in which the polymerase 'reads' an intact DNA strand as a template and uses it to synthesize the new strand. This process copies a piece of DNA. The newly polymerized molecule is complementary to the template strand and identical to the template's original partner strand.
Reporter gene	In molecular biology, a reporter gene is a gene that researchers attach to a regulatory sequence of another gene of interest in bacteria, cell culture, animals or plants.

Certain genes are chosen as reporters because the characteristics they confer on oanisms expressing them are easily identified and measured, or because they are selectable markers. Reporter genes are often used as an indication of whether a certain gene has been taken up by or expressed in the cell or oanism population.

Starvation response	Starvation response in animals is a set of adaptive biochemical and physiological changes that reduce metabolism in response to a lack of food.
Acetone	Acetone is the organic compound with the formula $(CH_3)_2CO$, a colorless, mobile, flammable liquid, the simplest example of the ketones. Acetone is miscible with water and serves as an important solvent in its own right, typically as the solvent of choice for cleaning purposes in the laboratory. About 6.7 million tonnes were produced worldwide in 2010, mainly for use as a solvent and production of methyl methacrylate and bisphenol A. It is a common building block in organic chemistry.
Clostridium acetobutylicum	Clostridium acetobutylicum, included in the genus Clostridium, is a commercially valuable bacterium. It is sometimes called the 'Weizmann Organism', after Chaim Weizmann, who in 1916 helped discover how C. acetobutylicum culture could be used to produce acetone, butanol, and ethanol from starch using the ABE process (Acetone Butanol Ethanol process) for industrial purposes such as gunpowder and Cordite (using acetone) production. The A.B.E. process was an industry standard until the late 1940s, when low oil costs drove more-efficient processes based on hydrocarbon cracking and petroleum distillation techniques.
Clostridium botulinum	Clostridium botulinum is a Gram-positive, rod-shaped bacterium that produces several toxins. The best known are its neurotoxins, subdivided in types A-G, that cause the flaccid muscular paralysis seen in botulism. It is also the main paralytic agent in botox.
Delta endotoxin	Delta endotoxin, N-terminal domain Delta endotoxins (δ-endotoxins, also called Cry and Cyt toxins) are pore-forming toxins produced by Bacillus thuringiensis species of bacteria. They are useful for their insecticidal action. During spore formation the bacteria produce crystals of this protein.
Insecticide	An insecticide is a pesticide used against insects. They include ovicides and larvicides used against the eggs and larvae of insects respectively. Insecticides are used in agriculture, medicine, industry and the household.
Clostridium difficile	Clostridium difficile (from the Greek kloster , spindle, and Latin difficile, difficult), also known as 'CDF/cdf', or 'C.

diff', is a species of Gram-positive bacteria of the genus Clostridium that causes severe diarrhea and other intestinal disease when competing bacteria in the gut flora have been wiped out by antibiotics.

Clostridia are anaerobic, spore-forming rods (bacilli). C. difficile is the most serious cause of antibiotic-associated diarrhea (AAD) and can lead to pseudomembranous colitis, a severe inflammation of the colon, often resulting from eradication of the normal gut flora by antibiotics.

Botulinum toxin	Botulinum toxin is a protein and neurotoxin produced by the bacterium Clostridium botulinum. Botulinum toxin can cause botulism, a serious and life-threatening illness in humans and animals. When introduced intravenously in monkeys, type A (Botox Cosmetic) of the toxin exhibits an LD_{50} of 40-56 ng, type C1 around 32 ng, type D 3200 ng, and type E 88 ng; these are some of the most potent neurotoxins known.
Photosystem I	Photosystem I is the second photosystem in the photosynthetic light reactions of algae, plants, and some bacteria. Photosystem I is so named because it was discovered before photosystem II. Aspects of PS I were discovered in the 1950s, but the significances of these discoveries was not yet known. Louis Duysens first proposed the concepts of photosystems I and II in 1960, and, in the same year, a proposal by Fay Bendall and Robert Hill assembled earlier discoveries into a cohesive theory of serial photosynthetic reactions.
Lactic acid	Lactic acid, is a chemical compound that plays a role in various biochemical processes and was first isolated in 1780 by the Swedish chemist Carl Wilhelm Scheele. Lactic acid is a carboxylic acid with the chemical formula $C_3H_6O_3$. It has a hydroxyl group adjacent to the carboxyl group, making it an alpha hydroxy acid (AHA).
Lactobacillales	The Lactobacillales are an order of Gram-positive bacteria that comprise the lactic acid bacteria. They are widespread in nature, and are found in soil, water, plants and animals. They are widely used in the production of fermented foods, including dairy products such as yogurt, cheese, butter, buttermilk, kefir and koumiss.
Lactobacillus	Lactobacillus, is a genus of Gram-positive facultative anaerobic or microaerophilic rod-shaped bacteria. They are a major part of the lactic acid bacteria group, named as such because most of its members convert lactose and other sugars to lactic acid. In humans they are present in the vagina and the gastrointestinal tract, where they are symbiotic and make up a small portion of the gut flora.
Lactococcus	Lactococcus is a genus of lactic acid bacteria that were formerly included in the genus Streptococcus Group N1. They are known as homofermentors meaning that they produce a single product, lactic acid in this case, as the major or only product of glucose fermentation.

	Their homofermentative character can be altered by adjusting cultural conditions like pH, glucose concentration, and nutrient limitation. They are gram-positive, catalase negative, non-motile cocci that are found singly, in pairs, or in chains.
Leuconostoc	Leuconostoc is a genus of Gram-positive bacteria, placed within the family of Leuconostocaceae. They are generally ovoid cocci often forming chains. Leuconostoc sp.
Listeria	Listeria is a bacterial genus containing six species. Named after the English pioneer of sterile surgery Joseph Lister, the genus was given its current name in 1940. Listeria species are Gram-positive bacilli and are typified by L. monocytogenes, the causative agent of listeriosis. Listeria ivanovii is a pathogen of ruminants, and can infect mice in the laboratory, although it is only rarely the cause of human disease.
Listeria monocytogenes	Listeria monocytogenes, a facultative anaerobe, intracellular bacterium, is the causative agent of listeriosis. It is one of the most virulent foodborne pathogens, with 20 to 30 percent of clinical infections resulting in death. Responsible for approximately 2,500 illnesses and 500 deaths in the United States (U.S).
Nosocomial infection	A nosocomial infection also known as a hospital-acquired infection or HAI, is an infection whose development is favoured by a hospital environment, such as one acquired by a patient during a hospital visit or one developing among hospital staff. Such infections include fungal and bacterial infections and are aggravated by the reduced resistance of individual patients. In the United States, the Centers for Disease Control and Prevention estimate that roughly 1.7 million hospital-associated infections, from all types of microorganisms, including bacteria, combined, cause or contribute to 99,000 deaths each year.
Probiotic	Probiotic are live microorganisms thought to be beneficial to the host organism. According to the currently adopted definition by FAOWHO, probiotics are: 'Live microorganisms which when administered in adequate amounts confer a health benefit on the host'. Lactic acid bacteria (LAB) and bifidobacteria are the most common types of microbes used as probiotics; but certain yeasts and bacilli may also be used.
Sauerkraut	Sauerkraut, French Choucroute, Polish Kiszona kapusta) directly translated: 'sour cabbage', is finely shredded cabbage that has been fermented by various lactic acid bacteria, including Leuconostoc, Lactobacillus, and Pediococcus. It has a long shelf-life and a distinctive sour flavor, both of which result from the lactic acid that forms when the bacteria ferment the sugars in the cabbage. It is not to be confused with coleslaw, which receives its acidic taste from vinegar.

Chapter 4. Part 4: Microbial Diversity and Ecology

Staphylococcus aureus	Staphylococcus aureus is a bacterial species named from Greek σταφυλ?κοκκος meaning the 'golden grape-cluster berry'. Also known as 'golden staph' and Oro staphira, it is a facultative anaerobic Gram-positive coccal bacterium. It is frequently found as part of the normal skin flora on the skin and nasal passages.
Staphylococcus epidermidis	Staphylococcus epidermidis is one of thirty-three known species belonging to the genus Staphylococcus. It is part of our skin flora, and consequently part of human flora. It can also be found in the mucous membranes and in animals.
Substrate-level phosphorylation	Substrate-level phosphorylation is a type of metabolism that results in the formation and creation of adenosine triphosphate (ATP) or guanosine triphosphate (GTP) by the direct transfer and donation of a phosphoryl (PO_3) group to adenosine diphosphate (ADP) or guanosine diphosphate (GDP) from a phosphorylated reactive intermediate. Note that the phosphate group does not have to directly come from the substrate. By convention, the phosphoryl group that is transferred is referred to as a phosphate group.
Cheese	Cheese is a generic term for a diverse group of milk-based food products. Cheese is produced throughout the world in wide-ranging flavors, textures, and forms. Cheese consists of proteins and fat from milk, usually the milk of cows, buffalo, goats, or sheep.
Toxic shock syndrome	Toxic shock syndrome is a potentially fatal illness caused by a bacterial toxin. Different bacterial toxins may cause toxic shock syndrome, depending on the situation. The causative bacteria include Staphylococcus aureus and Streptococcus pyogenes.
Yogurt	Yogurt, UK: /'j?g?t/) is a dairy product produced by bacterial fermentation of milk. The bacteria used to make yogurt are known as 'yogurt cultures'. Fermentation of lactose by these bacteria produces lactic acid, which acts on milk protein to give yogurt its texture and its characteristic tang.
Anaerobic respiration	Anaerobic respiration is a form of respiration using electron acceptors other than oxygen. Although oxygen is not used as the final electron acceptor, the process still uses a respiratory electron transport chain; it is respiration without oxygen. In order for the electron transport chain to function, an exogenous final electron acceptor must be present to allow electrons to pass through the system.
Enterococcus faecalis	Enterococcus faecalis - formerly classified as part of the Group D Streptococcus system - is a Gram-positive commensal bacterium inhabiting the gastrointestinal tracts of humans and other mammals. It is among the main constituents of some probiotic food supplements. A commensal organism like other species in the genus Enterococcus, E.

faecalis can cause life-threatening infections in humans, especially in the nosocomial (hospital) environment, where the naturally high levels of antibiotic resistance found in E. faecalis contribute to its pathogenicity.

Streptococcus	Streptococcus is a genus of spherical Gram-positive bacteria belonging to the phylum Firmicutes and the lactic acid bacteria group. Cellular division occurs along a single axis in these bacteria, and thus they grow in chains or pairs, hence the name -- from Greek στρεπτος streptos, meaning easily bent or twisted, like a chain (twisted chain). Contrast this with staphylococci, which divide along multiple axes and generate grape-like clusters of cells.
Mycoplasma	Mycoplasma refers to a genus of bacteria that lack a cell wall. Without a cell wall, they are unaffected by many common antibiotics such as penicillin or other beta-lactam antibiotics that target cell wall synthesis. They can be parasitic or saprotrophic.
Corynebacterium	Corynebacterium is a genus of Gram-positive rod-shaped bacteria. They are widely distributed in nature and are mostly innocuous. Some are useful in industrial settings such as C. glutamicum.
Geosmin	Geosmin, which literally translates to 'earth smell', is an organic compound with a distinct earthy flavor and aroma, and is responsible for the earthy taste of beets and a contributor to the strong scent (petrichor) that occurs in the air when rain falls after a dry spell of weather or when soil is disturbed.

Geosmin is produced by several classes of microbes, including cyanobacteria (blue-green algae) and actinobacteria (especially Streptomyces), and released when these microbes die. Communities whose water supplies depend on surface water can periodically experience episodes of unpleasant-tasting water when a sharp drop in the population of these bacteria releases geosmin into the local water supply. |
| Mycobacterium | Mycobacterium is a genus of Actinobacteria, given its own family, the Mycobacteriaceae. The genus includes pathogens known to cause serious diseases in mammals, including tuberculosis (Mycobacterium tuberculosis) and leprosy (Mycobacterium leprae). The Greek prefix 'myco--' means fungus, alluding to the way mycobacteria have been observed to grow in a mould-like fashion on the surface of liquids when cultured. |
| Ecology | Ecology is the scientific study of the relations that living organisms have with respect to each other and their natural environment. Variables of interest to ecologists include the composition, distribution, amount (biomass), number, and changing states of organisms within and among ecosystems. |

Actinomyces	Actinomyces from Greek 'actino' that means mukas and fungus, is a genus of the actinobacteria class of bacteria. They are all Gram-positive and are characterized by contiguous spread, suppurative and granulomatous inflammation, and formation of multiple abscesses and sinus tracts that may discharge sulfur granules.' They can be either anaerobic or facultatively anaerobic . Actinomyces species do not form endospores, and, while individual bacteria are rod-shaped, morphologically Actinomyces colonies form fungus-like branched networks of hyphae.
Frankia	Frankia is a genus of nitrogen fixing filamentous bacteria that live in symbiosis with actinorhizal plants, similar to Rhizobia. Bacteria of this genus form root nodules.
Mycobacterium tuberculosis	Mycobacterium tuberculosis is a pathogenic bacterial species in the genus Mycobacterium and the causative agent of most cases of tuberculosis. First discovered in 1882 by Robert Koch, M. tuberculosis has an unusual, waxy coating on the cell surface (primarily mycolic acid), which makes the cells impervious to Gram staining so acid-fast detection techniques are used instead. The physiology of M. tuberculosis is highly aerobic and requires high levels of oxygen.
Mycolic acid	Mycolic acids are long fatty acids found in the cell walls of the mycolata taxon, a group of bacteria that includes Mycobacterium tuberculosis, the causative agent of the disease tuberculosis. They form the major component of the cell wall of mycolata species. Despite their name, mycolic acids have no biological link to fungi; the name arises from the filamentous appearance their presence gives mycolata under high magnification.
Actinomycosis	Actinomycosis is an infectious bacterial disease caused by Actinomyces species such as Actinomyces israelii or A. gerencseriae. It can also be caused by Propionibacterium propionicus, and the condition is likely to be polymicrobial aerobic anaerobic infection. Actinomycosis occurs rarely in humans but rather frequently in cattle as a disease called lumpy jaw.
Cell envelope	The cell envelope is the cell membrane and cell wall plus an outer membrane, if one is present. Most bacterial cell envelopes fall into two major categories: Gram positive and Gram negative. These are differentiated by their Gram staining characteristics.
Chromosome	In genetic algorithms, a chromosome (also sometimes called a genome) is a set of parameters which define a proposed solution to the problem that the genetic algorithm is trying to solve.
Endophyte	An endophyte is an endosymbiont, often a bacterium or fungus, that lives within a plant for at least part of its life without causing apparent disease. Endophytes are ubiquitous and have been found in all the species of plants studied to date; however, most of these endophyte/plant relationships are not well understood.

Arthrobacter	Arthrobacter is a genus of bacteria that is commonly found in soil. All species in this genus are Gram-positive obligate aerobes that are rods during exponential growth and cocci in their stationary phase. Colonies of Arthrobacter have a greenish metallic center on mineral salts pyridone broth incubated at 20°C. This genus is distinctive because of its unusual habit of 'snapping division' in which the outer bacterial cell wall ruptures at a joint (hence its name).
Chromium	Chromium is a chemical element which has the symbol Cr and atomic number 24. It is the first element in Group 6. It is a steely-gray, lustrous, hard metal that takes a high polish and has a high melting point. It is also odorless, tasteless, and malleable. The name of the element is derived from the Greek word 'chroma' , meaning colour, because many of its compounds are intensely coloured.
Corynebacterium diphtheriae	Corynebacterium diphtheriae is a pathogenic bacterium that causes diphtheria. It is also known as the Klebs-Löffler bacillus, because it was discovered in 1884 by German bacteriologists Edwin Klebs (1834 - 1912) and Friedrich Löffler (1852 - 1915). Classification Four subspecies are recognized: C. diphtheriae mitis, C. diphtheriae intermedius, C. diphtheriae gravis, and C. diphtheriae belfanti.
Diphtheria	Diphtheria (Greek διφθ?ρα (diphthera) 'pair of leather scrolls') is an upper respiratory tract illness caused by Corynebacterium diphtheriae, a facultative anaerobic, Gram-positive bacterium. It is characterized by sore throat, low fever, and an adherent membrane (a pseudomembrane) on the tonsils, pharynx, and/or nasal cavity. A milder form of diphtheria can be restricted to the skin.
Metabolism	Metabolism is the set of chemical reactions that happen in the cells of living organisms to sustain life. These processes allow organisms to grow and reproduce, maintain their structures, and respond to their environments. The word metabolism can also refer to all chemical reactions that occur in living organisms, including digestion and the transport of substances into and between different cells, in which case the set of reactions within the cells is called intermediary metabolism or intermediate metabolism.
Alphaproteobacteria	Alphaproteobacteria is a class of Proteobacteria. Like all Proteobacteria, they are Gram-negative.

Chapter 4. Part 4: Microbial Diversity and Ecology

Proteorhodopsin	Proteorhodopsin is a photoactive retinylidene protein in marine planktonic prokaryotes and eukaryotes. Just like the homologous pigment bacteriorhodopsin found in some archaea, it consists of a transmembrane protein bound to a retinal molecule and functions as a light-driven proton pump. Some members of the family (of more than 800 types) are believed to have sensory functions.
Purple sulfur bacteria	The purple sulfur bacteria are a group of Proteobacteria capable of photosynthesis, collectively referred to as purple bacteria. They are anaerobic or microaerophilic, and are often found in hot springs or stagnant water. Unlike plants, algae, and cyanobacteria, they do not use water as their reducing agent, and so do not produce oxygen.
Retinal	Retinal, is one of the many forms of vitamin A (the number of which varies from species to species). Retinal is a polyene chromophore, and bound to proteins called opsins, is the chemical basis of animal vision. Bound to proteins called type 1 rhodopsins, retinal allows certain microorganisms to convert light into metabolic energy.
Rhodobacter sphaeroides	Rhodobacter sphaeroides is a kind of purple bacteria; a group of bacteria that can obtain energy through photosynthesis. Its best growth conditions are anaerobic phototrophy (photoheterotrophic and photoautotrophic) and aerobic chemoheterotrophy in the absence of light. R. sphaeroides is also able to fix nitrogen.
Photoheterotroph	Photoheterotrophs (Gk: photo = light, hetero = (an)other, troph = nourishment) are heterotrophic organisms that use light for energy, but cannot use carbon dioxide as their sole carbon source. Consequently, they use organic compounds from the environment to satisfy their carbon requirements. They use compounds such as carbohydrates, fatty acids and alcohols as their organic 'food'.
Arsenic	Arsenic is a chemical element with the symbol As, atomic number 33 and relative atomic mass 74.92. Arsenic occurs in many minerals, usually in conjunction with sulfur and metals, and also as a pure elemental crystal. It was first documented by Albertus Magnus in 1250. Arsenic is a metalloid. It can exist in various allotropes, although only the grey form has important use in industry.
Caulobacter crescentus	Caulobacter crescentus is a Gram-negative, oligotrophic bacterium widely distributed in fresh water lakes and streams. It plays an important role in the carbon cycle. Caulobacter is an important model for studying the regulation of the cell cycle and cellular differentiation.

Selenium	Selenium is a chemical element with atomic number 34, chemical symbol Se, and an atomic mass of 78.96. It is a nonmetal, whose properties are intermediate between those of adjacent chalcogen elements sulfur and tellurium. It rarely occurs in its elemental state in nature, but instead is obtained as a side-product in the refining of other elements. Selenium is found in sulfide ores such as pyrite, where it partially replaces the sulfur.
Cell division	Cell division is the process by which a parent cell divides into two or more daughter cells. Cell division is usually a small segment of a larger cell cycle. This type of cell division in eukaryotes is known as mitosis, and leaves the daughter cell capable of dividing again.
Bradyrhizobium	Bradyrhizobium is a genus of Gram-negative soil bacteria, many of which fix nitrogen. Nitrogen fixation is an important part of the nitrogen cycle. Plants cannot use atmospheric nitrogen (N_2) they must use nitrogen compounds such as nitrates.
Carbon monoxide	Carbon monoxide also called carbonous oxide, is a colorless, odorless, and tasteless gas that is slightly lighter than air. It can be toxic to humans and animals when encountered in higher concentrations, although it is also produced in normal animal metabolism in low quantities, and is thought to have some normal biological functions. In the atmosphere however, it is short lived and spatially variable, since it combines with oxygen to form carbon dioxide and ozone.
Roseobacter	In taxonomy, Roseobacter is a genus of the Rhodobacteraceae. The Roseobacter clade falls within the {alpha}-3 subclass of the class Proteobacteria. The first strain descriptions appeared in 1991 which described members Roseobacter litoralis and Roseobacter denitrificans, both pink-pigmented bacteriochlorophyll a-producing strains isolated from marine algae.
Rhizobium	Rhizobium is a genus of Gram-negative soil bacteria that fix nitrogen. Rhizobium forms an endosymbiotic nitrogen fixing association with roots of legumes and Parasponia. The bacteria colonize plant cells within root nodules; here the bacteria convert atmospheric nitrogen to ammonia and then provide organic nitrogenous compounds such as glutamine or ureides to the plant.
Sinorhizobium meliloti	Sinorhizobium meliloti is a Gram-negative nitrogen-fixing bacterium (rhizobium). It forms a symbiotic relationship with legumes from the genera Medicago, Melilotus and Trigonella, including the model legume Medicago truncatula. This symbiosis results in a new plant organ termed a root nodule.
Aphotic zone	The aphotic zone is the portion of a lake or ocean where there is little or no sunlight.

	It is formally defined as the depths beyond which less than 1% of sunlight penetrates. Consequently, bioluminescence is essentially the only light found in this zone.
Nitrogen fixation	Nitrogen fixation is a process by which nitrogen (N_2) in the atmosphere is converted into ammonium (NH_{4+}). Atmospheric nitrogen or elemental nitrogen (N_2) is relatively inert: it does not easily react with other chemicals to form new compounds. Fixation processes free up the nitrogen atoms from their diatomic form (N_2) to be used in other ways.
Photic zone	The photic zone is exposed to sufficient sunlight for photosynthesis to occur. The depth of the photic zone can be affected greatly by seasonal turbidity. It extends from the atmosphere-water interface downwards to a depth where light intensity falls to one percent of that at the surface, called the euphotic depth.
Filopodia	Filopodia are slender cytoplasmic projections that extend beyond the leading edge of lamellipodia in migrating cells. They contain actin filaments cross-linked into bundles by actin-binding proteins, e.g. fascin and fimbrin. Filopodia form focal adhesions with the substratum, linking it to the cell surface.
Leghemoglobin	Leghemoglobin is a nitrogen or oxygen carrier, because naturally occurring oxygen and nitrogen interact similarly with this protein; and a hemoprotein found in the nitrogen-fixing root nodules of leguminous plants. It is produced by legumes in response to the roots being infected by nitrogen-fixing bacteria, termed rhizobia, as part of the symbiotic interaction between plant and bacterium: roots uninfected with Rhizobium do not synthesise leghemoglobin. Leghemoglobin has close chemical and structural similarities to hemoglobin, and, like hemoglobin, is red in colour.
Marine habitats	Marine habitats can be divided into coastal and open ocean habitats. Coastal habitats are found in the area that extends from as far as the tide comes in on the shoreline out to the edge of the continental shelf. Most marine life is found in coastal habitats, even though the shelf area occupies only seven percent of the total ocean area.
Q fever	Q fever is a disease caused by infection with Coxiella burnetii, a bacterium that affects humans and other animals. This organism is uncommon, but may be found in cattle, sheep, goats and other domestic mammals, including cats and dogs. The infection results from inhalation of a spore-like small cell variant, and from contact with the milk, urine, feces, vaginal mucus, or semen of infected animals.

Rickettsia	Rickettsia is a genus of non-motile, Gram-negative, non-sporeforming, highly pleomorphic bacteria that can present as cocci (0.1 µm in diameter), rods (1-4 µm long) or thread-like (10 µm long). Being obligate intracellular parasites, the Rickettsia survival depends on entry, growth, and replication within the cytoplasm of eukaryotic host cells (typically endothelial cells). Because of this, Rickettsia cannot live in artificial nutrient environments and are grown either in tissue or embryo cultures (typically, chicken embryos are used).
Rickettsia rickettsii	Rickettsia rickettsii is native to the New World and causes the malady known as Rocky Mountain spotted fever (RMSF). RMSF is transmitted by the bite of an infected tick while feeding on warm-blooded animals, including humans. Humans are accidental hosts in the rickettsia-tick life cycle and are not required to maintain the rickettsiae in nature.
Biochemical oxygen demand	Biochemical oxygen demand or B.O.D. is the amount of dissolved oxygen needed by aerobic biological organisms in a body of water to break down organic material present in a given water sample at certain temperature over a specific time period. The term also refers to a chemical procedure for determining this amount. This is not a precise quantitative test, although it is widely used as an indication of the organic quality of water.
Betaproteobacteria	Betaproteobacteria is a class of Proteobacteria. Betaproteobacteria are, like all Proteobacteria, gram-negative. The Betaproteobacteria consist of several groups of aerobic or facultative bacteria that are often highly versatile in their degradation capacities, but also contain chemolithotrophic genera (e.g., the ammonia-oxidising genus Nitrosomonas) and some phototrophs (members of the genera Rhodocyclus and Rubrivivax).
Diplococcus	A diplococcus is a round bacterium (a coccus) that typically occurs in the form of two joined cells. Examples are Moraxella catarrhalis, Neisseria gonorrhoeae and Neisseria meningitidis. Of these, all are Gram-negative.
Neisseria gonorrhoeae	Neisseria gonorrhoeae, or gonococcus, is a species of Gram-negative coffee bean-shaped diplococci bacteria responsible for the sexually transmitted infection gonorrhea. N. gonorrhoea was first described by Albert Neisser in 1879. Microbiology Neisseria are fastidious Gram-negative cocci that require nutrient supplementation to grow in laboratory cultures.

Chapter 4. Part 4: Microbial Diversity and Ecology

CHAPTER HIGHLIGHTS & NOTES: KEY TERMS, PEOPLE, PLACES, CONCEPTS

Nitrogen cycle	The nitrogen cycle is the process by which nitrogen is converted between its various chemical forms. This transformation can be carried out to both biological and non-biological processes. Important processes in the nitrogen cycle include fixation, mineralization, nitrification, and denitrification.
Nitrosomonas	Nitrosomonas is a genus comprising rod shaped chemoautotrophic bacteria. This rare bacteria oxidizes ammonia into nitrite as a metabolic process. Nitrosomonas are useful in treatment of industrial and sewage waste and in the process of bioremediation.
Antibiotic resistance	Antibiotic resistance is a type of drug resistance where a microorganism is able to survive exposure to an antibiotic. While a spontaneous or induced genetic mutation in bacteria may confer resistance to antimicrobial drugs, genes that confer resistance can be transferred between bacteria in a horizontal fashion by conjugation, transduction, or transformation. Thus, a gene for antibiotic resistance that evolves via natural selection may be shared.
Asymptomatic carrier	An asymptomatic carrier is a person or other organism that has contracted an infectious disease, but who displays no symptoms. Although unaffected by the disease themselves, carriers can transmit it to others. A number of animal species can act as a vector of human disease.
Enterobacteriaceae	The Enterobacteriaceae are a large family of bacteria, including many of the more familiar pathogens, such as Salmonella and Escherichia coli. Genetic studies place them among the Proteobacteria, and they are given their own order (Enterobacteriales), though this is sometimes taken to include some related environmental samples. Characteristics Members of the Enterobacteriaceae are rod-shaped, and are typically 1-5 μm in length.
Gammaproteobacteria	Gammaproteobacteria is a class of several medically, ecologically and scientifically important groups of bacteria, such as the Enterobacteriaceae (Escherichia coli), Vibrionaceae and Pseudomonadaceae. An exceeding number of important pathogens belongs to this class, e.g. Salmonella (enteritis and typhoid fever), Yersinia (plague), Vibrio (cholera), Pseudomonas aeruginosa (lung infections in hospitalised or cystic fibrosis patients), Klebsiella pneumoniae responsible for causing pneumonia. Like all Proteobacteria, the Gammaproteobacteria are Gram-negative.
Enterobacter	Enterobacter is a genus of common Gram-negative, facultatively-anaerobic, rod-shaped bacteria of the family Enterobacteriaceae.

	Several strains of the these bacteria are pathogenic and cause opportunistic infections in immunocompromised (usually hospitalized) hosts and in those who are on mechanical ventilation. The urinary and respiratory tract are the most common sites of infection.
Erwinia	Erwinia is a genus of Enterobacteriaceae bacteria containing mostly plant pathogenic species which was named for the first phytobacteriologist, Erwin Smith. It is a gram negative bacterium related to E. coli, Shigella, Salmonella and Yersinia. It is primarily a rod-shaped bacteria.
Proteus mirabilis	Proteus mirabilis is a Gram-negative, facultatively anaerobic, rod shaped bacterium. It shows swarming motility, and urease activity. P. mirabilis causes 90% of all 'Proteus' infections in humans.
Proteus vulgaris	Proteus vulgaris is a rod-shaped, Gram negative bacterium that inhabits the intestinal tracts of humans and animals. It can be found in soil, water and fecal matter. It is grouped with the enterobacteriaceae and is an opportunistic pathogen of humans.
Pseudomonadaceae	The Pseudomonadaceae is a family of bacteria that includes the genera Azomonas, Azomonotrichon, Azorhizophilus, Azotobacter, Cellvibrio, Mesophilobacter, Pseudomonas (the type genus), Rhizobacter, Rugamonas, and Serpens . The Azotobacteriaceae were recently published as belonging in this family as well. History Pseudomonad literally means 'false unit', being derived from the Greek pseudo (ψευδο 'false') and monas (μονος 'a single unit').
Pseudomonas aeruginosa	Pseudomonas aeruginosa is a common bacterium which can cause disease in animals, including humans. It is found in soil, water, skin flora, and most man-made environments throughout the world. It thrives not only in normal atmospheres, but also in hypoxic atmospheres, and has thus colonized many natural and artificial environments.
Pseudomonas fluorescens	Pseudomonas fluorescens is a common Gram-negative, rod-shaped bacterium. It belongs to the Pseudomonas genus; 16S rRNA analysis has placed P. fluorescens in the P. fluorescens group within the genus, to which it lends its name. General characteristics P. fluorescens has multiple flagella.
Shewanella oneidensis	Shewanella oneidensis proteobacterium was first isolated from Lake Oneida, NY in 1988, thus the derivation of the name.

	This species is also sometimes referred to as Shewanella oneidensis MR-1, indicating 'metal reducing', a special feature of this particular organism. Shewanella oneidesnsis is a facultative bacterium, capable of surviving and proliferating in both aerobic and anaerobic conditions.
Deltaproteobacteria	Deltaproteobacteria is a class of Proteobacteria. All species of this group are, like all Proteobacteria, gram-negative. The Deltaproteobacteria comprise a branch of predominantly aerobic genera, the fruiting-body-forming Myxobacteria which release myxospores in unfavorable environments, and a branch of strictly anaerobic genera, which contains most of the known sulfate- (Desulfovibrio, Desulfobacter, Desulfococcus, Desulfonema, etc).
Myxobacteria	The myxobacteria are a group of bacteria that predominantly live in the soil. The myxobacteria have very large genomes, relative to other bacteria, e.g. 9-10 million nucleotides. Sorangium cellulosum has the largest known (as of 2008) bacterial genome, at 13.0 million nucleotides.
Myxococcus xanthus	Myxococcus xanthus colonies exist as a self-organized, predatory, saprotrophic, single-species biofilm called a swarm. Myxococcus xanthus, which can be found almost ubiquitously in soil, are thin rod shaped, gram-negative cells that exhibit self-organizing behavior as a response to environmental cues. The swarm, which has been compared to a 'wolf-pack,' modifies its environment through stigmergy.
Sorangium cellulosum	Sorangium cellulosum is a soil-dwelling Gram-negative bacterium of the group myxobacteria. It is motile and shows gliding motility. Under stressful conditions this motility, as in other myxobacteria, the cells congregate to form fruiting bodies and differentiate into myxospores.
Slime mold	Slime mold is a broad term describing protists that use spores to reproduce. Slime molds were formerly classified as fungi, but are no longer considered part of this kingdom. Their common name refers to part of some of these organisms' life cycles where they can appear as gelatinous 'slime'.
Agrobacterium	Agrobacterium is a genus of Gram-negative bacteria established by H. J. Conn that uses horizontal gene transfer to cause tumors in plants. Agrobacterium tumefaciens is the most commonly studied species in this genus. Agrobacterium is well known for its ability to transfer DNA between itself and plants, and for this reason it has become an important tool for genetic engineering.
Bdellovibrio	Bdellovibrio is a genus of Gram-negative, obligate aerobic bacteria.

	One of the more notable characteristics of this genus is that members parasitize other Gram-negative bacteria by entering into their periplasmic space and feeding on the biopolymers, e.g. proteins and nucleic acids, of their hosts. After entering the periplasmic space of its host the Bdellovibrio bacterium forms a structure called a bdelloplast, which consists of both predator and prey.
Epsilonproteobacteria	Epsilonproteobacteria is a class of Proteobacteria. All species of this class are, like all Proteobacteria, gram-negative.

The Epsilonproteobacteria consist of few known genera, mainly the curved to spirilloid Wolinella spp., Helicobacter spp., and Campylobacter spp. |
| Helicobacter | Helicobacter is a genus of Gram-negative bacteria possessing a characteristic helix shape. They were initially considered to be members of the Campylobacter genus, but since 1989 they have been grouped in their own genus. The Helicobacter genus belongs to class Epsilonproteobacteria, order Campylobacterales, family Helicobacteraceae and already involves >35 species. |
| Bacteroides | Bacteroides is a genus of Gram-negative, bacillus bacteria. Bacteroides species are non-endospore-forming, anaerobes, and may be either motile or non-motile, depending on the species. The DNA base composition is 40-48% GC. Unusual in bacterial organisms, Bacteroides membranes contain sphingolipids. |
| Bacteroides fragilis | Bacteroides fragilis is a Gram-negative bacillus bacterium species, and an obligate anaerobe of the gut.

B. fragilis group is the most commonly isolated bacteriodaceae in anaerobic infections especially those that originate from the gastrointestinal flora. B. fragilis is the most prevalent organism in the B. fragilis group, accounting for 41% to 78% of the isolates of the group. |
| Chlorobium | Chlorobium is a genus of green sulfur bacteria. They are photolithotrophic oxidizers of sulfur and most notably utilise a noncyclic electron transport chain to reduce NAD+. Hydrogen sulfide is used as an electron source and carbon dioxide its carbon source. |
| Flavobacterium | Flavobacterium is a genus of Gram-negative, non-motile and motile, rod-shaped bacteria that consists of ten recognized species, as well as three newly proposed species (F. gondwanense, F. salegens, and F. scophthalmum). Flavobacteria are found in soil and fresh water in a variety of environments. Several species are known to cause disease in freshwater fish. |

Chapter 4. Part 4: Microbial Diversity and Ecology

Obligate anaerobe	Obligate anaerobes are microorganisms that live and grow in the absence of molecular oxygen; some of these are killed by oxygen. Historically, it was widely accepted that obligate (strict) anaerobes die in presence of oxygen due to the absence of the enzymes superoxide dismutase and catalase, which would convert the lethal superoxide formed in their cells due to the presence of oxygen. While this is true in some cases, these enzyme activities have been identified in some obligate anaerobes, and genes for these enzymes and related proteins have been found in their genomes, such as Clostridium butyricum and Methanosarcina barkeri, among others.
Borrelia burgdorferi	Borrelia burgdorferi is a species of Gram negative bacteria of the spirochete class of the genus Borrelia. B. burgdorferi is predominant in North America, but also exists in Europe, and is the agent of Lyme disease. It is a zoonotic, vector-borne disease transmitted by ticks and is named after the researcher Willy Burgdorfer who first isolated the bacterium in 1982. B. burgdorferi is one of the few pathogenic bacteria that can survive without iron, having replaced all of its iron-sulfur cluster enzymes with enzymes that use manganese, thus avoiding the problem many pathogenic bacteria face in acquiring iron.
Chlamydia	Chlamydia is a genus of bacteria that are obligate intracellular parasites. Chlamydia infections are the most common bacterial sexually transmitted infections in humans and are the leading cause of infectious blindness worldwide. The three Chlamydia species include Chlamydia trachomatis (a human pathogen), Chlamydia suis (affects only swine), and Chlamydia muridarum (affects only mice and hamsters).
Chlamydia trachomatis	Chlamydia trachomatis, an obligate intracellular human pathogen, is one of three bacterial species in the genus Chlamydia. C. trachomatis is a Gram-negative bacteria, therefore its cell wall components retain the counter-stain safranin and appear pink under a light microscope. Identified in 1907, C. trachomatis was the first chlamydial agent discovered in humans.
Chlamydomonas reinhardtii	Chlamydomonas reinhardtii is a single celled green alga about 10 micrometres in diameter that swims with two flagella. It has a cell wall made of hydroxyproline-rich glycoproteins, a large cup-shaped chloroplast, a large pyrenoid, and an 'eyespot' that senses light. Although widely distributed worldwide in soil and fresh water, C. reinhardtii is primarily used as a model organism in biology in a wide range of subfields.
Leptospirosis	Leptospirosis is a disease caused by infection with bacteria of the genus Leptospira that affects humans as well as other mammals, birds, amphibians, and reptiles.

	The disease was first described by Adolf Weil in 1886 when he reported an 'acute infectious disease with enlargement of spleen, jaundice and nephritis'. Leptospira was first observed in 1907 from a post mortem renal tissue slice.
Confocal microscopy	Confocal microscopy is an optical imaging technique used to increase optical resolution and contrast of a micrograph by using point illumination and a spatial pinhole to eliminate out-of-focus light in specimens that are thicker than the focal plane. It enables the reconstruction of three-dimensional structures from the obtained images. This technique has gained popularity in the scientific and industrial communities and typical applications are in life sciences, semiconductor inspection and materials science.
Methanomicrobiales	In the taxonomy of microorganisms, the Methanomicrobiales are an order of the Methanomicrobia. Methanomicrobiales are strictly carbon dioxide reducing methanogens, using hydrogen or formate as the reducing agent. As seen from the phylogenetic tree based on the 'The All-Species Living Tree' Project the family Methanomicrobiaceae is highly polyphyletic within the Methanomicrobiales.
Extreme environment	An extreme environment exhibits extreme conditions which are challenging to most life forms. These may be extremely high or low ranges of temperature, radiation, pressure, acidity, alkalinity, air, water, salt, sugar, carbon dioxide, sulphur, petroleum and many others. An extreme environment is one place where humans generally do not live or could die there.
Periodontal disease	Periodontal disease is a type of disease that affects one or more of the periodontal tissues:•alveolar bone•periodontal ligament•cementum•gingiva While many different diseases affect the tooth-supporting structures, plaque-induced inflammatory lesions make up the vast majority of periodontal diseases and have traditionally been divided into two categories:•gingivitis or•periodontitis. While in some sites or individuals, gingivitis never progresses to periodontitis, data indicates that gingivitis always precedes periodontitis. Diagnosis In 1976, Page & Schroeder introduced an innovative new analysis of periodontal disease based on histopathologic and ultrastructural features of the diseased gingival tissue.

Chapter 4. Part 4: Microbial Diversity and Ecology

DNA gyrase	DNA gyrase, often referred to simply as gyrase, is an enzyme that relieves strain while double-stranded DNA is being unwound by helicase. This causes negative supercoiling of the DNA. Bacterial DNA gyrase is the target of many antibiotics, including nalidixic acid, novobiocin, and ciprofloxacin.

DNA gyrase is a type II topoisomerase (EC 5.99.1.3) that introduces negative supercoils into DNA by looping the template so as to form a crossing, then cutting one of the double helices and passing the other through it before releasing the break, changing the linking number by two in each enzymatic step. |
| Halobacterium | In taxonomy, Halobacterium is a genus of the Halobacteriaceae.

The genus Halobacterium consists of several species of archaea with an aerobic metabolism which require an environment with a high concentration of salt; many of their proteins will not function in low-salt environments. They grow on amino acids in their aerobic conditions. |
Halorhodopsin	Halorhodopsin is a light-driven ion pump, specific for chloride ions, and found in phylogenetically ancient archaea, known as halobacteria. It is a seven-transmembrane protein of the retinylidene protein family, homologous to the light-driven proton pump bacteriorhodopsin, and similar in tertiary structure (but not primary sequence structure) to vertebrate rhodopsins, the pigments that sense light in the retina. Halorhodopsin also shares sequence similarity to channelrhodopsin, a light-driven ion channel.
Pyrococcus furiosus	Pyrococcus furiosus is an extremophilic species of Archaea. It can be classified as a hyperthermophile because it thrives best under extremely high temperatures--higher than those preferred of a thermophile. It is notable for having an optimum growth temperature of 100°C (a temperature that would destroy most living organisms), and for being one of the few organisms identified as possessing enzymes containing tungsten, an element rarely found in biological molecules.
S-layer	An S-layer is a part of the cell envelope commonly found in bacteria, as well as among archaea . It consists of a monomolecular layer composed of identical proteins or glycoproteins. This two-dimensional structure is built via self-assembly and encloses the whole cell surface.
Cell membrane	The cell membrane is a biological membrane that separates the interior of all cells from the outside environment. The cell membrane is selectively permeable to ions and organic molecules and controls the movement of substances in and out of cells. It basically protects the cell from outside forces.
Phase variation	Phase variation is a method for dealing with rapidly varying environments without requiring random mutation employed by various types of bacteria, including Salmonella species.

	It involves the variation of protein expression, frequently in an on-off fashion, within different parts of a bacterial population. Although it has been most commonly studied in the context of immune evasion, it is observed in many other areas as well.
RNA polymerase	RNA polymerase also known as DNA-dependent RNA polymerase, is an enzyme that produces RNA. In cells, RNAP is necessary for constructing RNA chains using DNA genes as templates, a process called transcription. RNA polymerase enzymes are essential to life and are found in all organisms and many viruses. In chemical terms, RNAP is a nucleotidyl transferase that polymerizes ribonucleotides at the 3' end of an RNA transcript.
Sigma factor	A sigma factor is a protein needed only for initiation of RNA synthesis. It is a bacterial transcription initiation factor that enables specific binding of RNA polymerase to gene promoters. The specific sigma factor used to initiate transcription of a given gene will vary, depending on the gene and on the environmental signals needed to initiate transcription of that gene.
TATA-binding protein	The TATA-binding protein is a general transcription factor that binds specifically to a DNA sequence called the TATA box. This DNA sequence is found about 25 base pairs upstream of the transcription start site in some eukaryotic gene promoters. TBP, along with a variety of TBP-associated factors, make up the TFIID, a general transcription factor that in turn makes up part of the RNA polymerase II preinitiation complex.
Transcription factor	In molecular biology and genetics, a transcription factor is a protein that binds to specific DNA sequences, thereby controlling the flow of genetic information from DNA to mRNA. Transcription factors perform this function alone or with other proteins in a complex, by promoting (as an activator), or blocking (as a repressor) the recruitment of RNA polymerase (the enzyme that performs the transcription of genetic information from DNA to RNA) to specific genes.

A defining feature of transcription factors is that they contain one or more DNA-binding domains (DBDs), which attach to specific sequences of DNA adjacent to the genes that they regulate. Additional proteins such as coactivators, chromatin remodelers, histone acetylases, deacetylases, kinases, and methylases, while also playing crucial roles in gene regulation, lack DNA-binding domains, and, therefore, are not classified as transcription factors. |
| Transfer RNA | Transfer RNA is an adaptor molecule composed of RNA, typically 73 to 93 nucleotides in length, that is used in biology to bridge the four-letter genetic code (ACGU) in messenger RNA (mRNA) with the twenty-letter code of amino acids in proteins. The role of tRNA as an adaptor is best understood by considering its three-dimensional structure. One end of the tRNA carries the genetic code in a three-nucleotide sequence called the anticodon. |
| Archaea | The Archaea are a group of single-celled microorganisms. |

	A single individual or species from this domain is called an archaeon (sometimes spelled 'archeon'). They have no cell nucleus or any other membrane-bound organelles within their cells.
Crenarchaeota	In taxonomy, the Crenarchaeota has been classified as either a phylum of the Archaea kingdom or a kingdom of its own. Initially, the Crenarchaeota were thought to be extremophiles (e.g., thermophilic and psychrophilic organisms) but recent studies have identified them as the most abundant archaea in the marine environment. Originally, they were separated from the other archaea based on rRNA sequences; since then physiological features, such as lack of histones have supported this division.
Euryarchaeota	In the taxonomy of microorganisms, the Euryarchaeota are a phylum of the Archaea. The Euryarchaeota include the methanogens, which produce methane and are often found in intestines, the halobacteria, which survive extreme concentrations of salt, and some extremely thermophilic aerobes and anaerobes. They are separated from the other archaeans based mainly on rRNA sequences.
Histone	In biology, histones are highly alkaline proteins found in eukaryotic cell nuclei that package and order the DNA into structural units called nucleosomes. They are the chief protein components of chromatin, acting as spools around which DNA winds, and play a role in gene regulation. Without histones, the unwound DNA in chromosomes would be very long (a length to width ratio of more than 10 million to one in human DNA).
Archaeoglobus	Archaeoglobus is a genus of the phylum Euryarchaeota. Archaeoglobus can be found in high-temperature oil fields where they may contribute to oil field souring. Archaeoglobus grow at extremely high temperatures between 60 and 95 °C, with optimal growth at 83 °C (ssp.
Ferroplasma	In taxonomy, Ferroplasma is a genus of the Ferroplasmaceae. The genus Ferroplasma consists solely of F. acidophilum, an acidophilic iron-oxidizing member of the Euryarchaeota. Unlike other members of the Thermoplasmata F.acidophilum is a mesophile with a temperature optimum of approximately 35°C, at which grows optimally at pH of 1.7. F. acidophilum is generally found in acidic mine tailings, primarily those containing pyrite (FeS).
Thermococcales	In taxonomy, the Thermococcales are an order of the Thermococci.

High pressure	High pressure in science and engineering is studying the effects of high pressure on materials and the design and construction of devices, such as a diamond anvil cell, which can create high pressure. By high pressure it is usually meant pressures of thousands (kilobars) or millions (megabars) of times atmospheric pressure (about 1 bar or 100,000 Pa). Percy Williams Bridgman received a Nobel prize for advancing this area of physics by several magnitudes of pressure (400 MPa to 40,000 MPa).
Thermophile	A thermophile is an organism -- a type of extremophile -- that thrives at relatively high temperatures, between 45 and 80 °C (113 and 176 °F). Many thermophiles are archaea. It has been suggested that thermophilic eubacteria are among the earliest bacteria.
Korarchaeota	In taxonomy, the Korarchaeota are a phylum of the Archaea. Analysis of their 16S rRNA gene sequences suggests that they are a deeply-branching lineage that does not belong to the main archaeal groups, Crenarchaeota and Euryarchaeota. Analysis of the genome of one korarchaeote that was enriched from a mixed culture revealed a number of both Crenarchaeota- and Euryarchaeota-like features and supports the hypothesis of a deep-branching ancestry.
Thermoprotei	In taxonomy, the Thermoprotei are a class of the Crenarchaeota (eocytes). The currently accepted taxonomy is based on the List of Prokaryotic names with Standing in Nomenclature (LPSN) and National Center for Biotechnology Information (NCBI) and the phylogeny is based on 16S rRNA-based LTP release 106 by The All-Species Living Tree Project Notes:? Strain found at the National Center for Biotechnology Information (NCBI) but not listed in the List of Prokaryotic names with Standing in Nomenclature (LPSN)
Habitat	A habitat is an ecological or environmental area that is inhabited by a particular species of animal, plant, or other type of organism. It is the natural environment in which an organism lives, or the physical environment that surrounds (influences and is utilized by) a species population. The term 'population' is preferred to 'organism' because, while it is possible to describe the habitat of a single turtle, it is also possible that one may not find any particular or individual bear but the grouping of bears that constitute a breeding population and occupy a certain biogeographical area.
Desulfurococcales	In taxonomy, the Desulfurococcales are an order of the Thermoprotei.

Chapter 4. Part 4: Microbial Diversity and Ecology

Pyrodictium	In taxonomy, Pyrodictium is a genus of the Pyrodictiaceae.Pyrodictium is a genera of submarine hyperthermophilic archaea whose optimal growth temperature ranges is 80 to 105 °C. They have a unique cell structure involving a network of cannulae and flat, disk-shaped cells. Pyrodictium are found in the porous walls of deep-sea vents where the temperatures inside get as high as 400 °C, while the outside marine environment is typically 3 °C. Pyrodictium is apparently able to adapt morphologically to this type of hot-cold habitat.
	Much research has been done on the genetics of Pyrodictium in order to understand its ability to survive and even thrive in such extreme temperatures.
Sulfolobales	In taxonomy, the Sulfolobales are an order of the Thermoprotei.
Sulfolobus	Sulfolobus is a genus of microorganism in the family Sulfolobaceae. It belongs to the archaea domain.
	Sulfolobus species grow in volcanic springs with optimal growth occurring at pH 2-3 and temperatures of 75-80 °C, making them acidophiles and thermophiles respectively.
Sulfur metabolism	Sulfur metabolism is vital for all living organisms as it is a constituent of a number of essential organic molecules like cysteine, methionine, coenzyme A, and iron-sulfur clusters. These compounds are involved in a number of essential cellular processes such as protein biosynthesis or the transfer of electrons and acyl groups. Sulfur, therefore, is an essential component of all living cells.
Fusellovirus	Fusellovirus is a genus of dsDNA virus that infects the species of the clade Archaea. The Fuselloviridae are ubiquitous in high-temperature (≥70°C), acidic (pH ≤4) hot springs around the world. They are one of the few viruses to possess a lipid membrane and a protective inner capsid in the form of a core.
Sulfolobus turreted icosahedral virus	Sulfolobus turreted icosahedral virus is a virus that infects the archaeon Sulfolobus solfataricus.
	This virus was isolated from a hot spring in the Rabbit Creek thermal area which is located in the Midway Geyser Basin of Yellowstone National Park. Virology
	It is an icosahedrally symmetric virus with a unique triangulation number (T) of 31. At the 12 fivefold symmetrical positions of the icosahedron protrude 'turrets' that extend 13 nanometers (nm) above the capsid surface.
Caldisphaera	In taxonomy, Caldisphaera is a genus of the Caldisphaeraceae.

Cenarchaeum symbiosum	Cenarchaeum symbiosum is a species of Archaebacteria in the genus Cenarchaeum, and is one of the three species contained by the newly proposed phylum Thaumarchaeota in the domain Archaea. C. symbiosum is psychrophilic and is found inhabiting marine sponges. The genome of C. symbiosum is estimated to be 2.02 Million bp in length, with a predicted amount of 2011 genes.
Thermoproteales	In taxonomy, the Thermoproteales are an order of the Thermoprotei.
Thermoproteus	In taxonomy, Thermoproteus is a genus of the Thermoproteaceae. These prokaryotes are thermophilic sulphur-dependent organisms related to the genera Sulfolobus, Pyrodictium and Desulfurococcus. They are hydrogen-sulphur autotrophs and can grow at temperatures of up to 95 degrees Celsius. Thermoproteus is a genus of anaerobes that grow in the wild by autotrophic sulfur reduction.
Mesophile	A mesophile is an organism that grows best in moderate temperature, neither too hot nor too cold, typically between 20 and 45 °C (68 and 113 °F). The term is mainly applied to microorganisms. The habitats of these organisms include especially cheese, yogurt, and mesophile organisms are often included in the process of beer and wine making.
Benthos	Benthos refers to species living in or on the ocean bottom and represent the greatest proportion of marine species. In fact 98% of marine species are from the benthic community which comprises a wide range of bacteria, plants and animals, which are classified into three categories epifauna, infauna and demersal. Epifauna are organisms that live in the sediment, infauna are organisms which are either attach to the bottom or substrate (move within the sediment), or that live on the sediment surface and demersal which are fish that feed on the benthic epifauna and infauna.
Methanobacteriales	In taxonomy, the Methanobacteriales are an order of the Methanobacteria.
Methanogen	Methanogens are microorganisms that produce methane as a metabolic byproduct in anoxic conditions. They are classified as archaea, a group quite distinct from bacteria. They are common in wetlands, where they are responsible for marsh gas, and in the guts of animals such as ruminants and humans, where they are responsible for the methane content of belching in ruminants and flatulence in humans.
Methanobacterium	In taxonomy, Methanobacterium is a genus of the Methanobacteriaceae.

Chapter 4. Part 4: Microbial Diversity and Ecology

Methanosarcina	Methanosarcina are the only known anaerobic methanogens to produce methane using all three known metabolic pathways for methanogenesis. Most methanogens make methane from carbon dioxide and hydrogen gas. Some others utilize acetate in the acetoclastic pathway.
Methanococcus	In taxonomy, Methanococcus is a genus of the Methanococcaceae.
	Methanococcus is a genus of coccoid methanogens. They are all mesophiles, except the thermophilic M. thermolithotrophicus and the hyperthermophilic M. jannaschii.
Methanomicrobium	In taxonomy, Methanomicrobium is a genus of the Methanomicrobiaceae.
Methanosaeta	In taxonomy, Methanosaeta is a genus of the Methanosaetaceae.
Wetland	A wetland is a land area that is saturated with water, either permanently or seasonally, such that it takes on characteristics that distinguish it as a distinct ecosystem. The primary factor that distinguishes wetlands is the characteristic vegetation that is adapted to its unique soil conditions: Wetlands are made up primarily of hydric soil, which supports aquatic plants.
	The water found in wetlands can be saltwater, freshwater, or brackish.
Cattle	Cattle are the most common type of large domesticated ungulates. They are a prominent modern member of the subfamily Bovinae, are the most widespread species of the genus Bos, and are most commonly classified collectively as Bos primigenius. Cattle are raised as livestock for meat (beef and veal), as dairy animals for milk and other dairy products, and as draft animals (pulling carts, plows and the like).
Landfill	A landfill site (also known as tip, dump or rubbish dump and historically as a midden), is a site for the disposal of waste materials by burial and is the oldest form of waste treatment. Historically, landfills have been the most common methods of organized waste disposal and remain so in many places around the world.
	Landfills may include internal waste disposal sites (where a producer of waste carries out their own waste disposal at the place of production) as well as sites used by many producers.
Termite	Termites are a group of eusocial insects that, until recently, were classified at the taxonomic rank of order Isoptera , but are now accepted as the epifamily Termitoidae, of the cockroach order Blattodea. While termites are commonly known, especially in Australia, as 'white ants', they are only distantly related to the ants.

Sulfur-reducing bacteria	Sulfur-reducing bacteria get their energy by reducing elemental sulfur to hydrogen sulfide. They couple this reaction with the oxidation of acetate, succinate or other organic compounds. Several types of bacteria and many non-methanogenic archaea can reduce sulfur.
Global warming	Global warming is the continuing rise in the average temperature of Earth's atmosphere and oceans. Global warming is caused by increased concentrations of greenhouse gases in the atmosphere, resulting from human activities such as deforestation and burning of fossil fuels. This finding is recognized by the national science academies of all the major industrialized countries and is not disputed by any scientific body of national or international standing.
Molybdenum	Molybdenum is a Group 6 chemical element with the symbol Mo and atomic number 42. The name is from Neo-Latin Molybdaenum, from Ancient Greek M?λυβδος molybdos, meaning lead, since its ores were confused with lead ores. Molybdenum minerals have been known into prehistory, but the element was 'discovered' (in the sense of differentiating it as a new entity from the mineral salts of other metals) in 1778 by Carl Wilhelm Scheele. The metal was first isolated in 1781 by Peter Jacob Hjelm.
Halobacteriales	In taxonomy, the Halobacteriales are an order of the Halobacteria, found in water saturated or nearly saturated with salt. They are also called halophiles, though this name is also used for other organisms which live in somewhat less concentrated salt water. They are common in most environments where large amounts of salt, moisture, and organic material are available.
Halophile	Halophiles are extremophile organisms that thrive in environments with very high concentrations of salt. The name comes from the Greek for 'salt-loving'. While the term is perhaps most often applied to some halophiles classified into the Archaea domain, there are also bacterial halophiles and some eukaryota, such as the alga Dunaliella salina.
Carotenoid	Carotenoids are tetraterpenoid organic pigments that are naturally occurring in the chloroplasts and chromoplasts of plants and some other photosynthetic organisms like algae, some bacteria, and some types of fungus. Carotenoids can be synthesized fats and other basic organic metabolic building blocks by all these organisms. Carotenoids generally cannot be manufactured by species in the animal kingdom (although one species of aphid is known to have acquired the genes for synthesis of the carotenoid torulene from fungi by horizontal gene transfer).
Halocin	Halocins are bacteriocins produced by halophilic Archaea and a type of archaeocin.

Since their discovery in 1982, halocins have been demonstrated to be diverse in a similar ways as the other bacteriocins. Some are large proteins, some small polypeptides (microhalocins).

Taq polymerase	Taq polymerase, exonuclease

Taq polymerase is a thermostable DNA polymerase named after the thermophilic bacterium Thermus aquaticus from which it was originally isolated by Thomas D. Brock in 1965. It is often abbreviated to 'Taq Pol' (or simply 'Taq'), and is frequently used in polymerase chain reaction (PCR), a method for greatly amplifying short segments of DNA.

T. aquaticus is a bacterium that lives in hot springs and hydrothermal vents, and Taq polymerase was identified as an enzyme able to withstand the protein-denaturing conditions (high temperature) required during PCR. Therefore it replaced the DNA polymerase from E. coli originally used in PCR. Taq's optimum temperature for activity is 75-80°C, with a half-life of greater than 2 hours at 92.5°C, 40 minutes at 95°C and 9 minutes at 97.5°C, and can replicate a 1000 base pair strand of DNA in less than 10 seconds at 72°C.

One of Taq's drawbacks is its relatively low replication fidelity. It lacks a 3' to 5' exonuclease proofreading activity, and has an error rate measured at about 1 in 9,000 nucleotides. The remaining two domains however may act in coordination, via coupled domain motion.

Thermococcus

In taxonomy, Thermococcus is a genus of extreme thermophiles in the family the Thermococcaceae.

They consist of gram-negative spheres that move with flagella. They are organothrophic anaerobes, growing in temperatures above 70°C. Some species require elemental sulphur as electron acceptor, producing hydrogen sulphide (H_2S) and carbon dioxide (CO_2), while some species are also able to reduce H^+. Thermococcus coalescens is an example species which has cells that can fuse in culturing to produce monster single cells.

Chain reaction

A chain reaction is a sequence of reactions where a reactive product or by-product causes additional reactions to take place. In a chain reaction, positive feedback leads to a self-amplifying chain of events.

Chain reactions are one way in which systems which are in thermodynamic non-equilibrium can release energy or increase entropy in order to reach a state of higher entropy.

Polymerase chain reaction

The polymerase chain reaction is a scientific technique in molecular biology to amplify a single or a few copies of a piece of DNA across several orders of magnitude, generating thousands to millions of copies of a particular DNA sequence.

	Developed in 1983 by Kary Mullis, PCR is now a common and often indispensable technique used in medical and biological research labs for a variety of applications. These include DNA cloning for sequencing, DNA-based phylogeny, or functional analysis of genes; the diagnosis of hereditary diseases; the identification of genetic fingerprints (used in forensic sciences and paternity testing); and the detection and diagnosis of infectious diseases.
Archaeoglobales	In taxonomy, the Archaeoglobales are an order of the Archaeoglobi.
Okazaki fragment	An Okazaki fragment is a relatively short fragment of DNA (with no RNA primer at the 5' terminus) created on the lagging strand during DNA replication. The lengths of Okazaki fragments are between 1,000 to 2,000 nucleotides long in E. coli and are generally between 100 to 200 nucleotides long in eukaryotes. It was originally discovered in 1968 by Reiji Okazaki, Tsuneko Okazaki, and their colleagues while studying replication of bacteriophage DNA in Escherichia coli.
Thermococcus litoralis	Thermococcus litoralis is a species of Archaea that is found around deep-sea hydrothermal vents as well as shallow submarine thermal springs and oil wells. It is an anaerobic organotroph hyperthermophile that is between 0.5- 3.0 microns in diameter. Like the other species in the order thermococcales, T. litoralis is an irregular cocci hyperthermophile that grow between 55°C-100°C. T. litoralis stands out from the rest of the thermococci in that it lacks motility.
Thermoplasmatales	In taxonomy, the Thermoplasmatales are an order of the Thermoplasmata. All are acidophiles, growing optimally at pH below 2. Picrophilus is currently the most acidophilic of all known organisms growing at a minimum pH of 0.06. Many of these organisms do not contain a cell wall, although this is not true in the case of Picrophilus. Most of members of the Thermotoplasmata are thermophilic.
Dinoflagellate	The dinoflagellates are a large group of flagellate protists. Most are marine plankton, but they are common in fresh water habitats, as well. Their populations are distributed depending on temperature, salinity, or depth.
Eukaryote	A eukaryote is an organism whose cells contain complex structures enclosed within membranes. Eukaryotes may more formally be referred to as the taxon Eukarya or Eukaryota. The defining membrane-bound structure that sets eukaryotic cells apart from prokaryotic cells is the nucleus, or nuclear envelope, within which the genetic material is carried.
Vorticella	Vorticella is a genus of protozoa, with over 16 known species. They are stalked, inverted bell-shaped ciliates, placed among the peritrichs.

Chapter 4. Part 4: Microbial Diversity and Ecology

Red tide	Red tide is a common name for a phenomenon also known as an algal bloom (large concentrations of aquatic microorganisms), an event in which estuarine, marine, or fresh water algae accumulate rapidly in the water column and results in discoloration of the surface water. It is usually found in coastal areas. These algae, known as phytoplankton, are single-celled protists, plant-like organisms that can form dense, visible patches near the water's surface.
Convergent evolution	Convergent evolution describes the acquisition of the same biological trait in unrelated lineages. The wing is a classic example of convergent evolution in action. Flying insects, birds, and bats have all evolved the capacity of flight independently.
Mycology	Mycology is the branch of biology concerned with the study of fungi, including their genetic and biochemical properties, their taxonomy and their use to humans as a source for tinder, medicinals (e.g., penicillin), food (e.g., beer, wine, cheese, edible mushrooms) and entheogens, as well as their dangers, such as poisoning or infection. From mycology arose the field of phytopathology, the study of plant diseases, and the two disciplines remain closely related because the vast majority of 'plant' pathogens are fungi. A biologist who studies mycology is called a mycologist.
Fairy ring	A fairy ring, elf circle, elf ring or pixie ring, is a naturally occurring ring or arc of mushrooms. The rings may grow to over 10 metres (33 ft) in diameter, and they become stable over time as the fungus grows and seeks food underground. They are found mainly in forested areas, but also appear in grasslands or rangelands.
Giardia	Giardia is a genus of anaerobic flagellated protozoan parasites of the phylum Diplomonada in the supergroup 'Excavata' that colonise and reproduce in the small intestines of several vertebrates, causing giardiasis. Their life cycle alternates between an actively swimming trophozoite and an infective, resistant cyst. The genus was named after French zoologist Alfred Mathieu Giard.
Immune system	An immune system is a system of biological structures and processes within an organism that protects against disease. To function properly, an immune system must detect a wide variety of agents, from viruses to parasitic worms, and distinguish them from the organism's own healthy tissue.

Blood cell	A blood cell, is a cell of any type normally found in blood. In mammals, these fall into three general categories:•red blood cells -- Erythrocytes•white blood cells -- Leukocytes•platelets -- Thrombocytes Together, these three kinds of blood cells sum up for a total 45% of blood tissue by volume (and the remaining 55% is plasma). This is called the hematocrit and can be determined by centrifuge or flow cytometry.
Amoebozoa	The Amoebozoa are a major group of amoeboid protozoa, including the majority that move by means of internal cytoplasmic flow. Their pseudopodia are characteristically blunt and finger-like, called lobopodia. Most are unicellular, and are common in soils and aquatic habitats, with some found as symbiotes of other organisms, including several pathogens.
Elongation factor	Elongation factors are a set of proteins that facilitate the events of translational elongation, the steps in protein synthesis from the formation of the first peptide bond to the formation of the last one. Elongation is the most rapid step in translation:•in prokaryotes it proceeds at a rate of 15 to 20 amino acids added per second (about 60 nucleotides per second)•in eukaryotes the rate is about two amino acids per second. Elongation factors play a role in orchestrating the events of this process, and in ensuring the 99.99% accuracy of translation at this speed.
Heterokont	The heterokonts or stramenopiles are a major line of eukaryotes currently containing more than 100,000 known species, most of them diatoms. Most are algae, ranging from the giant multicellular kelp to the unicellular diatoms, which are a primary component of plankton. Other notable members of the Stramenopila include the (generally parasitic) oomycetes, including Phytophthora of Irish potato famine infamy and Pythium which causes seed rot and damping off.
Oomycete	Oömycota or oömycetes form a distinct phylogenetic lineage of fungus-like eukaryotic microorganisms. They are filamentous microscopic, absorptive organisms that reproduce both sexually and asexually. Oomycetes occupy both saprophytic and pathogenic lifestyles - and include some of the most notorious pathogens of plants, causing devastating diseases such as late blight of potato and sudden oak death.
Opisthokont	The opisthokonts (Greek: ?π?σθιος (opísthios) = 'rear, posterior' + κοντ?ς (kontós) = 'pole' i.e. 'flagellum') or 'Fungi/Metazoa group' are a broad group of eukaryotes, including both the animal and fungus kingdoms, together with the eukaryotic microorganisms that are sometimes grouped in the paraphyletic phylum Choanozoa (previously assigned to the protist 'kingdom').

Chapter 4. Part 4: Microbial Diversity and Ecology

	Both genetic and ultrastructural studies strongly support that opisthokonts form a monophyletic group. One common characteristic of opisthokonts is that flagellate cells, such as most animal sperm and chytrid spores, propel themselves with a single posterior flagellum.
Rhizaria	The Rhizaria are a species-rich supergroup of unicellular eukaryotes. This supergroup was proposed by Cavalier-Smith in 2002. They vary considerably in form, but for the most part they are amoeboids with filose, reticulose, or microtubule-supported pseudopods. Many produce shells or skeletons, which may be quite complex in structure, and these make up the vast majority of protozoan fossils.
Trypanosoma brucei	Trypanosoma brucei is a parasitic protist species that causes African trypanosomiasis (or sleeping sickness) in humans and nagana in animals in Africa. There are 3 sub-species of T. brucei: T. b. brucei, T. b. gambiense and T. b.
Zoospore	A zoospore is a motile asexual spore that uses a flagellum for locomotion. Also called a swarm spore, these spores are created by some algae, bacteria and fungi to propagate themselves. There are two types of flagellated zoospores, tinsel or 'decorated', and whiplash.
Chloroflexus aurantiacus	Chloroflexus aurantiacus is a photosynthetic bacterium isolated from hot springs, belonging to the green non-sulfur bacteria. This organism is thermophilic and can grow at temperatures from 35 °C to 70 °C. Chloroflexus aurantiacus can survive in the dark if oxygen is available. When grown in the dark, Chloroflexus aurantiacus has a dark orange color.
Chytridiomycota	Chytridiomycota is a division of the Fungi kingdom. The name is derived from the Greek chytridion, meaning 'little pot', describing the structure containing unreleased spores. In older classifications, chytrids (except the recently established order Spizellomycetales) were placed in the Class Phycomycetes under the subdivision Myxomycophyta of the Kingdom Fungi.
Alveolate	The alveolates ('with cavities') are a major line of protists. Phyla There are three phyla, which are very divergent in form, but are now known to be close relatives based on various ultrastructural and genetic similarities:•Ciliates, very common protozoa, with many short cilia arranged in rows•Apicomplexa, parasitic protozoa that lack axonemal locomotive structures except in gametes•Dinoflagellates, mostly marine flagellates, many of which have chloroplasts

	Characteristics
	The most notable shared characteristic is the presence of cortical alveoli, flattened vesicles packed into a continuous layer supporting the membrane, typically forming a flexible pellicle. In dinoflagellates they often form armor plates.
Green algae	The green algae are the large group of algae from which the embryophytes (higher plants) emerged. As such, they form a paraphyletic group, although the group including both green algae and embryophytes is monophyletic (and often just known as kingdom Plantae). The green algae include unicellular and colonial flagellates, most with two flagella per cell, as well as various colonial, coccoid, and filamentous forms, and macroscopic seaweeds.
Mixotroph	A mixotroph is a microorganism that can use a mix of different sources of energy and carbon. Possible are alternations between photo- and chemotrophy, between litho- and organotrophy, between auto- and heterotrophy or a combination of it. Mixotrophs can be either eukaryotic or prokaryotic. They can take advantage of different environmental conditions.
Phytoplankton	Phytoplankton are the autotrophic component of the plankton community. The name comes from the Greek words φυτ?ν (phyton), meaning 'plant', and πλαγκτ?ς (planktos), meaning 'wanderer' or 'drifter'. Most phytoplankton are too small to be individually seen with the unaided eye.
Red algae	The red algae, thus red plant), are one of the oldest groups of eukaryotic algae, and also one of the largest, with about 5,000-6,000 species of mostly multicellular, marine algae, including many notable seaweeds. Other references indicate as many as 10,000 species; more detailed counts indicate ~4,000 in ~600 genera (3,738 marine spp in 546 genera and 10 orders (plus the unclassifiable); 164 freshwater spp in 30 genera in 8 orders).
	The red algae form a distinct group characterized by the following attributes: eukaryotic cells without flagella and centrioles, using floridean starch as food reserve, with phycobiliproteins as accessory pigments (giving them their red color), and with chloroplasts lacking external endoplasmic reticulum and containing unstacked thylakoids. Most red algae are also multicellular, macroscopic, marine, and have sexual reproduction.
Zooplankton	Zooplankton are heterotrophic (sometimes detritivorous) plankton. Plankton are organisms drifting in oceans, seas, and bodies of fresh water. The word 'zooplankton' is derived from the Greek zoon , meaning 'animal', and planktos , meaning 'wanderer' or 'drifter'.
Kleptoplasty	Kleptoplasty are sequestered by host organisms. The alga is eaten normally and partially digested, leaving the plastid intact. The plastids are maintained within the host, temporarily retaining functional photosynthesis for use by the predator.

Chapter 4. Part 4: Microbial Diversity and Ecology

Ciliate	The ciliates are a group of protozoans characterized by the presence of hair-like organelles called cilia, which are identical in structure to eukaryotic flagella, but typically shorter and present in much larger numbers with a different undulating pattern than flagella. Cilia occur in all members of the group (although the peculiar Suctoria only have them for part of the life-cycle) and are variously used in swimming, crawling, attachment, feeding, and sensation.
	Ciliates are one of the most important groups of protists, common almost everywhere there is water -- in lakes, ponds, oceans, rivers, and soils.
Euglenozoa	The Euglenozoa are a large group of flagellate protozoa. They include a variety of common free-living species, as well as a few important parasites, some of which infect humans. There are two main subgroups, the euglenids and kinetoplastids.
Kelp	Kelps are large seaweeds (algae) belonging to the brown algae (Phaeophyceae) in the order Laminariales. There are about 30 different genera.
	Kelps grow in underwater 'forests' (kelp forests) in shallow oceans.
Ostreococcus tauri	Ostreococcus tauri is a unicellular species of marine green alga about 0.8 micrometres (μm) in diameter, the smallest free-living (non-symbiotic) eukaryote yet described. It has a very simple ultrastructure, and a compact genome.
	As a common member of global oceanic picoplankton populations, this organism has a major role in the carbon cycle in many areas.
Picoeukaryote	Picoeukaryotes are picoplanktonic eukaryotic organisms that range in size from 0.2 - 2.0 μm. They are distributed throughout the world's marine and freshwater ecosystems and constitute a significant contribution to autotrophic communities. Though the SI prefix pico- might imply an organism smaller than atomic size, the term was likely used to avoid confusion with existing size classifications of plankton.
Mycelium	Mycelium is the vegetative part of a fungus, consisting of a mass of branching, thread-like hyphae. The mass of hyphae is sometimes called shiro, especially within the fairy ring fungi. Fungal colonies composed of mycelia are found in soil and on or within many other substrates.
Nystatin	Nystatin is a polyene antifungal medication to which many molds and yeast infections are sensitive, including Candida. Due to its toxicity profile, there are currently no injectable formulations of this drug on the US market. However, nystatin may be safely given orally as well as applied topically due to its minimal absorption through mucocutaneous membranes such as the gut and the skin.

Blastomycosis	Blastomycosis is a fungal infection caused by the organism Blastomyces dermatitidis. Endemic to portions of North America, blastomycosis causes clinical symptoms similar to histoplasmosis. Blastomycosis can present in one of the following ways:•a flu-like illness with fever, chills, myalgia, headache, and a nonproductive cough which resolves within days.•an acute illness resembling bacterial pneumonia, with symptoms of high fever, chills, a productive cough, and pleuritic chest pain.•a chronic illness that mimics tuberculosis or lung cancer, with symptoms of low-grade fever, a productive cough, night sweats, and weight loss.•a fast, progressive, and severe disease that manifests as ARDS, with fever, shortness of breath, tachypnea, hypoxemia, and diffuse pulmonary infiltrates.•skin lesions, usually asymptomatic, appear as ulcerated lesions with small pustules at the margins•bone lytic lesions can cause bone or joint pain.•prostatitis may be asymptomatic or may cause pain on urinating.•laryngeal involvement causes hoarseness.Cause Infection occurs by inhalation of the fungus from its natural soil habitat.
Budding	Budding is a form of asexual reproduction in which a new organism develops from an outgrowth or bud on another one. The new organism remains attached as it grows, separating from the parent organism only when it is mature. Since the reproduction is asexual, the newly created organism is a clone and is genetically identical to the parent organism.
Candida albicans	Candida albicans is a diploid fungus that grows both as yeast and filamentous cells and a causal agent of opportunistic oral and genital infections in humans. Systemic fungal infections (fungemias) including those by C. albicans have emerged as important causes of morbidity and mortality in immunocompromised patients (e.g., AIDS, cancer chemotherapy, organ or bone marrow transplantation). C. albicans biofilms may form on the surface of implantable medical devices.
Meiosis	Meiosis is a special type of cell division necessary for sexual reproduction in eukaryotes. The cells produced by meiosis are gametes or spores. In many organisms, including all animals and land plants (but not some other groups such as fungi), gametes are called sperm and egg cells.
Pneumocystis jirovecii	Pneumocystis jirovecii is a yeast-like fungus of the genus Pneumocystis. The causative organism of Pneumocystis pneumonia, it is an important human pathogen, particularly among immunocompromised hosts. Prior to its discovery as a human-specific pathogen, P. jirovecii was known as P. carinii.
Saccharomyces cerevisiae	Saccharomyces cerevisiae is a species of yeast. It is perhaps the most useful yeast, having been instrumental to baking and brewing since ancient times.

Chapter 4. Part 4: Microbial Diversity and Ecology

Dimorphic fungi	Dimorphic fungi are fungi which can exist as moldhyphal/filamentous form or as yeast. An example is Penicillium marneffei:•At room temperature, it grows as a mold.•At body temperature, it grows as a yeast. Several species are potential pathogens, including Coccidioides immitis, Paracoccidioides brasiliensis, Candida albicans, Ustilago maydis, Blastomyces dermatitidis, Histoplasma capsulatum, and Sporothrix schenckii.
Life cycle	A life cycle is a period involving all different generations of a species succeeding each other through means of reproduction, whether through asexual reproduction or sexual reproduction (a period from one generation of organisms to the same identical). For example, a complex life cycle of Fasciola hepatica includes three different multicellular generations: 1) 'adult' hermaphroditic; 2) sporocyst; 3) redia.
Vaginal flora	The micro-organisms that colonize the vagina, collectively referred to as the vaginal microbiome or vaginal flora, were discovered by the German gynecologist Albert Döderlein in 1892. The amount and type of bacteria present have significant implications for a woman's overall health. The primary colonizing bacteria of a healthy individual are of the genus lactobacillus, such as L. acidophilus and L. doderlein, and the lactic acid they produce (some species produce hydrogen peroxide or antibiotic), in combination with fluids secreted during sexual arousal, are greatly responsible for the characteristic scent associated with the vaginal area. During menstruation, the concentration of vaginal microbiome is observed to decline.
Penicillium	Penicillium is a genus of ascomycetous fungi of major importance in the natural environment as well as food and drug production. Members of the genus produce penicillin, a molecule that is used as an antibiotic, which kills or stops the growth of certain kinds of bacteria inside the body. According to the Dictionary of the Fungi (10th edition, 2008), the widespread genus contains over 300 species.
Reproduction	Reproduction is the biological process by which new 'offspring' individual organisms are produced from their 'parents'. Reproduction is a fundamental feature of all known life; each individual organism exists as the result of reproduction. The known methods of reproduction are broadly grouped into two main types: sexual and asexual.
Batrachochytrium dendrobatidis	Batrachochytrium dendrobatidis is a chytrid fungus that causes the disease chytridiomycosis. In the decade after it was first discovered in amphibians in 1998, the disease devastated amphibian populations around the world, in a global decline towards multiple extinctions, part of the Holocene extinction.

Hydrogenosome	A hydrogenosome is a membrane-enclosed organelle of some anaerobic ciliates, trichomonads and fungi. The hydrogenosomes of trichomonads (the most studied of the hydrogenosome-containing microorganisms) produce molecular hydrogen, acetate, carbon dioxide and ATP by the combined actions of pyruvate:ferredoxin oxido-reductase, hydrogenase, acetate:succinate CoA transferase and succinate thiokinase. Superoxide dismutase, malate dehydrogenase, ferredoxin, adenylate kinase and NADH:ferredoxin oxido-reductase are also localized in the hydrogenosome.
Mucor	Mucor is a microbial genus of about 3000 species of moulds commonly found in soil, digestive systems, plant surfaces, and rotten vegetable matter. Colonies of this fungal genus are typically white to beige or grey and fast-growing. Colonies on culture medium may grow to several centimeters in height.
Rhizopus	Rhizopus is a genus of common saprobic fungi on plants and specialized parasites on animals. They are found on a wide variety of organic substrates, including 'mature fruits and vegetables', faeces, jellies, syrups, leather, bread, peanuts and tobacco. Some Rhizopus species are opportunistic agents of human zygomycosis (fungal infection) and can be fatal.
Sporangium	A sporangium (modern Latin, from Greek σπ?ρος (sporos) 'spore' + αγγε?ον (angeion) 'vessel') is an enclosure in which spores are formed. It can be composed of a single cell or can be multicellular. All plants, fungi, and many other lineages form sporangia at some point in their life cycle.
Ascospore	An ascospore is a spore contained in an ascus or that was produced inside an ascus. This kind of spore is specific to fungi classified as ascomycetes (Ascomycota). Typically, a single ascus will contain eight ascospores.
Aspergillus	Aspergillus is a genus consisting of several hundred mold species found in various climates worldwide. Aspergillus was first catalogued in 1729 by the Italian priest and biologist Pier Antonio Micheli. Viewing the fungi under a microscope, Micheli was reminded of the shape of an aspergillum (holy water sprinkler), from Latin spargere (to sprinkle), and named the genus accordingly.
Zygospore	A zygospore is a diploid reproductive stage in the life cycle of many fungi and protists. Zygospores are created by the nuclear fusion of haploid cells. In fungi, zygospores are termed chlamydospores and are formed after the fusion of hyphae of different mating types.

Chapter 4. Part 4: Microbial Diversity and Ecology

Mycotoxin	A mycotoxin (from Greek μ?κης (mykes, mukos) 'fungus' and Latin (toxicum) 'poison') is a toxic secondary metabolite produced by organisms of the fungus kingdom, commonly known as molds. The term 'mycotoxin' is usually reserved for the toxic chemical products produced by fungi that readily colonize crops. One mold species may produce many different mycotoxins, and the same mycotoxin may be produced by several species.
Truffle	A truffle is a fungi fruiting body that develops underground and relies on mycophagy for spore dispersal. Almost all truffles are ectomycorrhizal and are therefore usually found in close association with trees. There are hundreds of species of truffles that are big, but the fruiting body of some (mostly in the genus 'Tuber') are highly prized as a food.
Microsporum	Microsporum is a genus of fungi that causes tinea capitis, tinea corpus, ringworm, and other dermatophytoses (fungal infections of the skin). Microsporum forms both macroconidia (large asexual reproductive structures) and microconidia (smaller asexual reproductive structures) on short conidiophores. Macroconidia are hyaline, multiseptate, variable in form, fusiform, spindle-shaped to obovate, 7-20 by 30-160 um in size, with thin or thick echinulate to verrucose cell walls.
Trichophyton	The fungus genus Trichophyton is characterized by the development of both smooth-walled macro- and microconidia. Macroconidia are mostly borne laterally directly on the hyphae or on short pedicels, and are thin- or thick-walled, clavate to fusiform, and range from 4 to 8 by 8 to 50 μm in size. Macroconidia are few or absent in many species.
Amanita phalloides	Amanita phalloides commonly known as the death cap, is a deadly poisonous basidiomycete fungus, one of many in the genus Amanita. Widely distributed across Europe, A. phalloides forms ectomycorrhizas with various broadleaved trees. In some cases, death cap has been introduced to new regions with the cultivation of non-native species of oak, chestnut, and pine.
Basidiospore	A basidiospore is a reproductive spore produced by Basidiomycete fungi. Basidiospores typically each contain one haploid nucleus that is the product of meiosis, and they are produced by specialized fungal cells called basidia. In gills under a cap of one common species in the phylum of Basidiomycota, there exist millions of basidia.
Chlamydomonas	Chlamydomonas is a genus of green algae. They are unicellular flagellates. Chlamydomonas is used as a model organism for molecular biology, especially studies of flagellar motility and chloroplast dynamics, biogenesis, and genetics.
Encephalitozoon intestinalis	Encephalitozoon intestinalis is a parasite. It can cause microsporidiosis.

Phytophthora infestans	Phytophthora infestans is an oomycete that causes the serious potato disease known as late blight or potato blight. (Early blight, caused by Alternaria solani, is also often called 'potato blight'). Late blight was a major culprit in the 1840s European, the 1845 Irish and 1846 Highland potato famines.
Conidia	Conidia, are asexual, non-motile spores of a fungus; they are also called mitospores due to the way they are generated through the cellular process of mitosis. The two new haploid cells are genetically identical to the haploid parent, and can develop into new organisms if conditions are favorable, and serve in biological dispersal. Asexual reproduction in Ascomycetes (the Phylum Ascomycota) is by the formation of conidia, which are borne on specialized stalks called conidiophores.
Pyrenoid	In cell biology, pyrenoids are organelles, centers of carbon dioxide fixation within the chloroplasts of algae and hornworts. Pyrenoids are not membrane-bound, but specialized areas of the plastid that contain high levels of ribulose-1,5-bisphosphate carboxylase/oxygenase (RubisCO). RubisCO fixes carbon dioxide by adding it to the 5-carbon sugar-phosphate, ribulose-1,5-bisphosphate, yielding two molecules of the 3-carbon compound, 3-phosphoglycerate.
Spirogyra	Spirogyra is a genus of filamentous green algae of the order Zygnematales and there are more than 400 species of Spirogyra in the world. Spirogyra measures approximately 10 to 100µm in width and may stretch centimeters long. This particular algal species, commonly found in polluted water, is often referred to as 'pond scum'.
Caulerpa taxifolia	Caulerpa taxifolia is a species of seaweed, an alga of the genus Caulerpa. Native to the Indian Ocean, it has been widely used ornamentally in aquariums. The alga has a stem which spreads horizontally just above the seafloor, and from this stem grow vertical fern-like pinnae, whose blades are flat like yew, hence the species name 'taxifolia' (the genus of yew is 'Taxus').
Porphyra	Porphyra is a foliose red algal genus of laver, comprising approximately 70 species. It grows in the intertidal zone, typically between the upper intertidal zone and the splash zone in cold waters of temperate oceans. In East Asia, it is used to produce the sea vegetable products nori and gim (in Korea), the most commonly eaten seaweed.
Volvox	Volvox is a genus of chlorophytes, a type of green algae. It forms spherical colonies of up to 50,000 cells. They live in a variety of freshwater habitats, and were first reported by Antonie van Leeuwenhoek in 1700. Volvox developed its colonial lifestyle 200 million years ago.

Chapter 4. Part 4: Microbial Diversity and Ecology

Coral bleaching	Coral bleaching is the loss of intracellular endosymbionts (Symbiodinium, also known as zooxanthellae) through either expulsion or loss of algal pigmentation. The corals that form the structure of the great reef ecosystems of tropical seas depend upon a symbiotic relationship with unicellular flagellate protozoa that are photosynthetic and live within their tissues. Zooxanthellae give coral its coloration, with the specific color depending on the particular clade.
Coral reef	Coral reefs are underwater structures made from calcium carbonate secreted by corals. Coral reefs are colonies of tiny living animals found in marine waters that contain few nutrients. Most coral reefs are built from stony corals, which in turn consist of polyps that cluster in groups.
Frustule	A frustule is the hard and porous cell wall or external layer of diatoms. The frustule is composed almost purely of silica, made from silicic acid, and is coated with a layer of organic substance, which was referred to in the early literature on diatoms as pectin, a fiber most commonly found in cell walls of plants. This layer is actually composed of several types of polysaccharides.
Brown algae	The Phaeophyceae or brown algae, is a large group of mostly marine multicellular algae, including many seaweeds of colder Northern Hemisphere waters. They play an important role in marine environments, both as food and for the habitats they form. For instance Macrocystis, a kelp of the order Laminariales, may reach 60 m in length, and forms prominent underwater forests.
Kelp forest	Kelp forests are underwater areas with a high density of kelp. They are recognized as one of the most productive and dynamic ecosystems on Earth. Smaller areas of anchored kelp are called kelp beds.
Acanthamoeba	Acanthamoeba is a genus of amoebae, one of the most common protozoa in soil, and also frequently found in fresh water and other habitats. The cells are small, usually 15 to 35 μm in length and oval to triangular in shape when moving. Cysts are common.
Dictyostelium discoideum	Dictyostelium discoideum is a species of soil-living amoeba belonging to the phylum Mycetozoa. D. discoideum, commonly referred to as slime mold, is a eukaryote that transitions from a collection of unicellular amoebae into a multicellular slug and then into a fruiting body within its lifetime. D. discoideum has a unique asexual lifecycle that consists of four stages: vegetative, aggregation, migration, and culmination.
Radiolarian	Radiolarians (also Radiolaria) are amoeboid protozoa (diameter 0.1-0.2 mm) that produce intricate mineral skeletons, typically with a central capsule dividing the cell into inner and outer portions, called endoplasm and ectoplasm. They are found as zooplankton throughout the ocean, and their skeletal remains cover large portions of the ocean bottom as radiolarian ooze.

Contractile vacuole	A contractile vacuole is a sub-cellular structure (organelle) involved in osmoregulation. It is found predominantly in protists and in unicellular algae. It was previously known as pulsatile or pulsating vacuole.
Telomere	A telomere is a region of repetitive nucleotide sequences at the end of a chromosome, which protects the end of the chromosome from deterioration or from fusion with neighboring chromosomes. Its name is derived from the Greek nouns telos 'end' and meros 'part.' Telomere regions deter the degradation of genes near the ends of chromosomes by allowing chromosome ends to shorten, which necessarily occurs during chromosome replication. Over time, due to each cell division, the telomere ends become shorter.
Sea anemone	Sea Anemones are a group of water dwelling, predatory animals of the order Actiniaria; they are named after the anemone, a terrestrial flower. Sea Anemones are classified in the phylum Cnidaria, class Anthozoa, subclass Zoantharia. Anthozoa often have large polyps that allow for digestion of larger prey and also lack a medusa stage.
Toxoplasma gondii	Toxoplasma gondii is a species of parasitic protozoa in the genus Toxoplasma. The definitive host of T. gondii is the cat, but the parasite can be carried by many warm-blooded animals (birds or mammals, including humans). Toxoplasmosis, the disease of which T. gondii is the causative agent, is usually minor and self-limiting but can have serious or even fatal effects on a fetus whose mother first contracts the disease during pregnancy or on an immunocompromised human or cat.
Zooxanthella	Zooxanthellae ( ) are flagellate protozoa from the genus Symbiodinium that are golden-brown intracellular endosymbionts of various marine animals and protozoa, especially anthozoans such as the scleractinian corals and the tropical sea anemone, Aiptasia. Zooxanthellae live in other protozoa (foraminiferans and radiolarians) and in some invertebrates. Most are autotrophs and provide the host with energy in the form of translocated reduced carbon compounds, such as glucose, glycerol, and amino acids, which are the products of photosynthesis.
Plasmodium falciparum	Plasmodium falciparum is a protozoan parasite, one of the species of Plasmodium that cause malaria in humans. It is transmitted by the female Anopheles mosquito. P. falciparum is the most dangerous of these infections as P. falciparum (or malignant) malaria has the highest rates of complications and mortality.
Malaria	Malaria is a mosquito-borne infectious disease of humans and other animals caused by eukaryotic protists of the genus Plasmodium. The disease results from the multiplication of Plasmodium parasites within red blood cells, causing symptoms that typically include fever and headache, in severe cases progressing to coma or death.

Chapter 4. Part 4: Microbial Diversity and Ecology

Kinetoplast	A kinetoplast is a network of circular DNA (called kDNA) inside a large mitochondrion that contains many copies of the mitochondrial genome. The most common kinetoplast structure is that of a disk, but has been observed in other arrangements. Kinetoplasts are only found in protozoa of the class Kinetoplastida.
Giardia lamblia	Giardia lamblia is a flagellated protozoan parasite that colonizes and reproduces in the small intestine, causing giardiasis. The giardia parasite attaches to the epithelium by a ventral adhesive disc, and reproduces via binary fission. Giardiasis does not spread via the bloodstream, nor does it spread to other parts of the gastro-intestinal tract, but remains confined to the lumen of the small intestine.
Leishmania major	Leishmania major is a species of Leishmania, associated with zoonotic cutaneous leishmaniasis. Its genome has been sequenced.
Leishmaniasis	Leishmaniasis is a disease caused by protozoan parasites that belong to the genus Leishmania and is transmitted by the bite of certain species of sand fly (subfamily Phlebotominae). Although the majority of the literature mentions only one genus transmitting Leishmania to humans (Lutzomyia) in America, a 2003 study by Galati suggested a new classification for American sand flies, elevating several subgenera to the genus level. Elsewhere in the world, the genus Phlebotomus is considered the vector of leishmaniasis.
Trypanosoma cruzi	Trypanosoma cruzi is a species of parasitic euglenoid trypanosomes. This species causes the trypanosomiasis diseases in humans and animals in America. Transmission occurs when the reduviid bug deposits feces on the skin surface and subsequently bites; the human host then scratches the bite area, which facilitates penetration of the infected feces. Human American trypanosomiasis, or Chagas disease, is a potentially fatal disease of humans.
Microbial metabolism	Microbial metabolism is the means by which a microbe obtains the energy and nutrients (e.g. carbon) it needs to live and reproduce. Microbes use many different types of metabolic strategies and species can often be differentiated from each other based on metabolic characteristics. The specific metabolic properties of a microbe are the major factors in determining that microbe's ecological niche, and often allow for that microbe to be useful in industrial processes or responsible for biogeochemical cycles.
Protein	Proteins are biochemical compounds consisting of one or more polypeptides typically folded into a globular or fibrous form, facilitating a biological function. A polypeptide is a single linear polymer chain of amino acids bonded together by peptide bonds between the carboxyl and amino groups of adjacent amino acid residues.

Desulfovibrio	Desulfovibrio is a genus of Gram negative sulfate-reducing bacteria. Some species of Desulfovibrio are capable of transduction. Desulfovibrio species are commonly found in aquatic environments with high levels of organic material, as well as in water-logged soils, and form major community members of extreme oligotrophic habitats such as deep granitic fractured rock aquifers.
Copepod	Copepods (; meaning 'oar-feet') are a group of small crustaceans found in the sea and nearly every freshwater habitat. Some species are planktonic (drifting in sea waters), some are benthic (living on the ocean floor), and some continental species may live in limno-terrestrial habitats and other wet terrestrial places, such as swamps, under leaf fall in wet forests, bogs, springs, ephemeral ponds and puddles, damp moss, or water-filled recesses (phytotelmata) of plants such as bromeliads and pitcher plants. Many live underground in marine and freshwater caves, sinkholes, or stream beds.
Vibrio cholerae	Vibrio cholerae is a gram negative comma-shaped bacterium with a polar flagellum that causes cholera in humans. V. cholerae and other species of the genus Vibrio belong to the gamma subdivision of the Proteobacteria. There are two major biotypes of V. cholerae identified by hemagglutination testing, classical and El Tor, and numerous serogroups.
Assimilation	Biological assimilation, involves one of two different processes to supply animal cells with nutrients. The first is the process of absorbing vitamins, minerals, and other chemicals from food within the gastrointestinal tract. In humans this is done with a chemical breakdown (enzymes and acids) and physical breakdown (oral mastication and stomach churning).
Carbon cycle	The carbon cycle is the biogeochemical cycle by which carbon is exchanged among the biosphere, pedosphere, geosphere, hydrosphere, and atmosphere of the Earth. It is one of the most important cycles of the Earth and allows for carbon to be recycled and reused throughout the biosphere and all of its organisms. The global carbon budget is the balance of the exchanges (incomes and losses) of carbon between the carbon reservoirs or between one specific loop (e.g., atmosphere ↔ biosphere) of the carbon cycle.
Dissimilation	In phonology, particularly within historical linguistics, dissimilation is a phenomenon whereby similar consonant or vowel sounds in a word become less similar. For example, when one sound occurs before another in the middle of a word in rhotic dialects of English, the first tends to drop out, as in 'beserk' for berserk, 'supprise' for surprise, 'paticular' for particular, and 'govenor' for governor . There are several hypotheses as to what causes dissimilation.

Chapter 4. Part 4: Microbial Diversity and Ecology

Ecosystem	An ecosystem is a biological system consisting of all the living organisms or biotic components in a particular area and the nonliving or abiotic components with which the organisms interact, such as air, mineral soil, water, and sunlight. Key processes in ecosystems include the capture of light energy and carbon through photosynthesis, the transfer of carbon and energy through food webs, and the release of nutrients and carbon through decomposition. Biodiversity affects ecosystem functioning, as do the processes of disturbance and succession.
Consumers	Consumers are organisms of an ecological food chain that receive their energy by consuming other organisms. These organisms are formally referred to as heterotrophs, which includes animals, bacteria and fungus. Classification Consumers are typically viewed as predatory animals such as the wolf and hyena.
Decomposer	Decomposers are organisms that break down dead or decaying organisms, and in doing so carry out the natural process of decomposition. Like herbivores and predators, decomposers are heterotrophic, meaning that they use organic substrates to get their energy, carbon and nutrients for growth and development. Decomposers can break down cells of other organisms using biochemical reactions that convert the prey tissue into metabolically useful chemical products, without need for internal digestion.
Extremophile	An extremophile (from Latin extremus meaning 'extreme' and Greek philia meaning 'love') is an organism that thrives in physically or geochemically extreme conditions that are detrimental to most life on Earth. In contrast, organisms that live in more moderate environments may be termed mesophiles or neutrophiles. The category name is unfortunate as it calls for subjective judgements of two issues - firstly, the degree of deviation from 'normal' justifying the use of 'extreme', and secondly, whether the organism prefers the environment or merely tolerates it.
Detritus	In biology, detritus is non-living particulate organic material (as opposed to dissolved organic material). It typically includes the bodies or fragments of dead organisms as well as fecal material. Detritus is typically colonized by communities of microorganisms which act to decompose the material.
Electron acceptor	An electron acceptor is a chemical entity that accepts electrons transferred to it from another compound. It is an oxidizing agent that, by virtue of its accepting electrons, is itself reduced in the process. Typical oxidizing agents undergo permanent chemical alteration through covalent or ionic reaction chemistry, resulting in the complete and irreversible transfer of one or more electrons.

Lyme disease	Lyme disease, is an emerging infectious disease caused by at least three species of bacteria belonging to the genus Borrelia. Borrelia burgdorferi sensu stricto is the main cause of Lyme disease in the United States, whereas Borrelia afzelii and Borrelia garinii cause most European cases. he town of Lyme, Connecticut, USA, where a number of cases were identified in 1975. Although Allen Steere realized that Lyme disease was a tick-borne disease in 1978, the cause of the disease remained a mystery until 1981, when B. burgdorferi was identified by Willy Burgdorfer.
Commensalism	In ecology, commensalism is a class of relationship between two organisms where one organism benefits but the other is neutral (there is no harm or benefit). There are two other types of association: mutualism (where both organisms benefit) and parasitism (one organism benefits and the other one is harmed). Commensalism derives from the English word commensal, meaning 'sharing of food' in human social interaction, which in turn derives from the Latin cum mensa, meaning 'sharing a table'.
Neuston	Neuston is the collective term for the organisms that float on the top of water (epineuston) or live right under the surface (hyponeuston). Neustons are made up of some species of fish , beetles , protozoans, bacteria and spiders . A water strider is a common example that skips across water's surface tension.
Pelagic zone	Any water in a sea or lake that is not close to the bottom or near to the shore can be said to be in the pelagic zone. The word pelagic comes from the Greek π?λαγος or pélagos, which means 'open sea'. The pelagic zone can be thought of in terms of an imaginary cylinder or water column that goes from the surface of the sea almost to the bottom.
Gas gangrene	Gas gangrene is a bacterial infection that produces gas tissues in gangrene. It is a deadly form of gangrene usually caused by Clostridium perfringens bacteria. It is a medical emergency.
Oil spill	An oil spill is a release of a liquid petroleum hydrocarbon into the environment due to human activity, and is a form of pollution. The term often refers to marine oil spills, where oil is released into the ocean or coastal waters. Oil spills include releases of crude oil from tankers, offshore platforms, drilling rigs and wells, as well as spills of refined petroleum products (such as gasoline, diesel) and their by-products, and heavier fuels used by large ships such as bunker fuel, or the spill of any oily white substance refuse or waste oil.
Thermocline	A thermocline is a thin but distinct layer in a large body of fluid (e.g. water, such as an ocean or lake, or air, such as an atmosphere), in which temperature changes more rapidly with depth than it does in the layers above or below. In the ocean, the thermocline may be thought of as an invisible blanket which separates the upper mixed layer from the calm deep water below.

Chapter 4. Part 4: Microbial Diversity and Ecology

Picoplankton	Picoplankton is the fraction of plankton composed by cells between 0.2 and 2 μm that can be either :•photosynthetic •heterotrophic

Some species can also be mixotrophic. Picoplankton are responsible for the most primary productivity in oligotrophic gyres, and are distinguished from nanoplankton and microplankton. Because they are small, they have a greater surface to volume ratio, enabling them to obtain the scarce nutrients in these ecosystems. |
| Marine snow | In the deep ocean, marine snow is a continuous shower of mostly organic detritus falling from the upper layers of the water column. It is a significant means of exporting energy from the light-rich photic zone to the aphotic zone below. The term was first coined by the explorer William Beebe as he observed it from his bathysphere. |
| MHC restriction | MHC-restricted antigen recognition, or MHC restriction, refers to the fact that a given T cell will recognize a peptide antigen only when it is bound to a host body's own MHC molecule. Normally, as T cells are stimulated only in the presence of self-MHC molecules, antigen is recognized only as peptides bound to self-MHC molecules.

MHC restriction is particularly important when primary lymphocytes are developing and differentiating in the thymus or bone marrow. |
Psychrophile	Psychrophiles or cryophiles (adj. cryophilic) are extremophilic organisms that are capable of growth and reproduction in cold temperatures, ranging from −15°C to +10°C. Temperatures as low as −15°C are found in pockets of very salty water (brine) surrounded by sea ice. They can be contrasted with thermophiles, which thrive at unusually hot temperatures.
Drug resistance	Drug resistance is the reduction in effectiveness of a drug such as an antimicrobial or an antineoplastic in curing a disease or condition. When the drug is not intended to kill or inhibit a pathogen, then the term is equivalent to dosage failure or drug tolerance. More commonly, the term is used in the context of resistance acquired by pathogens.
Littoral zone	The littoral zone is the part of a sea, lake or river that is close to the shore. In coastal environments the littoral zone extends from the high water mark, which is rarely inundated, to shoreline areas that are permanently submerged. It always includes this intertidal zone and is often used to mean the same as the intertidal zone.
Epilimnion	The epilimnion is the top-most layer in a thermally stratified lake, occurring above the deeper hypolimnion. It is warmer and typically has a higher pH and high dissolved oxygen concentration than the hypolimnion.

Eutrophication	Eutrophication, is the ecosystem response to the addition of artificial or natural substances, such as nitrates and phosphates, through fertilizers or sewage, to an aquatic system. One example is the 'bloom' or great increase of phytoplankton in a water body as a response to increased levels of nutrients. Negative environmental effects include hypoxia, the depletion of oxygen in the water, which induces reductions in specific fish and other animal populations.
Soil microbiology	Soil microbiology is the study of organisms in soil, their functions, and how they affect soil properties. It is believed that between two to four billion years ago, the first ancient bacteria and microorganisms came about in Earth's primitive seas. These bacteria could fix nitrogen, in time multiplied and as a result released oxygen into the atmosphere.
Rhizosphere	The rhizosphere is the narrow region of soil that is directly influenced by root secretions and associated soil microorganisms. Soil which is not part of the rhizosphere is known as bulk soil. The rhizosphere contains many bacteria that feed on sloughed-off plant cells, termed rhizodeposition, and the proteins and sugars released by roots.
Radioactive decay	Radioactive decay is the process by which an atomic nucleus of an unstable atom loses energy by emitting ionizing particles (ionizing radiation). The emission is spontaneous, in that the atom decays without any interaction with another particle from outside the atom (i.e., without a nuclear reaction). Usually, radioactive decay happens due to a process confined to the nucleus of the unstable atom, but, on occasion (as with the different processes of electron capture and internal conversion), an inner electron of the radioactive atom is also necessary to the process.
Ribosomal protein	Mitochondrial ribosomal protein L31 A ribosomal protein is any of the proteins that, in conjunction with rRNA, make up the ribosomal subunits involved in the cellular process of translation. A large part of the knowledge about these organic molecules has come from the study of E. coli ribosomes. Most ribosomic proteins have been isolated and specific anti-bodies have been produced.
Soil food web	The soil food web is the community of organisms living all or part of their lives in the soil. It describes a complex living system in the soil and how it interacts with the environment, plants, and animals. Food webs describe the transfer of energy between species in an ecosystem.
Lignin	Lignin is a complex chemical compound most commonly derived from wood, and an integral part of the secondary cell walls of plants and some algae. The term was introduced in 1819 by de Candolle and is derived from the Latin word lignum, meaning wood.

Chapter 4. Part 4: Microbial Diversity and Ecology

Hydric soil	A hydric soil is a soil that formed under conditions of saturation, flooding, or ponding long enough during the growing season to develop anaerobic conditions in the upper part. This term is part of the legal definition of a wetland included in the US Food Security Act of 1985 (P.L. 99-198). The US Natural Resources Conservation Service maintains the official list of hydric soils.
Atmospheric methane	Atmospheric methane levels are of interest due to its impact on climate change. Atmospheric methane is one of the most potent and influential greenhouse gases on Earth. The 100-year global warming potential of methane is 25, i.e. it traps 25 times more heat per mass unit than carbon dioxide.
Flavonoid	Flavonoids , are a class of plant secondary metabolites. Flavonoids were referred to as Vitamin P (probably due to the effect they had on the permeability of vascular capillaries) from the mid-1930s to early 50s, but the term has since fallen out of use. According to the IUPAC nomenclature, they can be classified into:•flavonoids, derived from 2-phenylchromen-4-one (2-phenyl-1,4-benzopyrone) structure (examples: quercetin, rutin).•isoflavonoids, derived from 3-phenylchromen-4-one (3-phenyl-1,4-benzopyrone) structure•neoflavonoids, derived from 4-phenylcoumarine (4-phenyl-1,2-benzopyrone) structure. The three flavonoid classes above are all ketone-containing compounds, and as such, are flavonoids and flavonols.
Haber process	The Haber process, is the nitrogen fixation reaction of nitrogen gas and hydrogen gas, over an enriched iron or ruthenium catalyst, which is used to industrially produce ammonia. Despite the fact that 78.1% of the air we breathe is nitrogen, the gas is relatively unavailable because it is so unreactive: nitrogen molecules are held together by strong triple bonds. It was not until the early 20th century that the Haber process was developed to harness the atmospheric abundance of nitrogen to create ammonia, which can then be oxidized to make the nitrates and nitrites essential for the production of nitrate fertilizer and explosives.
Neotyphodium coenophialum	Neotyphodium coenophialum is a systemic and seed-transmissible symbiont (endophyte) of Schedonorus arundinaceus (=Festuca arundinacea; tall fescue), a grass endemic to Eurasia and North Africa, but widely naturalized in North America, Australia and New Zealand / Aotearoa. The endophyte has been identified as the cause of the 'fescue toxicosis' syndrome sometimes suffered by livestock that graze the N. coenophialum-infected grass.

Rhizobia	Rhizobia are soil bacteria that fix nitrogen (diazotrophs) after becoming established inside root nodules of legumes (Fabaceae). Rhizobia require a plant host; they cannot independently fix nitrogen. In general, they are Gram-negative, motile, non-sporulating rods.
Salmonella enterica	Salmonella enterica is a rod-shaped flagellated, facultative anaerobic, Gram-negative bacterium, and a member of the genus Salmonella.
	Most cases of salmonellosis are caused by food infected with S. enterica, which often infects cattle and poultry, though also other animals such as domestic cats and hamsters have also been shown to be sources of infection to humans. However, investigations of vacuum cleaner bags have shown that households can act as a reservoir of the bacterium; this is more likely if the household has contact with an infection source, for example members working with cattle or in a veterinary clinic.
Soybean	The soybean. or soya bean (UK) (Glycine max) is a species of legume native to East Asia, widely grown for its edible bean which has numerous uses. The plant is classed as an oilseed rather than a pulse by the Food and Agricultural Organization (FAO).
Stenotrophomonas	Stenotrophomonas is a genus of Gram-negative bacteria. With species ranging from common soil organisms (S. nitritireducens) to opportunistic human pathogens (S. maltophilia), the molecular taxonomy of the genus is still somewhat unclear.
B cell	B cells are lymphocytes that play a large role in the humoral immune response (as opposed to the cell-mediated immune response, which is governed by T cells). B cells are an essential component of the adaptive immune system. Their principal functions are to make antibodies against antigens, perform the role of antigen-presenting cells (APCs) and eventually develop into memory B cells after activation by antigen interaction.
Fistula	In medicine, a fistula is an abnormal connection or passageway between two epithelium-lined organs or vessels that normally do not connect. It is generally a disease condition, but a fistula may be surgically created for therapeutic reasons.
Prevotella	Prevotella is a genus of bacteria.
	Bacteroides melaninogenicus has recently been reclassified and split into Prevotella melaninogenica and Prevotella intermedia.
	Prevotella spp.
Chlorophyll	Chlorophyll is a green pigment found in almost all plants, algae, and cyanobacteria. Its name is derived from the Greek words χλωρος, chloros ('green') and φ?λλον, phyllon ('leaf').

Chapter 4. Part 4: Microbial Diversity and Ecology

Fossil fuel	Fossil fuels are fuels formed by natural resources such as anaerobic decomposition of buried dead organisms. The age of the organisms and their resulting fossil fuels is typically millions of years, and sometimes exceeds 650 million years. The fossil fuels, which contain high percentages of carbon, include coal, petroleum, and natural gas.
Iron-sulfur cluster	Iron-sulfur clusters are ensembles of iron and sulfide centres. Fe-S clusters are most often discussed in the context of the biological role for iron-sulfur proteins. Many Fe-S clusters are known in the area of organometallic chemistry and as precursors to synthetic analogues of the biological clusters .
Biogeochemistry	Biogeochemistry is the scientific discipline that involves the study of the chemical, physical, geological, and biological processes and reactions that govern the composition of the natural environment (including the biosphere, the hydrosphere, the pedosphere, the atmosphere, and the lithosphere). In particular, biogeochemistry is the study of the cycles of chemical elements, such as carbon and nitrogen, and their interactions with and incorporation into living things transported through earth scale biological systems in space through time. The field focuses on chemical cycles which are either driven by or have an impact on biological activity.
Micronutrient	Micronutrients are nutrients required by humans and other living things throughout life in small quantities to orchestrate a whole range of physiological functions, but which the organism itself cannot produce. For people, they include dietary trace minerals in amounts generally less than 100 milligrams/day - as opposed to macrominerals which are required in larger quantities. The microminerals or trace elements include at least iron, cobalt, chromium, copper, iodine, manganese, selenium, zinc and molybdenum.
Sulfur cycle	The sulfur cycle is the collection of processes by which sulfur moves to and from minerals (including the waterways) and living systems. Such biogeochemical cycles are important in geology because they affect many minerals. Biogeochemical cycles are also important for life because sulfur is an essential element, being a constituent of many proteins and cofactors.
Oxidation state	In chemistry, the oxidation state is an indicator of the degree of oxidation of an atom in a chemical compound. The formal oxidation state is the hypothetical charge that an atom would have if all bonds to atoms of different elements were 100% ionic. Oxidation states are typically represented by integers, which can be positive, negative, or zero.
Mesocosm	A mesocosm is an experimental tool that brings a small part of the natural environment under controlled conditions. In this way mesocosms provide a link between observational field studies that take place in natural environments, but without replication, and controlled laboratory experiments that may take place under somewhat unnatural conditions .

| Stable isotope | Stable isotopes are chemical isotopes that may or may not be radioactive, but if radioactive, have half-lives too long to be measured.

Only 90 nuclides from the first 40 elements are energetically stable to any kind of decay save proton decay, in theory . An additional 165 are theoretically unstable to known types of decay, but no evidence of decay has ever been observed, for a total of 255 nuclides for which there is no evidence of radioactivity. |
|---|---|
| Reservoir | A reservoir is used to store water. Reservoirs may be created in river valleys by the construction of a dam or may be built by excavation in the ground or by conventional construction techniques such a brickwork or cast concrete.

The term reservoir may also be used to describe underground reservoirs such as an oil or water well. |
| Deforestation | Deforestation is the removal of a forest or stand of trees where the land is thereafter converted to a nonforest use. Examples of deforestation include conversion of forestland to farms, ranches, or urban use.

The term deforestation is often misused to describe any activity where all trees in an area are removed. |
Nitrosopumilus	Nitrosopumilus maritimus is an extremely common archaeon living in seawater. It is the first member of the Group 1a Crenarchaeota to be isolated in pure culture. Gene sequences suggest that the Group 1a Crenarchaeota are ubiquitous with the oligotrophic surface ocean and can be found in most non-coastal marine waters around the planet.
Carbon sink	A carbon sink is a natural or artificial reservoir that accumulates and stores some carbon-containing chemical compound for an indefinite period. The process by which carbon sinks remove carbon dioxide (CO_2) from the atmosphere is known as carbon sequestration. Public awareness of the significance of CO_2 sinks has grown since passage of the Kyoto Protocol, which promotes their use as a form of carbon offset.
Water cycle	The water cycle, describes the continuous movement of water on, above and below the surface of the Earth. Water can change states among liquid, vapor, and ice at various places in the water cycle. Although the balance of water on Earth remains fairly constant over time, individual water molecules can come and go, in and out of the atmosphere.
Dead zone	Dead zones are hypoxic (low-oxygen) areas in the world's oceans, the observed incidences of which have been increasing since oceanographers began noting them in the 1970s. These occur near inhabited coastlines, where aquatic life is most concentrated.

Chapter 4. Part 4: Microbial Diversity and Ecology

Hypoxia	Hypoxia, is a phenomenon that occurs in aquatic environments as dissolved oxygen (DO; molecular oxygen dissolved in the water) becomes reduced in concentration to a point where it becomes detrimental to aquatic organisms living in the system. Dissolved oxygen is typically expressed as a percentage of the oxygen that would dissolve in the water at the prevailing temperature and salinity . An aquatic system lacking dissolved oxygen (0% saturation) is termed anaerobic, reducing, or anoxic; a system with low concentration--in the range between 1 and 30% saturation--is called hypoxic or dysoxic.
Sludge	Sludge refers to the residual, semi-solid material left from industrial wastewater, or sewage treatment processes. It can also refer to the settled suspension obtained from conventional drinking water treatment, and numerous other industrial processes. The term is also sometimes used as a generic term for solids separated from suspension in a liquid; this 'soupy' material usually contains significant quantities of 'interstitial' water (between the solid particles).
Treatment wetland	A Treatment wetland is an engineered sequence of water bodies designed to filter and treat waterborne pollutants found in storm water runoff or effluent.
	In treatment wetlands aerobic and anaerobic biological processes can neutralize and capture most of the dissolved nutrients and toxic pollutants from the water, resulting in the discharge of clean water. Types
	Types of treatment wetlands include:•Subsurface wetlands•Surface wetlands•Sewage treatment - tertiary, secondary, and primary treatmentBest management practices
	Many regulatory agencies list treatment wetlands as one of their recommended 'best management practices' for controlling urban runoff.
Nitrification	Nitrification is the biological oxidation of ammonia with oxygen into nitrite followed by the oxidation of these nitrites into nitrates. Degradation of ammonia to nitrite is usually the rate limiting step of nitrification. Nitrification is an important step in the nitrogen cycle in soil.
Nitrogenase	Nitrogenases (EC 1.18.6.1EC 1.19.6.1) are enzymes used by some organisms to fix atmospheric nitrogen gas (N_2). It is the only known family of enzymes that accomplish this process. Dinitrogen is quite inert because of the strength of its N-N triple bond.
Methemoglobinemia	Methemoglobinemia is a disorder characterized by the presence of a higher than normal level of methemoglobin (metHb) in the blood. Methemoglobin is an oxidized form of hemoglobin that has almost no affinity for oxygen, resulting in almost no oxygen delivery to the tissues. When its concentration is elevated in red blood cells, tissue hypoxia can occur.
Nitrobacter	Nitrobacter is genus of mostly rod-shaped, gram-negative, and chemoautotrophic bacteria.

Nitrobacter plays an important role in the nitrogen cycle by oxidizing nitrite into nitrate in soil. Unlike plants, where electron transfer in photosynthesis provides the energy for carbon fixation, Nitrobacter use energy from the oxidation of nitrite ions, NO_2^-, into nitrate ions, NO_3^- to fulfill their carbon requirements.

Denitrification	Denitrification is a microbially facilitated process of nitrate reduction that may ultimately produce molecular nitrogen (N_2) through a series of intermediate gaseous nitrogen oxide products.

This respiratory process reduces oxidized forms of nitrogen in response to the oxidation of an electron donor such as organic matter. The preferred nitrogen electron acceptors in order of most to least thermodynamically favorable include nitrate (NO_3^-), nitrite (NO_2^-), nitric oxide (NO), and nitrous oxide (N_2O)and dinitrigen [N2].

Nitrous oxide

Nitrous oxide, commonly known as laughing gas or sweet air, is a chemical compound with the formula N_2O. It is an oxide of nitrogen. At room temperature, it is a colorless non-flammable gas, with a slightly sweet odor and taste.

Iron cycle

In ecology or geoscience, the iron cycle is the biogeochemical cycle of iron through landforms, the atmosphere, and oceans. The iron cycle affects dust deposition and aerosol iron bioavailability..

Siderophore

Siderophores (compound from the Ancient Greek nouns síderos and phoros (φορος) meaning 'iron carrier') are small, high-affinity iron chelating compounds secreted by grasses and microorganisms such as bacteria and fungi. Siderophores are amongst the strongest soluble Fe^{3+} binding agents known.

Iron is essential for almost all life, essential for processes such as respiration and DNA synthesis.

Iron fertilization

Iron fertilization is the intentional introduction of iron to the upper ocean to stimulate a phytoplankton bloom. This is intended to enhance biological productivity, which can benefit the marine food chain and remove carbon dioxide from the atmosphere. Iron is a trace element necessary for photosynthesis in all plants.

Astrobiology

Astrobiology is the study of the origin, evolution, distribution, and future of extraterrestrial life.

Chapter 4. Part 4: Microbial Diversity and Ecology

This interdisciplinary field encompasses the search for habitable environments in our Solar System and habitable planets outside our Solar System, the search for evidence of prebiotic chemistry, laboratory and field research into the origins and early evolution of life on Earth, and studies of the potential for life to adapt to challenges on Earth and in outer space. Astrobiology addresses the question of whether life exists beyond Earth, and how humans can detect it if it does.

Extrasolar planet

An extrasolar planet, is a planet outside the Solar System. A total of 770 such planets (in 616 planetary systems and 102 multiple planetary systems) have been identified as of May 28, 2012. Estimates of the frequency of systems strongly suggest that more than 50% of Sun-like stars harbor at least one planet. In a 2012 study, each star of the 100 billion or so in our Milky Way galaxy is estimated to host 'on average ... at least 1.6 planets.' Accordingly, at least 160 billion star-bound planets may exist in the Milky Way Galaxy alone.

Terraforming

Terraforming of a planet, moon, or other body is the hypothetical process of deliberately modifying its atmosphere, temperature, surface topography or ecology to be similar to the biosphere of Earth, in order to make it habitable by humans.

The term is sometimes used more generally as a synonym for planetary engineering, although some consider this more general usage an error. The concept of terraforming developed from both science fiction and actual science.

1. _____ is a genus of filamentous green algae of the order Zygnematales and there are more than 400 species of _____ in the world. _____ measures approximately 10 to 100μm in width and may stretch centimeters long. This particular algal species, commonly found in polluted water, is often referred to as 'pond scum'.

   a. Spirogyra
   b. S/MARt
   c. Sarcomere
   d. Secondary cell wall

2. . _____ is a genus of bacteria that is commonly found in soil. All species in this genus are Gram-positive obligate aerobes that are rods during exponential growth and cocci in their stationary phase.

   Colonies of _____ have a greenish metallic center on mineral salts pyridone broth incubated at 20°C. This genus is distinctive because of its unusual habit of 'snapping division' in which the outer bacterial cell wall ruptures at a joint (hence its name).

   a. Arthrobacter

b. Atopobium

c. Edwardsiella

d. Ehrlichia ewingii

3.

_____s are ensembles of iron and sulfide centres. Fe-S clusters are most often discussed in the context of the biological role for iron-sulfur proteins. Many Fe-S clusters are known in the area of organometallic chemistry and as precursors to synthetic analogues of the biological clusters .

a. Iron-sulfur protein

b. Octahedral cluster

c. Iron-sulfur cluster

d. Low-carbon economy

4. _____ is a genus of Gram-negative, non-motile and motile, rod-shaped bacteria that consists of ten recognized species, as well as three newly proposed species (F. gondwanense, F. salegens, and F. scophthalmum). Flavobacteria are found in soil and fresh water in a variety of environments. Several species are known to cause disease in freshwater fish.

a. Flavobacterium columnare

b. Flexibacter

c. Flavobacterium

d. Myroidaceae

5. _____ is a pathogenic bacterium that causes diphtheria. It is also known as the Klebs-Löffler bacillus, because it was discovered in 1884 by German bacteriologists Edwin Klebs (1834 - 1912) and Friedrich Löffler (1852 - 1915).

Classification

Four subspecies are recognized: C. diphtheriae mitis, C. diphtheriae intermedius, C. diphtheriae gravis, and C. diphtheriae belfanti.

a. Stephen Switzer

b. Fluorine

c. Corynebacterium diphtheriae

d. Human iron metabolism

1. a
2. a
3. c
4. c
5. c

You can take the complete Chapter Practice Test

for Chapter 4. Part 4: Microbial Diversity and Ecology
on all key terms, persons, places, and concepts.

Online 99 Cents

http://www.epub13.5.20451.4.cram101.com/

Use www.Cram101.com for all your study needs

including Cram101's online interactive problem solving labs in

chemistry, statistics, mathematics, and more.

Chapter 5. Part 5: Medicine and Immunology

CHAPTER OUTLINE: KEY TERMS, PEOPLE, PLACES, CONCEPTS

Immunoprecipitation

Salmonella

Clostridium difficile

Nitric oxide

Borrelia burgdorferi

Toll-like receptor

Confocal microscopy

Microbiome

Microbiota

Vibrio vulnificus

Acne

Epidermis

Propionibacterium acnes

Sebaceous gland

Staphylococcus epidermidis

Actinomyces

Blackhead

Fusobacterium

Keratin

	Moraxella catarrhalis
	Nasopharynx
	Oropharynx
	Prevotella
	Streptococcus oralis
	Streptococcus salivarius
	Tonsil
	Bacteremia
	Cholera
	Mitral valve
	Respiratory tract
	Vibrio cholerae
	Cholera toxin
	Cholera vaccine
	Vaccine trial
	Vegetation
	Helicobacter pylori
	Bacteroidetes
	Candida albicans

Clostridium

Enterobacteriaceae

Lactobacillus

Peptostreptococcus

Probiotic

Biosynthesis

Gas gangrene

Enterococcus

Urinary tract infection

Vaginal flora

Bacteroides fragilis

Cytokine

Enterotoxin

Lactococcus

Cell-mediated immunity

Dendritic cell

Gnotobiotic animal

Immune system

Mast cell

Chapter 5. Part 5: Medicine and Immunology

CHAPTER OUTLINE: KEY TERMS, PEOPLE, PLACES, CONCEPTS

- Neisseria gonorrhoeae
- Antibiotic resistance
- Blood cell
- Stem cell
- Macrophage
- Monocyte
- Phagosome
- Antigen-presenting cell
- B cell
- Lymphocyte
- Spleen
- T cell
- Gut-associated lymphoid tissue
- Lymph node
- Langerhans cell
- M cell
- Mucous membrane
- Infectious disease
- Alveolar macrophage

_____ Cystic fibrosis

_____ Cystic fibrosis transmembrane conductance regulator

_____ Defensin

_____ Pseudomonas aeruginosa

_____ Lung

_____ Salmonella enterica

_____ Extravasation

_____ Interleukin

_____ Prostaglandin

_____ Selectin

_____ Staphylococcus aureus

_____ Aspirin

_____ Bradykinin

_____ Histamine

_____ Chemokine

_____ Granulocyte

_____ Mycobacterium tuberculosis

_____ Granuloma

_____ Phagocytosis

CHAPTER OUTLINE: KEY TERMS, PEOPLE, PLACES, CONCEPTS

Mycobacterium

Streptococcus pneumoniae

Autophagy

Interferon

Major histocompatibility complex

Natural killer cell

Antibody-dependent cell-mediated cytotoxicity

Perforin

Anaphylatoxin

Lectin

Opsonin

C-reactive protein

Cardiovascular diseases

Severe combined immunodeficiency

Combined immunodeficiencies

Antibodies

Epitope

Immunogenicity

Smallpox

_____ | Bubonic plague _____

_____ | ABO blood group system _____

_____ | Cowpox _____

_____ | Vaccination _____

_____ | Polio vaccine _____

_____ | Smallpox vaccine _____

_____ | Hapten _____

_____ | Heavy chain _____

_____ | Immunoglobulin superfamily _____

_____ | ATP synthase _____

_____ | Pepsin _____

_____ | Idiotype _____

_____ | Anaphylaxis _____

_____ | B-cell receptor _____

_____ | Memory B cell _____

_____ | Plasma cell _____

_____ | Serum _____

_____ | Clonal selection _____

_____ | Recombination signal sequences _____

Chapter 5. Part 5: Medicine and Immunology

CHAPTER OUTLINE: KEY TERMS, PEOPLE, PLACES, CONCEPTS

- Cytotoxic T cell
- MHC restriction
- T cell receptor
- Superantigen
- Granzyme
- Cellular differentiation
- Yersinia enterocolitica
- Yersinia pestis
- Decay-accelerating factor
- Factor H
- Adenylate cyclase
- Histamine antagonist
- Edema
- Type IV hypersensitivity
- Vaccine
- Contact dermatitis
- Type II hypersensitivity
- Regulatory T cell
- Autoimmune disease

Autoimmunity

Rheumatic fever

Pathogenesis

Rickettsia rickettsii

Rocky Mountain spotted fever

Corynebacterium diphtheriae

Filariasis

Infection

Pathogen

Viral pathogenesis

Wuchereria bancrofti

Diphtheria

Epstein-Barr virus

Herpes simplex

Marburg virus

Pathogenicity

Pneumocystis jirovecii

Rickettsia prowazekii

Shigella flexneri

	Virulence
	Infectious dose
	Pathogenicity island
	Bacillus thuringiensis
	Baculovirus
	Fomite
	Horizontal transmission
	Reservoir
	West Nile virus
	Yellow fever
	Equine encephalitis
	Insecticide
	Transovarial transmission
	Vertical transmission
	Rotavirus
	Virulence factor
	Genomic island
	Streptococcus agalactiae
	Clostridium tetani

Neisseria meningitidis

Pertactin

Biofilm

Group B

Membrane protein

Exotoxin

Endotoxin

AB toxin

Shiga toxin

Alpha toxin

Toxin

G factor

Anthrax

Bacillus anthracis

Pertussis toxin

Anthrax toxin

Delta endotoxin

Agrobacterium tumefaciens

Pseudomonas syringae

Chapter 5. Part 5: Medicine and Immunology

CHAPTER OUTLINE: KEY TERMS, PEOPLE, PLACES, CONCEPTS

| | Ti plasmid |

Genetic analysis

Genomics

Legionella pneumophila

Listeria monocytogenes

Q fever

Shigella dysenteriae

Parasitism

Ames test

Holliday junction

Protein A

Phase variation

Site-specific recombination

Quorum sensing

Antigenic shift

Coronavirus

Palmaria palmata

Severe acute respiratory syndrome

Common cold

Molecular biology

Superinfection

Chemokine receptor

Haemophilus influenzae

RNA polymerase

Influenza

Pandemic

Symptom

TATA-binding protein

Cancer vaccine

Human papillomavirus

Protein

Wart

Typhoid fever

Toxic shock syndrome

DNA Research

Francisella tularensis

Livestock

Coagulase

Chapter 5. Part 5: Medicine and Immunology

CHAPTER OUTLINE: KEY TERMS, PEOPLE, PLACES, CONCEPTS

Nosocomial infection

Soft tissue

Gentamicin

Glomerulonephritis

Necrotizing fasciitis

Cryptococcus

Generation time

Measles

Rubella vaccine

Bordetella pertussis

Pneumonia

Bacterial pneumonia

Respiratory tract infection

Macrolide

Disseminated disease

Meningitis

Pneumococcal polysaccharide vaccine

Amphotericin B

Blastomycosis

	Coccidioidomycosis
	Cryptococcosis
	Histoplasmosis
	Conidia
	Ethambutol
	Isoniazid
	Drug resistance
	Entamoeba histolytica
	Slime mold
	Dysentery
	John Snow
	Epidemiology
	Urease
	Cryptosporidiosis
	Cryptosporidium parvum
	Naegleria
	Acanthamoeba
	Giardia lamblia
	Metronidazole

Chapter 5. Part 5: Medicine and Immunology

CHAPTER OUTLINE: KEY TERMS, PEOPLE, PLACES, CONCEPTS

- ______ Lumbar puncture
- ______ Sexually transmitted disease
- ______ Syphilis
- ______ Substrate-level phosphorylation
- ______ Chancre
- ______ Chlamydia
- ______ Chlamydia trachomatis
- ______ Congenital syphilis
- ______ Brugia malayi
- ______ Chocolate agar
- ______ Prevalence
- ______ Trichomoniasis
- ______ Opportunistic infection
- ______ Nervous system
- ______ Viral meningitis
- ______ Clostridium botulinum
- ______ Tetanospasmin
- ______ Tetanus
- ______ Transcytosis

Botulinum toxin

Botulism

Prion

Encephalitozoon intestinalis

Serum albumin

Atherosclerosis

Endocarditis

Lyme disease

Pericarditis

Viremia

Myocarditis

Plasmodium falciparum

Life cycle

Malaria

Red blood cell

Chloroquine

Genome

Pneumonic plague

Septicemic plague

Chapter 5. Part 5: Medicine and Immunology

CHAPTER OUTLINE: KEY TERMS, PEOPLE, PLACES, CONCEPTS

- Borrelia afzelii
- Francisella
- Hepatitis
- Systemic disease
- Hepatitis A
- Hepatitis B
- Hepatitis C
- Ebola virus
- Herd immunity
- Booster dose
- Drug discovery
- Penicillin
- Penicillium
- Folic acid
- Arsenic
- Amp resistance
- Aplastic anemia
- Side effect
- Cell wall

Bacillus licheniformis

Bacillus subtilis

Cycloserine

Gramicidin

Polymyxin

Vancomycin

Cell membrane

Fusobacterium necrophorum

Quinolone

Myxopyronin

Erysipelas

Protein synthesis inhibitor

Aminoglycoside

Chloramphenicol

Streptogramin

Secondary metabolite

Enterococcus faecalis

Integron

Transposable element

Clavulanic acid

Antibiotic tolerance

Platensimycin

Amantadine

Neuraminidase inhibitor

Zanamivir

Hemagglutinin

Neuraminidase

Protease inhibitor

Zidovudine

Clotrimazole

Fluconazole

Griseofulvin

Miconazole

Nystatin

Index case

Fecal-oral route

Osteomyelitis

Chain reaction

Chapter 5. Part 5: Medicine and Immunology
CHAPTER OUTLINE: KEY TERMS, PEOPLE, PLACES, CONCEPTS

Oxidase test

Gram-positive bacteria

Catabolite repression

Restriction fragment

Restriction fragment length polymorphism

Anaerobic respiration

Anaerobic infection

Obligate anaerobe

Blood culture

Disease surveillance

DNA polymerase

Pantoea agglomerans

Nitrogen fixation

Lactose permease

Long terminal repeat

Energy carrier

Glutamate decarboxylase

Haber process

Insertion sequence

CHAPTER OUTLINE: KEY TERMS, PEOPLE, PLACES, CONCEPTS

Tuberculosis

EcoRI

Human genome

Retrotransposon

Downstream processing

Phylogenetic tree

Betaproteobacteria

Amoeba proteus

Marine habitats

Algal bloom

Biochemical oxygen demand

Brown algae

Bacterial toxin

Chapter 5. Part 5: Medicine and Immunology

Immunoprecipitation	Immunoprecipitation is the technique of precipitating a protein antigen out of solution using an antibody that specifically binds to that particular protein. This process can be used to isolate and concentrate a particular protein from a sample containing many thousands of different proteins. Immunoprecipitation requires that the antibody be coupled to a solid substrate at some point in the procedure.
Salmonella	Salmonella is a genus of rod-shaped, Gram-negative, non-spore-forming, predominantly motile enterobacteria with diameters around 0.7 to 1.5 μm, lengths from 2 to 5 μm, and flagella which grade in all directions (i.e. peritrichous). They are chemoorganotrophs, obtaining their energy from oxidation and reduction reactions using organic sources, and are facultative anaerobes. Most species produce hydrogen sulfide, which can readily be detected by growing them on media containing ferrous sulfate, such as TSI. Most isolates exist in two phases: a motile phase I and a nonmotile phase II. Cultures that are nonmotile upon primary culture may be switched to the motile phase using a Cragie tube.
Clostridium difficile	Clostridium difficile (from the Greek kloster , spindle, and Latin difficile, difficult), also known as 'CDF/cdf', or 'C. diff', is a species of Gram-positive bacteria of the genus Clostridium that causes severe diarrhea and other intestinal disease when competing bacteria in the gut flora have been wiped out by antibiotics. Clostridia are anaerobic, spore-forming rods (bacilli). C. difficile is the most serious cause of antibiotic-associated diarrhea (AAD) and can lead to pseudomembranous colitis, a severe inflammation of the colon, often resulting from eradication of the normal gut flora by antibiotics.
Nitric oxide	Nitric oxide, is a molecule with chemical formula NO. It is a free radical and is an important intermediate in the chemical industry. Nitric oxide is a by-product of combustion of substances in the air, as in automobile engines, fossil fuel power plants, and is produced naturally during the electrical discharges of lightning in thunderstorms. In mammals including humans, NO is an important cellular signaling molecule involved in many physiological and pathological processes.
Borrelia burgdorferi	Borrelia burgdorferi is a species of Gram negative bacteria of the spirochete class of the genus Borrelia. B. burgdorferi is predominant in North America, but also exists in Europe, and is the agent of Lyme disease. It is a zoonotic, vector-borne disease transmitted by ticks and is named after the researcher Willy Burgdorfer who first isolated the bacterium in 1982. B.

Toll-like receptor	Toll-like receptors are a class of proteins that play a key role in the innate immune system. They are single, membrane-spanning, non-catalytic receptors that recognize structurally conserved molecules derived from microbes. Once these microbes have breached physical barriers such as the skin or intestinal tract mucosa, they are recognized by Toll like receptors, which activate immune cell responses.
Confocal microscopy	Confocal microscopy is an optical imaging technique used to increase optical resolution and contrast of a micrograph by using point illumination and a spatial pinhole to eliminate out-of-focus light in specimens that are thicker than the focal plane. It enables the reconstruction of three-dimensional structures from the obtained images. This technique has gained popularity in the scientific and industrial communities and typical applications are in life sciences, semiconductor inspection and materials science.
Microbiome	A microbiome is the totality of microbes, their genetic elements (genomes), and environmental interactions in a particular environment. The term 'microbiome' was coined by Joshua Lederberg, who argued that microorganisms inhabiting the human body should be included as part of the human genome, because of their influence on human physiology. The human body contains over 10 times more microbial cells than human cells.
Microbiota	Microbiota is a monotypic] genus of evergreen coniferous shrub in the cypress family Cupressaceae, containing only one species, Microbiota decussata. The plant is native and endemic to a limited area of the Sikhote-Alin mountains in Primorsky Krai - in the Russian Far East region of Western Siberia in Northeast Asia. Microbiota decussata has never acquired a vernacular or common name in English, though Siberian Cypress and Russian Arborvitae have been proposed.
Vibrio vulnificus	Vibrio vulnificus is a species of Gram-negative, motile, curved, rod-shaped bacteria of the genus Vibrio. Present in marine environments such as estuaries, brackish ponds, or coastal areas, V. vulnificus is related to V. cholerae, the causative agent of cholera. Infection with V. vulnificus leads to rapidly expanding cellulitis or septicemia. It was first isolated in 1976.
Acne	Acne is a general term used for acneiform eruptions. It is usually used as a synonym for acne vulgaris, but may also refer to:•Acne aestivalis•Acne conglobata•Acne cosmetica•Acne fulminans•Acne keloidalis nuchae•Acne mechanica•Acne medicamentosa (drug-induced acne) (e.g., steroid acne)•Acne miliaris necrotica•Acne necrotica•Blackheads•Chloracne•Excoriated acne•Halogen acne•Infantile acne/Neonatal acne•Lupus miliaris disseminatus faciei•Occupational acne•Oil acne•Pomade acne•Tar acne•Tropical acne.
Epidermis	The Epidermis (zoology) is an epithelium (sheet of cells) that covers the body of an eumetazoan (animal more complex than a sponge).

Chapter 5. Part 5: Medicine and Immunology

	Eumetazoa have a cavity lined with a similar epithelium, the gastrodermis, which forms a boundary with the epidermis at the mouth. Sponges have no epithelium, and therefore no epidermis or gastrodermis.
Propionibacterium acnes	Propionibacterium acnes is a relatively slow growing, typically aerotolerant anaerobic gram positive bacterium (rod) that is linked to the skin condition acne; it can also cause chronic blepharitis and endophthalmitis, the latter particularly following intraocular surgery. The genome of the bacterium has been sequenced and a study has shown several genes that can generate enzymes for degrading skin and proteins that may be immunogenic (activating the immune system). This bacterium is largely commensal and part of the skin flora present on most healthy adult human skin.
Sebaceous gland	The sebaceous glands are microscopic glands in the skin that secrete an oily/waxy matter, called sebum, to lubricate the skin and hair of mammals. In humans, they are found in greatest abundance on the face and scalp, though they are distributed throughout all skin sites except the palms and soles. In the eyelids, meibomian sebaceous glands secrete a special type of sebum into tears. There are several related medical conditions, including acne, sebaceous cysts, hyperplasia, sebaceous adenoma and sebaceous gland carcinoma.
Staphylococcus epidermidis	Staphylococcus epidermidis is one of thirty-three known species belonging to the genus Staphylococcus. It is part of our skin flora, and consequently part of human flora. It can also be found in the mucous membranes and in animals.
Actinomyces	Actinomyces from Greek 'actino' that means mukas and fungus, is a genus of the actinobacteria class of bacteria. They are all Gram-positive and are characterized by contiguous spread, suppurative and granulomatous inflammation, and formation of multiple abscesses and sinus tracts that may discharge sulfur granules.' They can be either anaerobic or facultatively anaerobic . Actinomyces species do not form endospores, and, while individual bacteria are rod-shaped, morphologically Actinomyces colonies form fungus-like branched networks of hyphae.
Blackhead	A blackhead is a yellow or blackish bump or plug on the skin. Blackheads are one of the common findings in acne vulgaris. Contrary to the common belief that it is caused by poor hygiene, blackheads are caused by excess oils that have accumulated in the sebaceous gland's duct.
Fusobacterium	Fusobacterium is a genus of rod shaped baccilli with pointed ends, anaerobic, Gram-negative bacteria, similar to Bacteroides. Fusobacterium contribute to several human diseases, including periodontal diseases, Lemierre's syndrome, and topical skin ulcers.

Keratin	Keratin refers to a family of fibrous structural proteins. Keratin is the key of structural material making up the outer layer of human skin. It is also the key structural component of hair and nails.
Moraxella catarrhalis	Moraxella catarrhalis is a fastidious, nonmotile, Gram-negative, aerobic, oxidase-positive diplococcus that can cause infections of the respiratory system, middle ear, eye, central nervous system and joints of humans. History M. catarrhalis was previously placed in a separate genus named Branhamella. The rationale for this was that other members of the genus Moraxella are rod-shaped and rarely caused infections in humans.
Nasopharynx	The nasopharynx is the uppermost part of the pharynx. It extends from the base of the skull to the upper surface of the soft palate; it differs from the oral and laryngeal parts of the pharynx in that its cavity always remains patent (open). In front it communicates through the choanae with the nasal cavities.
Oropharynx	The Oropharynx reaches from the Uvula to the level of the hyoid bone. It opens anteriorly, through the isthmus faucium, into the mouth, while in its lateral wall, between the two palatine arches, is the palatine tonsil. Normal oropharyngeal flora Fusobacterium Although older resources have stated that Fusobacterium is a common occurrence in the human oropharynx, the current consensus is that Fusobacterium should always be treated as a pathogen.
Prevotella	Prevotella is a genus of bacteria. Bacteroides melaninogenicus has recently been reclassified and split into Prevotella melaninogenica and Prevotella intermedia. Prevotella spp.
Streptococcus oralis	Streptococcus oralis is a Gram positive bacterium that grows characteristically in chains. It forms small white colonies on a Wilkins-Chalgren agar plate. It is found in high numbers in the oral cavity.

Chapter 5. Part 5: Medicine and Immunology

Streptococcus salivarius	Streptococcus salivarius is a species of spherical, Gram-positive bacteria which colonize the mouth and upper respiratory tract of humans a few hours after birth, making further exposure to the bacteria harmless. The bacteria is considered an opportunistic pathogen, rarely finding its way into the bloodstream, where it has been implicated in septicemia cases in people with neutropenia. S. salivarius has distinct characteristics when exposed to different environmental nutrients.
Tonsil	The human palatine tonsils and the nasopharyngeal tonsil are lymphoepithelial tissues located in strategic areas of the oropharynx and nasopharynx, although most commonly the term 'tonsils' refers to the palatine tonsils [that can be seen in the back of the throat]. These immunocompetent tissues represent the defense mechanism of first line against ingested or inhaled foreign pathogens. However, the fundamental immunological roles of tonsils have yet to be addressed.
Bacteremia	Bacteremia is the presence of bacteria in the blood. The blood is normally a sterile environment, so the detection of bacteria in the blood (most commonly with blood cultures) is always abnormal. Bacteria can enter the bloodstream as a severe complication of infections (like pneumonia or meningitis), during surgery (especially when involving mucous membranes such as the gastrointestinal tract), or due to catheters and other foreign bodies entering the arteries or veins (including intravenous drug abuse).
Cholera	Cholera is an infection in the small intestine caused by the bacterium Vibrio cholerae. The main symptoms are profuse, watery diarrhea and vomiting. Transmission occurs primarily by drinking water or eating food that has been contaminated by the feces of an infected person, including one with no apparent symptoms.
Mitral valve	The mitral valve is a dual-flap valve in the heart that lies between the left atrium (LA) and the left ventricle (LV). The mitral valve and the tricuspid valve are known collectively as the atrioventricular valves because they lie between the atria and the ventricles of the heart and control the flow of blood. Overview A normally-functioning mitral valve opens secondary to increased pressure from the left atrium as it fills with blood.
Respiratory tract	In humans the respiratory tract is the part of the anatomy involved with the process of respiration.

	The respiratory tract is divided into 3 segments:•Upper respiratory tract: nose and nasal passages, paranasal sinuses, and throat or pharynx•Respiratory airways: voice box or larynx, trachea, bronchi, and bronchioles•Lungs: respiratory bronchioles, alveolar ducts, alveolar sacs, and alveoli
	The respiratory tract is a common site for infections. Upper respiratory tract infections are probably the most common infections in the world.
Vibrio cholerae	Vibrio cholerae is a gram negative comma-shaped bacterium with a polar flagellum that causes cholera in humans. V. cholerae and other species of the genus Vibrio belong to the gamma subdivision of the Proteobacteria. There are two major biotypes of V. cholerae identified by hemagglutination testing, classical and El Tor, and numerous serogroups.
Cholera toxin	Cholera toxin is a protein complex secreted by the bacterium Vibrio cholerae. CTX is responsible for the massive, watery diarrhea characteristic of cholera infection.
	The cholera toxin is an oligomeric complex made up of six protein subunits: a single copy of the A subunit (part A, enzymatic), and five copies of the B subunit (part B, receptor binding).
Cholera vaccine	Cholera vaccine is a vaccine used against cholera and traveler's diarrhea.
	The first vaccines against cholera were developed in the late nineteenth century. These injected whole cell vaccine became increasingly popular until they were replaced by oral vaccines starting in the 1980s.
Vaccine trial	A vaccine trial is a clinical trial that aims at establishing the safety and efficacy of a vaccine prior to it being licensed.
	A vaccine trial might involve forming two groups from the target population. For example, from the set of trail subjects, each subject may be randomly assigned to receive either a new vaccine or a 'control' treatment: The control treatment may be a placebo, or an adjuvant-containing cocktail, or an established vaccine (which might be intended to protect against a different pathogen).
Vegetation	Vegetation is a general term for the plant life of a region; it refers to the ground cover provided by plants. It is a general term, without specific reference to particular taxa, life forms, structure, spatial extent, or any other specific botanical or geographic characteristics. It is broader than the term flora which refers exclusively to species composition.
Helicobacter pylori	Helicobacter pylori previously named Campylobacter pyloridis, is a Gram-negative, microaerophilic bacterium found in the stomach.

	It was identified in 1982 by Barry Marshall and Robin Warren, who found that it was present in patients with chronic gastritis and gastric ulcers, conditions that were not previously believed to have a microbial cause. It is also linked to the development of duodenal ulcers and stomach cancer.
Bacteroidetes	The phylum Bacteroidetes is composed of three large classes of gram-negative, nonsporeforming, anaerobic, and rod-shaped bacteria that are widely distributed in the environment, including in soil, in sediments, sea water and in the guts and on the skin of animals. By far, the ones in the Bacteroidia class are the most well-studied, including the genus Bacteroides (an abundant organism in the feces of warm-blooded animals including humans), and Porphyromonas, a group of organisms inhabiting the human oral cavity. The class Bacteroidia was formally called Bacteroidetes as it was until recently the only class in the phylum, the name was changed in the fourth volume of Bergey's Manual of Systematic Bacteriology.
Candida albicans	Candida albicans is a diploid fungus that grows both as yeast and filamentous cells and a causal agent of opportunistic oral and genital infections in humans. Systemic fungal infections (fungemias) including those by C. albicans have emerged as important causes of morbidity and mortality in immunocompromised patients (e.g., AIDS, cancer chemotherapy, organ or bone marrow transplantation). C. albicans biofilms may form on the surface of implantable medical devices.
Clostridium	Clostridium is a genus of Gram-positive bacteria, belonging to the Firmicutes. They are obligate anaerobes capable of producing endospores. Individual cells are rod-shaped, which gives them their name, from the Greek kloster or spindle.
Enterobacteriaceae	The Enterobacteriaceae are a large family of bacteria, including many of the more familiar pathogens, such as Salmonella and Escherichia coli. Genetic studies place them among the Proteobacteria, and they are given their own order (Enterobacteriales), though this is sometimes taken to include some related environmental samples. Characteristics Members of the Enterobacteriaceae are rod-shaped, and are typically 1-5 μm in length.
Lactobacillus	Lactobacillus, is a genus of Gram-positive facultative anaerobic or microaerophilic rod-shaped bacteria. They are a major part of the lactic acid bacteria group, named as such because most of its members convert lactose and other sugars to lactic acid.

Peptostreptococcus	Peptostreptococcus is a genus of anaerobic, Gram-positive, non-spore forming bacteria. The cells are small, spherical, and can occur in short chains, in pairs or individually. Peptostreptococcus are slow-growing bacteria with increasing resistance to antimicrobial drugs.
Probiotic	Probiotic are live microorganisms thought to be beneficial to the host organism. According to the currently adopted definition by FAOWHO, probiotics are: 'Live microorganisms which when administered in adequate amounts confer a health benefit on the host'. Lactic acid bacteria (LAB) and bifidobacteria are the most common types of microbes used as probiotics; but certain yeasts and bacilli may also be used.
Biosynthesis	Biosynthesis is an enzyme-catalyzed process in cells of living organisms by which substrates are converted to more complex products. The biosynthesis process often consists of several enzymatic steps in which the product of one step is used as substrate in the following step. Examples for such multi-step biosynthetic pathways are those for the production of amino acids, fatty acids, and natural products.
Gas gangrene	Gas gangrene is a bacterial infection that produces gas tissues in gangrene. It is a deadly form of gangrene usually caused by Clostridium perfringens bacteria. It is a medical emergency.
Enterococcus	Enterococcus is a genus of lactic acid bacteria of the phylum Firmicutes. Enterococci are Gram-positive cocci that often occur in pairs (diplococci) or short chains and are difficult to distinguish from Streptococci on physical characteristics alone. Two species are common commensal organisms in the intestines of humans: E. faecalis (90-95%) and E. faecium (5-10%).
Urinary tract infection	A urinary tract infection is a bacterial infection that affects part of the urinary tract. When it affects the lower urinary tract it is known as a simple cystitis (a bladder infection) and when it affects the upper urinary tract it is known as pyelonephritis (a kidney infection). Symptoms from a lower urinary tract include painful urination and either frequent urination or urge to urinate (or both), while those of pyelonephritis include fever and flank pain in addition to the symptoms of a lower UTI. In the elderly and the very young, symptoms may be vague.
Vaginal flora	The micro-organisms that colonize the vagina, collectively referred to as the vaginal microbiome or vaginal flora, were discovered by the German gynecologist Albert Döderlein in 1892. The amount and type of bacteria present have significant implications for a woman's overall health. The primary colonizing bacteria of a healthy individual are of the genus lactobacillus, such as L. acidophilus and L. doderlein, and the lactic acid they produce (some species produce hydrogen peroxide or antibiotic), in combination with fluids secreted during sexual arousal, are greatly responsible for the characteristic scent associated with the vaginal area.

Chapter 5. Part 5: Medicine and Immunology

Bacteroides fragilis	Bacteroides fragilis is a Gram-negative bacillus bacterium species, and an obligate anaerobe of the gut. B. fragilis group is the most commonly isolated bacteriodaceae in anaerobic infections especially those that originate from the gastrointestinal flora. B. fragilis is the most prevalent organism in the B. fragilis group, accounting for 41% to 78% of the isolates of the group.
Cytokine	Cytokines are small cell-signaling protein molecules that are secreted by numerous cells and are a category of signaling molecules used extensively in intercellular communication. Cytokines can be classified as proteins, peptides, or glycoproteins; the term 'cytokine' encompasses a large and diverse family of regulators produced throughout the body by cells of diverse embryological origin. The term 'cytokine' has been used to refer to the immunomodulating agents, such as interleukins and interferons.
Enterotoxin	An enterotoxin is a protein toxin released by a microorganism in the intestine. Enterotoxins are chromosomally encoded exotoxins that are produced and secreted from several bacterial organisms. They are often heat stable, of low molecular weight and are water-soluble.
Lactococcus	Lactococcus is a genus of lactic acid bacteria that were formerly included in the genus Streptococcus Group N1. They are known as homofermentors meaning that they produce a single product, lactic acid in this case, as the major or only product of glucose fermentation. Their homofermentative character can be altered by adjusting cultural conditions like pH, glucose concentration, and nutrient limitation. They are gram-positive, catalase negative, non-motile cocci that are found singly, in pairs, or in chains.
Cell-mediated immunity	Cell-mediated immunity is an immune response that does not involve antibodies or complement but rather involves the activation of macrophages, natural killer cells (NK), antigen-specific cytotoxic T-lymphocytes, and the release of various cytokines in response to an antigen. Historically, the immune system was separated into two branches: humoral immunity, for which the protective function of immunization could be found in the humor (cell-free bodily fluid or serum) and cellular immunity, for which the protective function of immunization was associated with cells. CD4 cells or helper T cells provide protection against different pathogens.
Dendritic cell	Dendritic cells (DCs) are immune cells forming part of the mammalian immune system. Their main function is to process antigen material and present it on the surface to other cells of the immune system. That is, dendritic cells function as antigen-presenting cells.

Gnotobiotic animal	A gnotobiotic animal is an animal in which only certain known strains of bacteria and other microorganisms are present. Technically, the term also includes germ-free animals, as the status of their microbial communities is also known. However, the term gnotobiotic is often incorrectly contrasted with germ-free.
Immune system	An immune system is a system of biological structures and processes within an organism that protects against disease. To function properly, an immune system must detect a wide variety of agents, from viruses to parasitic worms, and distinguish them from the organism's own healthy tissue. Pathogens can rapidly evolve and adapt to avoid detection and neutralization by the immune system.
Mast cell	A mast cell is a resident cell of several types of tissues and contains many granules rich in histamine and heparin. Although best known for their role in allergy and anaphylaxis, mast cells play an important protective role as well, being intimately involved in wound healing and defense against pathogens. Mast cells were first described by Paul Ehrlich in his 1878 doctoral thesis on the basis of their unique staining characteristics and large granules.
Neisseria gonorrhoeae	Neisseria gonorrhoeae, or gonococcus, is a species of Gram-negative coffee bean-shaped diplococci bacteria responsible for the sexually transmitted infection gonorrhea. N. gonorrhoea was first described by Albert Neisser in 1879. Microbiology Neisseria are fastidious Gram-negative cocci that require nutrient supplementation to grow in laboratory cultures.
Antibiotic resistance	Antibiotic resistance is a type of drug resistance where a microorganism is able to survive exposure to an antibiotic. While a spontaneous or induced genetic mutation in bacteria may confer resistance to antimicrobial drugs, genes that confer resistance can be transferred between bacteria in a horizontal fashion by conjugation, transduction, or transformation. Thus, a gene for antibiotic resistance that evolves via natural selection may be shared.
Blood cell	A blood cell, is a cell of any type normally found in blood. In mammals, these fall into three general categories:•red blood cells -- Erythrocytes•white blood cells -- Leukocytes•platelets -- Thrombocytes

	Together, these three kinds of blood cells sum up for a total 45% of blood tissue by volume (and the remaining 55% is plasma). This is called the hematocrit and can be determined by centrifuge or flow cytometry.
Stem cell	Stem cells are biological cells found in all multicellular organisms, that can divide (through mitosis) and differentiate into diverse specialized cell types and can self-renew to produce more stem cells. In mammals, there are two broad types of stem cells: embryonic stem cells, which are isolated from the inner cell mass of blastocysts, and adult stem cells, which are found in various tissues. In adult organisms, stem cells and progenitor cells act as a repair system for the body, replenishing adult tissues.
Macrophage	Macrophages are cells produced by the differentiation of monocytes in tissues. Human macrophages are about 21 micrometres (0.00083 in) in diameter. Monocytes and macrophages are phagocytes.
Monocyte	Monocytes are a type of white blood cell and are part of the innate immune system of vertebrates including all mammals (including humans), birds, reptiles, and fish. Monocytes play multiple roles in immune function. Such roles include: (1) replenish resident macrophages and dendritic cells under normal states, and (2) in response to inflammation signals, monocytes can move quickly (approx. 8-12 hours) to sites of infection in the tissues and divide/differentiate into macrophages and dendritic cells to elicit an immune response.
Phagosome	In cell biology, a phagosome is a vesicle formed around a particle absorbed by phagocytosis. The vacuole is formed by the fusion of the cell membrane around the particle. A phagosome is a cellular compartment in which pathogenic microorganisms can be killed and digested.
Antigen-presenting cell	An antigen-presenting cell or accessory cell is a cell that displays foreign antigen complexes with major histocompatibility complex (MHC) on their surfaces. T-cells may recognize these complexes using their T-cell receptors (TCRs). These cells process antigens and present them to T-cells.
B cell	B cells are lymphocytes that play a large role in the humoral immune response (as opposed to the cell-mediated immune response, which is governed by T cells). B cells are an essential component of the adaptive immune system. Their principal functions are to make antibodies against antigens, perform the role of antigen-presenting cells (APCs) and eventually develop into memory B cells after activation by antigen interaction.
Lymphocyte	A lymphocytes is a type of white blood cell in the vertebrate immune system.

	Under the microscope, lymphocytes can be divided into large granular lymphocytes and small lymphocytes. Large granular lymphocytes include natural killer cells (NK cells).
Spleen	The spleen is an organ found in virtually all vertebrate animals. Similar in structure to a large lymph node, the spleen acts primarily as a blood filter. As such, it is a non-vital organ, with a healthy life possible after removal.
T cell	T cells or T lymphocytes belong to a group of white blood cells known as lymphocytes, and play a central role in cell-mediated immunity. They can be distinguished from other lymphocytes, such as B cells and natural killer cells (NK cells), by the presence of a T cell receptor (TCR) on the cell surface. They are called T cells because they mature in the thymus.
Gut-associated lymphoid tissue	The digestive tract's immune system is often referred to as gut-associated lymphoid tissue and works to protect the body from invasion. GALT is an example of mucosa-associated lymphoid tissue. The digestive tract is an important component of the body's immune system.
Lymph node	A lymph node is a small ball or an oval-shaped organ of the immune system, distributed widely throughout the body including the armpit and stomach and linked by lymphatic vessels. Lymph nodes are garrisons of B, T and other immune cells. Lymph nodes act as filters or traps for foreign particles and are important in the proper functioning of the immune system.
Langerhans cell	Langerhans cells are dendritic cells of the epidermis, containing large granules called Birbeck granules. They are also normally present in lymph nodes and other organs, including the stratum spinosum layer of the epidermis. They can be found elsewhere, particularly in association in the condition histiocytosis.
M cell	M cells (or microfold cells) are cells found in the follicle-associated epithelium of the Peyer's patch. They transport organisms and particles from the gut lumen to immune cells across the epithelial barrier, and thus are important in stimulating mucosal immunity. Unlike their neighbouring cells, they have the unique ability to take up antigen from the lumen of the small intestine via endocytosis or phagocytosis, and then deliver it via transcytosis to dendritic cells (an antigen presenting cell) and lymphocytes (namely T cells) located in a unique pocket-like structure on their basolateral side.
Mucous membrane	The mucous membranes are linings of mostly endodermal origin, covered in epithelium, which are involved in absorption and secretion. They line cavities that are exposed to the external environment and internal organs.

Chapter 5. Part 5: Medicine and Immunology

Infectious disease	Infectious diseases, also known as transmissible diseases or communicable diseases comprise clinically evident illness (i.e., characteristic medical signs and/or symptoms of disease) resulting from the infection, presence and growth of pathogenic biological agents in an individual host organism. In certain cases, infectious diseases may be asymptomatic for much or even all of their course in a given host. In the latter case, the disease may only be defined as a 'disease' (which by definition means an illness) in hosts who secondarily become ill after contact with an asymptomatic carrier.
Alveolar macrophage	An alveolar macrophage is a type of macrophage found in the pulmonary alveolus, near the pneumocytes, but separated from the wall.
	Activity of the alveolar macrophage is relatively high, because they are located at one of the major boundaries between the body and the outside world.
	Dust cells are another name for monocyte derivatives in the lungs that reside on respiratory surfaces and clean off particles such as dust or microorganisms.
Cystic fibrosis	
	Cystic fibrosis is an autosomal recessive genetic disorder affecting most critically the lungs, and also the pancreas, liver, and intestine. It is characterized by abnormal transport of chloride and sodium across an epithelium, leading to thick, viscous secretions.
	The name cystic fibrosis refers to the characteristic scarring (fibrosis) and cyst formation within the pancreas, first recognized in the 1930s.
Cystic fibrosis transmembrane conductance regulator	Cystic fibrosis transmembrane conductance regulator is a protein that in humans is encoded by the CFTR gene.
	CFTR is a ABC transporter-class ion channel that transports chloride and thiocyanate ions across epithelial cell membranes. Mutations of the CFTR gene affect functioning of the chloride ion channels in these cell membranes, leading to cystic fibrosis and congenital absence of the vas deferens.
Defensin	Defensins are small arginine rich cationic proteins found in both vertebrates and invertebrates. They have also been reported in plants. They are, and function as, host defense peptides.
Pseudomonas aeruginosa	Pseudomonas aeruginosa is a common bacterium which can cause disease in animals, including humans. It is found in soil, water, skin flora, and most man-made environments throughout the world.

Lung	The lung is the essential respiration organ in many air-breathing animals, including most tetrapods, a few fish and a few snails. In mammals and the more complex life forms, the two lungs are located near the backbone on either side of the heart. Their principal function is to transport oxygen from the atmosphere into the bloodstream, and to release carbon dioxide from the bloodstream into the atmosphere.
Salmonella enterica	Salmonella enterica is a rod-shaped flagellated, facultative anaerobic, Gram-negative bacterium, and a member of the genus Salmonella. Most cases of salmonellosis are caused by food infected with S. enterica, which often infects cattle and poultry, though also other animals such as domestic cats and hamsters have also been shown to be sources of infection to humans. However, investigations of vacuum cleaner bags have shown that households can act as a reservoir of the bacterium; this is more likely if the household has contact with an infection source, for example members working with cattle or in a veterinary clinic.
Extravasation	Extravasation is the leakage of a fluid out of its container. In the case of inflammation, it refers to the movement of white blood cells from the capillaries to the tissues surrounding them (diapedesis). In the case of malignant cancer metastasis it refers to cancer cells exiting the capillaries and entering organs.
Interleukin	Interleukins are a group of cytokines (secreted proteinssignaling molecules) that were first seen to be expressed by white blood cells (leukocytes). The term interleukin derives from (inter-) 'as a means of communication', and (-leukin) 'deriving from the fact that many of these proteins are produced by leukocytes and act on leukocytes'. The name is something of a relic, though (the term was coined by Dr. Vern Paetkau, University of Victoria); it has since been found that interleukins are produced by a wide variety of body cells.
Prostaglandin	A prostaglandin is any member of a group of lipid compounds that are derived enzymatically from fatty acids and have important functions in the animal body. Every prostaglandin contains 20 carbon atoms, including a 5-carbon ring. They are mediators and have a variety of strong physiological effects, such as regulating the contraction and relaxation of smooth muscle tissue.
Selectin	Selectins (cluster of differentiation 62 or CD62) are a family of cell adhesion molecules . All selectins are single-chain transmembrane glycoproteins that share similar properties to C-type lectins due to a related amino terminus and calcium-dependent binding. Selectins bind to sugar moieties and so are considered to be a type of lectin, cell adhesion proteins that bind sugar polymers.

Chapter 5. Part 5: Medicine and Immunology

Staphylococcus aureus	Staphylococcus aureus is a bacterial species named from Greek σταφυλ?κοκκος meaning the 'golden grape-cluster berry'. Also known as 'golden staph' and Oro staphira, it is a facultative anaerobic Gram-positive coccal bacterium. It is frequently found as part of the normal skin flora on the skin and nasal passages.
Aspirin	Aspirin also known as acetylsalicylic acid , is a salicylate drug, often used as an analgesic to relieve minor aches and pains, as an antipyretic to reduce fever, and as an anti-inflammatory medication. Salicylic acid, the main metabolite of aspirin, is an integral part of human and animal metabolism. While much of it is attributable to diet, a substantial part is synthesized endogenously.
Bradykinin	Bradykinin Bradykinin is a peptide that causes blood vessels to dilate (enlarge), and therefore causes blood pressure to lower. A class of drugs called ACE inhibitors, which are used to lower blood pressure, increase bradykinin further lowering blood pressure. Bradykinin works on blood vessels through the release of prostacyclin, nitric oxide, and Endothelium-Derived Hyperpolarizing Factor.
Histamine	Histamine is an organic nitrogen compound involved in local immune responses as well as regulating physiological function in the gut and acting as a neurotransmitter. Histamine triggers the inflammatory response. As part of an immune response to foreign pathogens, histamine is produced by basophils and by mast cells found in nearby connective tissues.
Chemokine	Small cytokines (intecrine/chemokine), interleukin-8 like Chemokines are a family of small cytokines, or proteins secreted by cells. Their name is derived from their ability to induce directed chemotaxis in nearby responsive cells; they are chemotactic cytokines. Proteins are classified as chemokines according to shared structural characteristics such as small size (they are all approximately 8-10 kilodaltons in size), and the presence of four cysteine residues in conserved locations that are key to forming their 3-dimensional shape.
Granulocyte	Granulocytes are a category of white blood cells characterized by the presence of granules in their cytoplasm. They are also called polymorphonuclear leukocytes (PMN or PML) because of the varying shapes of the nucleus, which is usually lobed into three segments. In common parlance, the term polymorphonuclear leukocyte often refers specifically to neutrophil granulocytes, the most abundant of the granulocytes.

Mycobacterium tuberculosis	Mycobacterium tuberculosis is a pathogenic bacterial species in the genus Mycobacterium and the causative agent of most cases of tuberculosis. First discovered in 1882 by Robert Koch, M. tuberculosis has an unusual, waxy coating on the cell surface (primarily mycolic acid), which makes the cells impervious to Gram staining so acid-fast detection techniques are used instead. The physiology of M. tuberculosis is highly aerobic and requires high levels of oxygen.
Granuloma	Granuloma is a medical term for a tiny collection of immune cells known as macrophages. Granulomas form when the immune system attempts to wall off substances that it perceives as foreign but is unable to eliminate. Such substances include infectious organisms such as bacteria and fungi as well as other materials such as keratin and suture fragments.
Phagocytosis	Phagocytosis (from Ancient Greek φαγε?ν (phagein), meaning 'to devour', κ?τος, (kytos), meaning ' cell', and -osis, meaning 'process') is the cellular process of engulfing solid particles by the cell membrane to form an internal phagosome by phagocytes and protists. Phagocytosis is a specific form of endocytosis involving the vesicular internalization of solids such as bacteria, and is, therefore, distinct from other forms of endocytosis such as the vesicular internalization of various liquids. Phagocytosis is involved in the acquisition of nutrients for some cells, and, in the immune system, it is a major mechanism used to remove pathogens and cell debris.
Mycobacterium	Mycobacterium is a genus of Actinobacteria, given its own family, the Mycobacteriaceae. The genus includes pathogens known to cause serious diseases in mammals, including tuberculosis (Mycobacterium tuberculosis) and leprosy (Mycobacterium leprae). The Greek prefix 'myco--' means fungus, alluding to the way mycobacteria have been observed to grow in a mould-like fashion on the surface of liquids when cultured.
Streptococcus pneumoniae	Streptococcus pneumoniae, is a Gram-positive, alpha-hemolytic, aerotolerant anaerobic member of the genus Streptococcus. A significant human pathogenic bacterium, S. pneumoniae was recognized as a major cause of pneumonia in the late 19th century, and is the subject of many humoral immunity studies.

Despite the name, the organism causes many types of pneumococcal infections other than pneumonia. |
| Autophagy | In cell biology, autophagy, is a catabolic process involving the degradation of a cell's own components through the lysosomal machinery. It is a tightly regulated process that plays a normal part in cell growth, development, and homeostasis, helping to maintain a balance between the synthesis, degradation, and subsequent recycling of cellular products. It is a major mechanism by which a starving cell reallocates nutrients from unnecessary processes to more-essential processes. |
| Interferon | Interferon alpha/beta domain |

Interferons (IFNs) are proteins made and released by host cells in response to the presence of pathogens such as viruses, bacteria, parasites or tumor cells. They allow for communication between cells to trigger the protective defenses of the immune system that eradicate pathogens or tumors.

IFNs belong to the large class of glycoproteins known as cytokines.

Major histocompatibility complex	Major histocompatibility complex is a cell surface molecule encoded by a large gene family in all vertebrates. MHC molecules mediate interactions of leukocytes, also called white blood cells (WBCs), which are immune cells, with other leukocytes or body cells. MHC determines compatibility of donors for organ transplant as well as one's susceptibility to an autoimmune disease via crossreacting immunization.
Natural killer cell	Natural killer cells are a type of cytotoxic lymphocyte critical to the innate immune system. The role NK cells play is analogous to that of cytotoxic T cells in the vertebrate adaptive immune response. NK cells provide rapid responses to virally infected cells and respond to tumor formation, acting at around 3 days after infection.
Antibody-dependent cell-mediated cytotoxicity	Antibody-Dependent Cell-Mediated Cytotoxicity is a mechanism of cell-mediated immunity whereby an effector cell of the immune system actively lyses a target cell that has been bound by specific antibodies. It is one of the mechanisms through which antibodies, as part of the humoral immune response, can act to limit and contain infection. Classical ADCC is mediated by natural killer (NK) cells; neutrophils and eosinophils can also mediate ADCC. For example, eosinophils can kill certain parasitic worms known as helminths through ADCC. ADCC is part of the adaptive immune response due to its dependence on a prior antibody response.
Perforin	Perforin-1 is a protein that in humans is encoded by the PRF1 gene. Perforin is a cytolytic protein found in the granules of CD8 T-cells and NK cells. Upon degranulation, perforin inserts itself into the target cell's plasma membrane, forming a pore.
Anaphylatoxin	Anaphylotoxin-like domain Anaphylatoxins, or anaphylotoxins, are fragments (C3a, C4a and C5a) that are produced as part of the activation of the complement system.. Complement components C3, C4 and C5 are large glycoproteins that have important functions in the immune response and host defense. They have a wide variety of biological activities and are proteolytically activated by cleavage at a specific site, forming a- and b-fragments.

Lectin	Lectins are sugar-binding proteins (not to be confused with glycoproteins, which are proteins containing sugar chains or residues) that are highly specific for their sugar moieties. They play a role in biological recognition phenomena involving cells and proteins. For example, some viruses use lectins to attach themselves to the cells of the host organism during infection.
Opsonin	An opsonin is any molecule that targets an antigen for an immune response. However, the term is usually used in reference to molecules that act as a binding enhancer for the process of phagocytosis, especially antibodies, which coat the negatively-charged molecules on the membrane. Molecules that activate the complement system are also considered opsonins.
C-reactive protein	C-reactive protein is a protein found in the blood, the levels of which rise in response to inflammation (i.e. C-reactive protein is an acute-phase protein). Its physiological role is to bind to phosphocholine expressed on the surface of dead or dying cells (and some types of bacteria) in order to activate the complement system via the C1Q complex. CRP is synthesized by the liver in response to factors released by fat cells (adipocytes).
Cardiovascular diseases	Cardiovascular diseases is the class of diseases that involve the heart or blood vessels (arteries and veins). While the term technically refers to any disease that affects the cardiovascular system (as used in MeSH C14), it is usually used to refer to those related to atherosclerosis (arterial disease). These conditions have similar causes, mechanisms, and treatments.
Severe combined immunodeficiency	Severe Combined Immunodeficiency (SCID) is a severe immunodeficiency genetic disorder that is characterized by the complete inability of the adaptive immune system to mount, coordinate, and sustain an appropriate immune response, usually due to absent or atypical T and B lymphocytes. In humans, SCID is colloquially known as 'bubble boy' disease, as victims may require complete clinical isolation to prevent lethal infection from environmental microbes. Several forms of SCID occur in animal species.
Combined immunodeficiencies	Combined immunodeficiencies are immunodeficiency disorders that involve multiple components of the immune system, including both humoral immunity and cell-mediated immunity.
Antibodies	Antibodies are gamma globulin proteins that are found in blood or other bodily fluids of vertebrates, and are used by the immune system to identify and neutralize foreign objects, such as bacteria and viruses. They are typically made of basic structural units--each with two large heavy chains and two small light chains--to form, for example, monomers with one unit, dimers with two units or pentamers with five units.

Chapter 5. Part 5: Medicine and Immunology

Epitope	An epitope, is the part of an antigen that is recognized by the immune system, specifically by antibodies, B cells, or T cells. The part of an antibody that recognizes the epitope is called a paratope. Although epitopes are usually thought to be derived from non-self proteins, sequences derived from the host that can be recognized are also classified as epitopes.
Immunogenicity	Immunogenicity is the ability of a particular substance, such as an antigen or epitope, to provoke an immune response in the body of a human or animal. The ability to induce humoral and/or cell-mediated immune responses. The ability of antigen to elicit immune response is called 'immunogenicity'.
Smallpox	Smallpox is an infectious disease unique to humans, caused by either of two virus variants, Variola major and Variola minor. The disease is also known by the Latin names Variola or Variola vera, which is a derivative of the Latin varius, meaning 'spotted', or varus, meaning 'pimple'. The term 'smallpox' was first used in Europe in the 15th century to distinguish variola from the 'great pox' (syphilis).
Bubonic plague	Bubonic plague is a zoonotic disease, circulating mainly among small rodents and their fleas, and is one of three types of infections caused by Yersinia pestis (formerly known as Pasteurella pestis), which belongs to the family Enterobacteriaceae. Without treatment, the bubonic plague kills about two out of three infected humans within 4 days. The term bubonic plague is derived from the Greek word βουβ?v, meaning 'groin.' Swollen lymph nodes (buboes) especially occur in the armpit and groin in persons suffering from bubonic plague.
ABO blood group system	The ABO blood group system is the most important blood type system (or blood group system) in human blood transfusion. The associated anti-A antibodies and anti-B antibodies are usually IgM antibodies, which are usually produced in the first years of life by sensitization to environmental substances such as food, bacteria, and viruses. ABO blood types are also present in some animals, for example apes such as chimpanzees, bonobos, and gorillas.
Cowpox	Cowpox is a skin disease caused by a virus known as the Cowpox virus. The pox is related to the vaccinia virus and got its name from the distribution of the disease when dairymaids touched the udders of infected cows. The ailment manifests itself in the form of red blisters and is transmitted by touch from infected animals to humans.
Vaccination	Vaccination is the administration of antigenic material (a vaccine) to stimulate the immune system of an individual to develop adaptive immunity to a disease. Vaccines can prevent or ameliorate the effects of infection by many pathogens.

Polio vaccine	Two polio vaccines are used throughout the world to combat poliomyelitis (or polio). The first was developed by Jonas Salk and first tested in 1952. Announced to the world by Salk on April 12, 1955, it consists of an injected dose of inactivated (dead) poliovirus. An oral vaccine was developed by Albert Sabin using attenuated poliovirus.
Smallpox vaccine	The smallpox vaccine was the first successful vaccine to be developed. The process of vaccination was first publicised by Edward Jenner in 1796, who acted upon his observation that milkmaids who caught the cowpox virus did not catch smallpox. Prior to widespread vaccination, mortality rates in individuals with smallpox were high--up to 35% in some cases.
Hapten	A hapten is a small molecule that can elicit an immune response only when attached to a large carrier such as a protein; the carrier may be one that also does not elicit an immune response by itself. (In general, only large molecules, infectious agents, or insoluble foreign matter can elicit an immune response in the body). Once the body has generated antibodies to a hapten-carrier adduct, the small-molecule hapten may also be able to bind to the antibody, but it will usually not initiate an immune response; usually only the hapten-carrier adduct can do this.
Heavy chain	A heavy chain is the large polypeptide subunit of a protein complex, such as a motor protein (e.g. myosin, kinesin, or dynein) or antibody .
	It commonly refers to the immunoglobulin heavy chain. The heavy (H) chain is the larger of the two types of chains that comprise a normal immunoglobulin or antibody molecule.
Immunoglobulin superfamily	Immunoglobulin superfamily
	The immunoglobulin superfamily is a large group of cell surface and soluble proteins that are involved in the recognition, binding, or adhesion processes of cells. Molecules are categorized as members of this superfamily based on shared structural features with immunoglobulins (also known as antibodies); they all possess a domain known as an immunoglobulin domain or fold. Members of the IgSF include cell surface antigen receptors, co-receptors and co-stimulatory molecules of the immune system, molecules involved in antigen presentation to lymphocytes, cell adhesion molecules, certain cytokine receptors and intracellular muscle proteins.
ATP synthase	ATP synthase is an important enzyme that provides energy for the cell to use through the synthesis of adenosine triphosphate (ATP). ATP is the most commonly used 'energy currency' of cells from most organisms. It is formed from adenosine diphosphate (ADP) and inorganic phosphate (P_i), and needs energy.
Pepsin	Pepsin is an enzyme whose precursor form (pepsinogen) is released by the chief cells in the stomach and that degrades food proteins into peptides.

Chapter 5. Part 5: Medicine and Immunology

It was discovered in 1836 by Theodor Schwann who also coined its name from the Greek word pepsis, meaning digestion (peptein: to digest). It was the first animal enzyme to be discovered, and, in 1929, it became one of the first enzymes to be crystallized, by John H. Northrop.

Idiotype	In immunology, an idiotype is a shared characteristic between a group of immunoglobulin or T cell receptor (TCR) molecules based upon the antigen binding specificity and therefore structure of their variable region. The variable region of antigen receptors of T cells (TCRs) and B cells (immunoglobulins) contains a complementarity determining region (CDR) with a unique amino acid structure that determines the antigen specificity of the receptor. The structure formed by the CDR is known as the idiotope.
Anaphylaxis	Anaphylaxis is an acute multi-system severe type I hypersensitivity reaction. The term comes from the Greek words ?v? ana (against) and φ?λαξις phylaxis (protection). Due in part to the variety of definitions, between 1% and 15% of the population of the United States can be considered 'at risk' for having an anaphylactic reaction if they are exposed to one or more allergens.
B-cell receptor	The B-cell receptor is a transmembrane receptor protein located on the outer surface of B-cells. The receptor's binding moiety is composed of a membrane-bound antibody that, like all antibodies, has a unique and randomly-determined antigen-binding site. When a B-cell is activated by its first encounter with an antigen that binds to its receptor (its 'cognate antigen'), the cell proliferates and differentiates to generate a population of antibody-secreting plasma B cells and memory B cells.
Memory B cell	Memory B cells are a B cell sub-type that are formed following primary infection. Primary response, paratopes, and epitopes In wake of first (primary response) infection involving a particular antigen, the responding naïve cells (ones which have never been exposed to the antigen) proliferate to produce a colony of cells, most of which differentiate into the plasma cells, also called effector B cells (which produce the antibodies) and clear away with the resolution of infection, and the rest persist as the memory cells that can survive for years, or even a lifetime. To understand the events taking place, it is important to appreciate that the antibody molecules present on a clone (a group of genetically identical cells) of B cells have a unique paratope (the sequence of amino acids that binds to the epitope on an antigen).
Plasma cell	Plasma cells are white blood cells which produce large volumes of antibodies. They are transported by the blood plasma and the lymphatic system.

Serum	In blood, the serum is the component that is neither a blood cell (serum does not contain white or red blood cells) nor a clotting factor; it is the blood plasma with the fibrinogens removed. Serum includes all proteins not used in blood clotting (coagulation) and all the electrolytes, antibodies, antigens, hormones, and any exogenous substances (e.g., drugs and microorganisms).
	The study of serum is serology, and may also include proteomics.
Clonal selection	The clonal selection hypothesis has become a widely accepted model for how the immune system responds to infection and how certain types of B and T lymphocytes are selected for destruction of specific antigens invading the body. •Each lymphocyte bears a single type of receptor with a unique specificity (by V(D)J recombination).•Receptor occupation is required for cell activation.•The differentiated effector cells derived from an activated lymphocyte will bear receptors of identical specificity as the parental cell.•Those lymphocytes bearing receptors for self molecules will be deleted at an early stage.Early work
	In 1954, Danish immunologist Niels Jerne put forward a hypothesis which stated that there is already a vast array of lymphocytes in the body prior to any infection. The entrance of an antigen into the body results in the selection of only one type of lymphocyte to match it and produce a corresponding antibody to destroy the antigen.
Recombination signal sequences	The regional genes (V, D, J), used to generate T-cell receptors and Immunoglobulin molecules, are flanked by Recombination Signal Sequences that are recognized by a group of enzymes known collectively as the VDJ recombinase. RSSs are composed of seven conserved nucleotides (a heptamer) that reside next to the gene encoding sequence followed by a spacer (containing either 12 or 23 unconserved nucleotides) followed by a conserved nonamer (9 base pairs). The RSSs are present on the 3' side (downstream) of a V region and the 5' side (upstream) of the J region.
Cytotoxic T cell	A cytotoxic T cell belongs to a sub-group of T lymphocytes (a type of white blood cell) that are capable of inducing the death of infected somatic or tumor cells; they kill cells that are infected with viruses , or are otherwise damaged or dysfunctional. Most cytotoxic T cells express T-cell receptors (TCRs) that can recognize a specific antigenic peptide bound to Class I MHC molecules, present on all nucleated cells, and a glycoprotein called CD8, which is attracted to non-variable portions of the Class I MHC molecule. The affinity between CD8 and the MHC molecule keeps the T_C cell and the target cell bound closely together during antigen-specific activation.
MHC restriction	MHC-restricted antigen recognition, or MHC restriction, refers to the fact that a given T cell will recognize a peptide antigen only when it is bound to a host body's own MHC molecule.

Chapter 5. Part 5: Medicine and Immunology

	Normally, as T cells are stimulated only in the presence of self-MHC molecules, antigen is recognized only as peptides bound to self-MHC molecules.
	MHC restriction is particularly important when primary lymphocytes are developing and differentiating in the thymus or bone marrow.
T cell receptor	The T cell receptor is a molecule found on the surface of T lymphocytes (or T cells) that is, in general, responsible for recognizing antigens bound to major histocompatibility complex (MHC) molecules. The binding between T cell receptor and antigen is of relatively low affinity and is degenerate: that is, many T cell receptor recognize the same antigen and many antigen are recognized by the same T cell receptor.
	The T cell receptor is composed of two different protein chains (that is, it is a heterodimer). In 95% of T cells, this consists of an alpha (α) and beta (β) chain, whereas in 5% of T cells this consists of gamma and delta (γ/δ) chains.
Superantigen	Superantigens (SAgs) are a class of antigens which cause non-specific activation of T-cells resulting in polyclonal T cell activation and massive cytokine release. SAgs can be produced by pathogenic microbes (including viruses, mycoplasma, and bacteria) as a defense mechanism against the immune system. Compared to a normal antigen-induced T-cell response where .001-.0001% of the body's T-cells are activated, these SAgs are capable of activating up to 20% of the body's T-cells.
Granzyme	Granzymes are serine proteases that are released by cytoplasmic granules within cytotoxic T cells and natural killer cells. Their purpose is to induce apoptosis within virus-infected cells, thus destroying them.
	Cytotoxic T cells and natural killer cells release a protein called perforin, which attacks the target cells.
Cellular differentiation	In developmental biology, cellular differentiation is the process by which a less specialized cell becomes a more specialized cell type. Differentiation occurs numerous times during the development of a multicellular organism as the organism changes from a simple zygote to a complex system of tissues and cell types. Differentiation is a common process in adults as well: adult stem cells divide and create fully differentiated daughter cells during tissue repair and during normal cell turnover.
Yersinia enterocolitica	Yersinia enterocolitica is a species of gram-negative coccobacillus-shaped bacterium, belonging to the family Enterobacteriaceae. Primarily a zoonotic disease (cattle, deer, pigs, and birds), animals that recover frequently become asymptomatic carriers of the disease.

	Signs and symptoms
	Acute Y. enterocolitica infections produce severe diarrhea in humans, along with Peyer's patch necrosis, chronic lymphadenopathy, and hepatic or splenic abscesses.
Yersinia pestis	Yersinia pestis is a Gram-negative rod-shaped bacterium. It is a facultative anaerobe that can infect humans and other animals. Human Y. pestis infection takes three main forms: pneumonic, septicemic, and the notorious bubonic plagues.
Decay-accelerating factor	Complement decay-accelerating factor is a protein that, in humans, is encoded by the CD55 gene. Decay accelerating factor is a 70 kDa membrane protein that regulates the complement system on the cell surface. It prevents the assembly of the C3bBb complex (the C3-convertase of the alternative pathway) or accelerates the disassembly of preformed convertase, thus blocking the formation of the membrane attack complex.
Factor H	Factor H is a member of the regulators of complement activation family and is a complement control protein. It is a large (155 kilodaltons), soluble glycoprotein that circulates in human plasma (at a concentration of 500-800 micrograms per milliliter). Its main job is to regulate the Alternative Pathway of the complement system, ensuring that the complement system is directed towards pathogens and does not damage host tissue.
Adenylate cyclase	Adenylate cyclase is part of the G protein signaling cascade, which transmits chemical signals from outside the cell across the membrane to the inside of the cell (cytoplasm). The outside signal binds to a receptor, which transmits a signal to the G protein, which transmits a signal to adenylate cyclase, which transmits a signal by converting adenosine triphosphate to cyclic adenosine monophosphate (cAMP). cAMP is known as a second messenger.
Histamine antagonist	A histamine antagonist, commonly referred to as antihistamine, is a pharmaceutical drug that inhibits action of histamine by blocking it from attaching to histamine receptors. H_1 antihistamines are used to treat symptoms of allergy, such as runny nose and watery eyes. Allergies are caused by a body's excessive type I hypersensitivity response to allergens, such as pollen.
Edema	Edema is an abnormal accumulation of fluid beneath the skin or in one or more cavities of the body.

Chapter 5. Part 5: Medicine and Immunology

Generally, the amount of interstitial fluid is determined by the balance of fluid homeostasis, and increased secretion of fluid into the interstitium or impaired removal of this fluid may cause edema.

Cutaneous edema is referred to as 'pitting' when, after pressure is applied to a small area, the indentation persists for some time after the release of the pressure. Peripheral pitting edema, as shown in the illustration, is the more common type, results from water retention. It can be caused by systemic diseases, pregnancy in some women, either directly or as a result of heart failure, or local conditions such as varicose veins, thrombophlebitis, insect bites, and dermatitis.

Type IV hypersensitivity	Type IV hypersensitivity is often called delayed type hypersensitivity as the reaction takes two to three days to develop. Unlike the other types, it is not antibody mediated but rather is a type of cell-mediated response.

CD8+ cytotoxic T cells and CD4+ helper T cells recognize antigen in a complex with either type 1 or 2 major histocompatibility complex. |
| Vaccine | A vaccine is a biological preparation that improves immunity to a particular disease. A vaccine typically contains an agent that resembles a disease-causing microorganism, and is often made from weakened or killed forms of the microbe, its toxins or one of its surface proteins. The agent stimulates the body's immune system to recognize the agent as foreign, destroy it, and 'remember' it, so that the immune system can more easily recognize and destroy any of these microorganisms that it later encounters. |
| Contact dermatitis | Contact dermatitis is a term for a skin reaction (dermatitis) resulting from exposure to allergens (allergic contact dermatitis) or irritants (irritant contact dermatitis). Phototoxic dermatitis occurs when the allergen or irritant is activated by sunlight.

Symptoms

Contact dermatitis is a localized rash or irritation of the skin caused by contact with a foreign substance. |
| Type II hypersensitivity | In type II hypersensitivity the antibodies produced by the immune response bind to antigens on the patient's own cell surfaces. The antigens recognized in this way may either be intrinsic ('self' antigen, innately part of the patient's cells) or extrinsic (absorbed onto the cells during exposure to some foreign antigen, possibly as part of infection with a pathogen). These cells are recognized by macrophages or dendritic cells which act as antigen presenting cells, this causes a B cell response where antibodies are produced against the foreign antigen. |

Regulatory T cell	Regulatory T cells (T_{reg}), formerly known as suppressor T cells, are a subpopulation of T cells which downregulates the immune system, maintains tolerance to self-antigens, and downregulates autoimmune disease. Mouse models have suggested that modulation of T_{regs} can treat autoimmune disease and cancer, and facilitate organ transplantation. T regulatory cells are a component of the immune system that suppress immune responses of other cells.
Autoimmune disease	Autoimmune diseases arise from an overactive immune response of the body against substances and tissues normally present in the body. In other words, the body actually attacks its own cells. The immune system mistakes some part of the body as a pathogen and attacks it.
Autoimmunity	Autoimmunity is the failure of an organism in recognizing its own constituent parts as self, which allows an immune response against its own cells and tissues. Any disease that results from such an aberrant immune response is termed an autoimmune disease. Autoimmunity is often caused by a lack of germ development of a target body and as such the immune response acts against its own cells and tissues.
Rheumatic fever	Rheumatic fever is an inflammatory disease that occurs following a Streptococcus pyogenes infection, such as streptococcal pharyngitis or scarlet fever. Believed to be caused by antibody cross-reactivity that can involve the heart, joints, skin, and brain, the illness typically develops two to three weeks after a streptococcal infection. Acute rheumatic fever commonly appears in children between the ages of 6 and 15, with only 20% of first-time attacks occurring in adults.
Pathogenesis	The pathogenesis of a disease is the mechanism by which the disease is caused. The term can also be used to describe the origin and development of the disease and whether it is acute, chronic or recurrent. The word comes from the Greek pathos, 'disease', and genesis, 'creation'.
Rickettsia rickettsii	Rickettsia rickettsii is native to the New World and causes the malady known as Rocky Mountain spotted fever (RMSF). RMSF is transmitted by the bite of an infected tick while feeding on warm-blooded animals, including humans. Humans are accidental hosts in the rickettsia-tick life cycle and are not required to maintain the rickettsiae in nature.
Rocky Mountain spotted fever	Rocky Mountain spotted fever is the most lethal and most frequently reported rickettsial illness in the United States. It has been diagnosed throughout the Americas. Some synonyms for Rocky Mountain spotted fever in other countries include 'tick typhus,' 'Tobia fever' (Colombia), 'São Paulo fever' or 'febre maculosa' (Brazil), and 'fiebre manchada' (Mexico).
Corynebacterium diphtheriae	Corynebacterium diphtheriae is a pathogenic bacterium that causes diphtheria. It is also known as the Klebs-Löffler bacillus, because it was discovered in 1884 by German bacteriologists Edwin Klebs (1834 - 1912) and Friedrich Löffler (1852 - 1915).

	Classification
	Four subspecies are recognized: C. diphtheriae mitis, C. diphtheriae intermedius, C. diphtheriae gravis, and C. diphtheriae belfanti.
Filariasis	Filariasis is a parasitic disease and is considered an infectious tropical disease, that is caused by thread-like filarial nematodes (roundworms) in the superfamily Filarioidea, also known as 'filariae'.
	There are 9 known filarial nematodes which use humans as their definitive host. These are divided into 3 groups according to the niche within the body that they occupy: 'lymphatic filariasis', 'subcutaneous filariasis', and 'serous cavity filariasis'.
Infection	An infection is the invasion of body tissues by disease-causing microorganisms, their multiplication and the reaction of body tissues to these microorganisms and the toxins that they produce. Infections are caused by microorganisms such as viruses, prions, bacteria, and viroids, though larger organisms like macroparasites and fungi can also infect.
	Hosts normally fight infections themselves via their immune system.
Pathogen	A pathogen (Greek: π?θος pathos, 'suffering, passion' and γεν?ς genes (-gen) 'producer of') or infectious agent -- in colloquial terms, a germ -- is a microorganism such as a virus, bacterium, prion, or fungus, that causes disease in its animal or plant host. There are several substrates including pathways wherein pathogens can invade a host; the principal pathways have different episodic time frames, but soil contamination has the longest or most persistent potential for harboring a pathogen.
	Not all pathogens are negative.
Viral pathogenesis	Viral pathogenesis is the study of how biological viruses cause diseases in their target hosts, usually carried out at the cellular or molecular level. It is a specialized field of study in virology.
Wuchereria bancrofti	Wuchereria bancrofti is a parasitic filarial nematode (roundworm) spread by a mosquito vector. It is one of the three parasites that cause lymphatic filariasis, an infection of the lymphatic system by filarial worms. It affects over 120 million people, primarily in Africa, South America, and other tropical and subtropical countries.
Diphtheria	Diphtheria (Greek διφθ?ρα (diphthera) 'pair of leather scrolls') is an upper respiratory tract illness caused by Corynebacterium diphtheriae, a facultative anaerobic, Gram-positive bacterium.

It is characterized by sore throat, low fever, and an adherent membrane (a pseudomembrane) on the tonsils, pharynx, and/or nasal cavity. A milder form of diphtheria can be restricted to the skin.

Epstein-Barr virus	The Epstein-Barr virus , also called human herpesvirus 4 (HHV-4), is a virus of the herpes family, which includes herpes simplex virus 1 and 2, and is one of the most common viruses in humans. It is best known as the cause of infectious mononucleosis. It is also associated with particular forms of cancer, particularly Hodgkin's lymphoma, Burkitt's lymphoma, nasopharyngeal carcinoma, and central nervous system lymphomas associated with HIV. Finally, there is evidence that infection with the virus is associated with a higher risk of certain autoimmune diseases, especially dermatomyositis, systemic lupus erythematosus, rheumatoid arthritis, Sjögren's syndrome, and multiple sclerosis.
Herpes simplex	Herpes simplex is a viral disease caused by both Herpes simplex virus type 1 (HSV-1) and type 2 (HSV-2). Infection with the herpes virus is categorized into one of several distinct disorders based on the site of infection. Oral herpes, the visible symptoms of which are colloquially called cold sores or fever blisters, infects the face and mouth.
Marburg virus	Marburg virus was first noticed and described during small epidemics in the German cities Marburg and Frankfurt and the Yugoslavian capital Belgrade. Workers were accidentally exposed to tissues of infected grivets (Chlorocebus aethiops) at the city's former main industrial plant, the Behringwerke, then part of Hoechst, and today of CSL Behring. During these outbreaks, 31 people became infected and seven of them died.
Pathogenicity	Pathogenicity is the ability of a pathogen to produce an infectious disease in an organism. It is often used interchangeably with the term 'virulence', although virulence is used more specifically to describe the relative degree of damage done by a pathogen, or the degree of pathogenicity caused by an organism. A pathogen is either pathogenic or not, and is determined by the pathogen's ability to produce toxins, its ability to enter tissue and colonize and its ability to spread from host to host.
Pneumocystis jirovecii	Pneumocystis jirovecii is a yeast-like fungus of the genus Pneumocystis. The causative organism of Pneumocystis pneumonia, it is an important human pathogen, particularly among immunocompromised hosts. Prior to its discovery as a human-specific pathogen, P. jirovecii was known as P. carinii.
Rickettsia prowazekii	Rickettsia prowazekii is a species of gram negative, Alpha Proteobacteria, obligate intracellular parasitic, aerobic bacteria that is the etiologic agent of epidemic typhus, transmitted in the faeces of lice. In North America, the main reservoir for R. prowazekii is the flying squirrel. R.

Chapter 5. Part 5: Medicine and Immunology

Shigella flexneri	Shigella flexneri is a species of Gram-negative bacteria in the genus Shigella that can cause diarrhea in humans. There are several different serogroups of Shigella; S. flexneri belongs to group B. S. flexneri infections can usually be treated with antibiotics although some strains have become resistant. Less severe cases are not usually treated because they become more resistant in the future.
Virulence	Virulence is by MeSH definition the degree of pathogenicity within a group or species of parasites as indicated by case fatality rates and/or the ability of the organism to invade the tissues of the host. The pathogenicity of an organism - its ability to cause disease - is determined by its virulence factors. The noun virulence derives from the adjective virulent.
Infectious dose	Infectious dose is the amount of pathogen (measured in number of microorganisms) required to cause an infection in the host. Usually it varies according to the pathogenic agent and the consumer's age and overall health. Infectious doses for some known microorganisms •Escherichia coli : very large (10^6 - 10^8 of organisms)•Salmonella : quite large (> 10^5 of organisms)•Cholera : relatively large (10^4 - 10^6 of organisms)•Bacillus anthracis : relatively large (10^4 spores)•Campylobacter jejuni: low (500 organisms)•Shigella : very low (10s of organisms)•C.parvum : very low (10 to 30 oocysts)•Escherichia coli O157:H7 : very low (< 10 organisms)•Entamoeba coli : extremely low (from 1 cyst).
Pathogenicity island	Pathogenicity islands (PAIs) are a distinct class of genomic islands acquired by microorganisms through horizontal gene transfer. They are incorporated in the genome of pathogenic organisms, but are usually absent from those nonpathogenic organisms of the same or closely related species. These mobile genetic elements may range from 10-200 kb and encode genes which contribute to the virulence of the respective pathogen.
Bacillus thuringiensis	Bacillus thuringiensis is a Gram-positive, soil-dwelling bacterium, commonly used as a biological pesticide; alternatively, the Cry toxin may be extracted and used as a pesticide. B. thuringiensis also occurs naturally in the gut of caterpillars of various types of moths and butterflies, as well as on the dark surface of plants. During sporulation many Bt strains produce crystal proteins (proteinaceous inclusions), called δ-endotoxins, that have insecticidal action.
Baculovirus	The baculoviruses are a family of large rod-shaped viruses that can be divided to two genera: nucleopolyhedroviruses (NPV) and granuloviruses (GV). While GVs contain only one nucleocapsid per envelope, NPVs contain either single (SNPV) or multiple (MNPV) nucleocapsids per envelope.

Fomite	A fomite is any inanimate object or substance capable of carrying infectious organisms (such as germs or parasites) and hence transferring them from one individual to another. A fomite can be anything (such as a cloth or mop head). Skin cells, hair, clothing, and bedding are common hospital sources of contamination.
Horizontal transmission	Horizontal disease transmission is the transmission of an infectious agent, such as bacterial, fungal, or viral infection, between members of the same species that are not in a parent-child relationship. Horizontal transmission tends to evolve virulence. It is therefore a critical concept for evolutionary medicine.
Reservoir	A reservoir is used to store water. Reservoirs may be created in river valleys by the construction of a dam or may be built by excavation in the ground or by conventional construction techniques such a brickwork or cast concrete. The term reservoir may also be used to describe underground reservoirs such as an oil or water well.
West Nile virus	West Nile virus is a virus of the family Flaviviridae. Part of the Japanese encephalitis (JE) antigenic complex of viruses, it is found in both tropical and temperate regions. It mainly infects birds, but is known to infect humans, horses, dogs, cats, bats, chipmunks, skunks, squirrels, domestic rabbits, crows, robins, crocodiles and alligators.
Yellow fever	Yellow fever is an acute viral hemorrhagic disease. The virus is a 40 to 50 nm enveloped RNA virus with positive sense of the Flaviviridae family. The yellow fever virus is transmitted by the bite of female mosquitoes (the yellow fever mosquito, Aedes aegypti, and other species) and is found in tropical and subtropical areas in South America and Africa, but not in Asia.
Equine encephalitis	Equine encephalitis may be caused by several viruses:•Eastern equine encephalitis virus•Western equine encephalitis virus•West Nile virus•Venezuelan equine encephalitis virus.
Insecticide	An insecticide is a pesticide used against insects. They include ovicides and larvicides used against the eggs and larvae of insects respectively. Insecticides are used in agriculture, medicine, industry and the household.
Transovarial transmission	Transovarial or transovarian transmission occurs in certain arthropod vectors as they transmit disease-causing bacteria from parent arthropod to offspring arthropod.

Chapter 5. Part 5: Medicine and Immunology

For instance, Rickettsia rickettsii, carried within ticks, is passed on from parent to offspring tick by transovarial transmission. In contrast, Rickettsia prowazekii is not passed on by transovarian transmission because it kills the vector that carries it (human louse).

Vertical transmission	Vertical transmission, is the transmission of an infection or other disease from mother to child immediately before and after birth during the perinatal period. A pathogen's transmissibility refers to its capacity for vertical transmission. The concept of vertical transmission is also used in population genetics to describe inheritance of an allele or condition from either the father or mother.
Rotavirus	Rotavirus is the most common cause of severe diarrhoea among infants and young children. It is a genus of double-stranded RNA virus in the family Reoviridae. By the age of five, nearly every child in the world has been infected with rotavirus at least once.
Virulence factor	Virulence factors are molecules expressed and secreted by pathogens (bacteria, viruses, fungi and protozoa) that enable them to achieve the following:•colonization of a niche in the host (this includes adhesion to cells)•Immunoevasion, evasion of the host's immune response•Immunosuppression, inhibition of the host's immune response•entry into and exit out of cells (if the pathogen is an intracellular one)•obtain nutrition from the host.

Virulence factors are very often responsible for causing disease in the host as they inhibit certain host functions.

Pathogens possess a wide array of virulence factors. Some are intrinsic to the bacteria (e.g. capsules and endotoxin) whereas others are obtained from plasmids (e.g. some toxins). |
| Genomic island | A Genomic island is part of a genome that has evidence of horizontal origins. The term is usually used in microbiology, especially with regard to bacteria. A GI can code for many functions, can be involved in symbiosis or pathogenesis, and may help an organism's adaptation. |
| Streptococcus agalactiae | Streptococcus agalactiae is a beta-hemolytic Gram-positive streptococcus.

The CAMP test is an important test for identification. GBS (group B Streptococcus species) are screened through this test. |
| Clostridium tetani | Clostridium tetani is a rod-shaped, anaerobic bacterium of the genus Clostridium. Like other Clostridium species, it is Gram-positive, and its appearance on a gram stain resembles tennis rackets or drumsticks. C. tetani is found as spores in soil or in the gastrointestinal tract of animals. |

Neisseria meningitidis	Neisseria meningitidis is a heterotrophic gram-negative diplococcal bacterium best known for its role in meningitis and other forms of meningococcal disease such as meningococcemia. N. meningitidis is a major cause of morbidity and mortality during childhood in industrialized countries and is responsible for epidemics in Africa and in Asia. Approximately 2500 to 3500 cases of N meningitidis infection occur annually in the United States, with a case rate of about 1 in 100,000. Children younger than 5 years are at greatest risk, followed by teenagers of high school age.
Pertactin	Pertactin Pertactin is a highly immunogenic virulence factor of Bordetella pertussis, a bacterium that causes pertussis. Specifically, it is an outer membrane protein that promotes adhesion to tracheal epithelial cells. PRN is purified from Bordetella pertussis and is used for the vaccine production as one of the important components of acellular pertussis vaccine.
Biofilm	A biofilm is an aggregate of microorganisms in which cells adhere to each other on a surface. These adherent cells are frequently embedded within a self-produced matrix of extracellular polymeric substance (EPS). Biofilm EPS, which is also referred to as slime (although not everything described as slime is a biofilm), is a polymeric conglomeration generally composed of extracellular DNA, proteins, and polysaccharides.
Group B	The Group B referred to a set of regulations introduced in 1982 for competition vehicles in sportscar racing and rallying regulated by the FIA. The Group B regulations fostered some of the quickest, most powerful and sophisticated rally cars ever built. However, a series of major accidents, some fatal, were blamed on their outright speed. After the death of Henri Toivonen and his co-driver in the 1986 Tour de Corse, the FIA disestablished the class, which was replaced as the top-line formula by Group A. The short-lived Group B era has acquired legendary status among rally fans.
Membrane protein	A membrane protein is a protein molecule that is attached to, or associated with the membrane of a cell or an organelle. More than half of all proteins interact with membranes. Biological membranes consist of a phospholipid bilayer and a variety of proteins that accomplish vital biological functions.
Exotoxin	An exotoxin is a toxin secreted by a microorganism, like bacteria, fungi, algae, and protozoa. An exotoxin can cause damage to the host by destroying cells or disrupting normal cellular metabolism. They are highly potent and can cause major damage to the host.

Chapter 5. Part 5: Medicine and Immunology

Endotoxin	The term endotoxin was coined by Richard Friedrich Johannes Pfeiffer, who distinguished between exotoxin, which he classified as a toxin that is released by bacteria into the environment, and endotoxin, which he considered to be a toxin kept 'within' the bacterial cell and to be released only after destruction of the bacterial cell wall. Today, the term 'endotoxin' is used synonymously to the term lipopolysaccharide, which is a major constituent of the outer cell wall of Gram-negative bacteria. Larger amounts of endotoxins can be mobilized if Gram-negative bacteria are killed or destroyed by detergents.
AB toxin	ADPrib_exo_Tox The AB toxins are two-component protein complexes secreted by a number of pathogenic bacteria. They can be classified as Type III toxins because they interfere with internal cell function. They are named AB toxins due to their components: the 'A' component is usually the 'active' portion, and the 'B' component is usually the 'binding' portion.
Shiga toxin	Shiga toxins are a family of related toxins with two major groups, Stx1 and Stx2, whose genes are considered to be part of the genome of lambdoid prophages. The toxins are named for Kiyoshi Shiga, who first described the bacterial origin of dysentery caused by Shigella dysenteriae. The most common sources for Shiga toxin are the bacteria S. dysenteriae and the Shigatoxigenic group of Escherichia coli (STEC), which includes serotypes O157:H7, O104:H4, and other enterohemorrhagic E. coli (EHEC).
Alpha toxin	Alpha Toxin or alpha-toxin refers to several different protein toxins produced by bacteria. Alpha toxin may be:•Staphylococcus aureus alpha toxin, a membrane-disrupting toxin that creates pores causing hemolysis and tissue damage.•Clostridium perfringens alpha toxin, a membrane-disrupting toxin with phospholipase C activity, which is directly responsible for gas gangrene and myonecrosis.•Pseudomonas aeruginosa alpha toxin..
Toxin	A toxin is a poisonous substance produced within living cells or organisms; man-made substances created by artificial processes are thus excluded. The term was first used by organic chemist Ludwig Brieger (1849-1919). For a toxic substance not produced within living organisms, 'toxicant' and 'toxics' are also sometimes used..
G factor	The g factor (short for 'general factor') is a construct developed in psychometric investigations of cognitive abilities. It is a variable that summarizes the all-positive correlations that empirical research has consistently found to exist between mental tests, regardless of the tests' contents. The g factor typically accounts for 40 to 50 percent of the variance in IQ test performance.
Anthrax	Anthrax is an acute disease caused by the bacterium Bacillus anthracis.

	Most forms of the disease are lethal, and it affects both humans and other animals. There are effective vaccines against anthrax, and some forms of the disease respond well to antibiotic treatment.
Bacillus anthracis	Bacillus anthracis is a Gram-positive spore-forming, rod-shaped bacterium, with a width of 1-1.2μm and a length of 3-5μm. It can be grown in an ordinary nutrient medium under aerobic or anaerobic conditions. It is the only bacterium known to synthesize a protein capsule (D-glutamate), and the only pathogenic bacterium to carry its own adenylyl cyclase virulence factor (edema factor).
Pertussis toxin	Pertussis toxin, subunit 1 Pertussis toxin is a protein-based AB_5-type exotoxin produced by the bacterium Bordetella pertussis, which causes whooping cough. PT is involved in the colonization of the respiratory tract and the establishment of infection. Research suggests PT may have a therapeutic role in treating a number of common human ailments, including hypertension, viral inhibition, and autoimmune inhibition.
Anthrax toxin	Anthrax toxin lethal factor middle domain Anthrax toxin is a three-protein exotoxin secreted by virulent strains of the bacterium, Bacillus anthracis--the causative agent of anthrax. The toxin was first discovered by Harry Smith in 1954. Anthrax toxin is composed of a cell-binding protein, known as protective antigen (PA), and two enzyme components, called edema factor (EF) and lethal factor (LF). These three protein components act together to impart their physiological effects.
Delta endotoxin	Delta endotoxin, N-terminal domain Delta endotoxins (δ-endotoxins, also called Cry and Cyt toxins) are pore-forming toxins produced by Bacillus thuringiensis species of bacteria. They are useful for their insecticidal action. During spore formation the bacteria produce crystals of this protein.
Agrobacterium tumefaciens	Agrobacterium tumefaciens is the causal agent of crown gall disease (the formation of tumours) in over 140 species of dicot. It is a rod shaped, Gram negative soil bacterium (Smith et al., 1907). Symptoms are caused by the insertion of a small segment of DNA (known as the T-DNA, for 'transfer DNA'), from a plasmid, into the plant cell, which is incorporated at a semi-random location into the plant genome.

Chapter 5. Part 5: Medicine and Immunology

Pseudomonas syringae	Pseudomonas syringae is a rod shaped, Gram-negative bacterium with polar flagella. It is a plant pathogen which can infect a wide range of plant species, and exists as over 50 different pathovars, all of which are available to legitimate researches via international culture collections such as the NCPPB, ICMP, and others. Many of these pathovars were once considered to be individual species within the Pseudomonas genus, but molecular biology techniques such as DNA hybridization have shown these to in fact all be part of the P. syringae species.
Ti plasmid	Ti plasmid is a circular plasmid that often, but not always, is a part of the genetic equipment that Agrobacterium tumefaciens and Agrobacterium rhizogenes use to transduce its genetic material to plants. Ti stands for tumor inducing. The Ti plasmid is lost when Agrobacterium is grown above 28° C. Such cured bacteria do not induce crown galls, i.e. they become avirulent.
Genetic analysis	Genetic analysis can be used generally to describe methods both used in and resulting from the sciences of genetics and molecular biology, or to applications resulting from this research. Genetic analysis may be done to identify genetic/inherited disorders and also to make a differential diagnosis in certain somatic diseases such as cancer. Genetic analyses of cancer include detection of mutations, fusion genes, and DNA copy number changes.
Genomics	Genomics is a discipline in genetics concerned with the study of the genomes of organisms. The field includes efforts to determine the entire DNA sequence of organisms and fine-scale genetic mapping. The field also includes studies of intragenomic phenomena such as heterosis, epistasis, pleiotropy and other interactions between loci and alleles within the genome.
Legionella pneumophila	Legionella pneumophila is a thin, pleomorphic, flagellated Gram-negative bacterium of the genus Legionella. L. pneumophila is the primary human pathogenic bacterium in this group and is the causative agent of legionellosis or Legionnaires' disease. Characterization L. pneumophila is non-acid-fast, non-sporulating, and morphologically a non-capsulated rod-like bacteria.
Listeria monocytogenes	Listeria monocytogenes, a facultative anaerobe, intracellular bacterium, is the causative agent of listeriosis. It is one of the most virulent foodborne pathogens, with 20 to 30 percent of clinical infections resulting in death. Responsible for approximately 2,500 illnesses and 500 deaths in the United States (U.S).
Q fever	Q fever is a disease caused by infection with Coxiella burnetii, a bacterium that affects humans and other animals.

	This organism is uncommon, but may be found in cattle, sheep, goats and other domestic mammals, including cats and dogs. The infection results from inhalation of a spore-like small cell variant, and from contact with the milk, urine, feces, vaginal mucus, or semen of infected animals.
Shigella dysenteriae	Shigella dysenteriae is a species of the rod-shaped bacterial genus Shigella. Shigella can cause shigellosis (bacillary dysentery). Shigellae are Gram-negative, non-spore-forming, facultatively anaerobic, non-motile bacteria.
Parasitism	Parasitism is a type of non mutual relationship between organisms of different species where one organism, the parasite, benefits at the expense of the other, the host. Traditionally parasite referred to organisms with lifestages that needed more than one host (e.g. Taenia solium). These are now called macroparasites (typically protozoa and helminths).
Ames test	The Ames test is a biological assay to assess the mutagenic potential of chemical compounds. A positive test indicates that the chemical is mutagenic and therefore may act as a carcinogen, since cancer is often linked to mutation. However, a number of false-positives and false-negatives are known.
Holliday junction	A Holliday junction is a mobile junction between four strands of DNA. The structure is named after Robin Holliday, who proposed it in 1964 to account for a particular type of exchange of genetic information he observed in yeast known as homologous recombination. Holliday junctions are highly conserved structures, from prokaryotes to mammals. Because these junctions are between homologous sequences, they can slide up and down the DNA. In bacteria, this sliding is facilitated by the RuvABC complex or RecG protein, molecular motors that use the energy of ATP hydrolysis to push the junction around.
Protein A	Protein A is a 40-60 kDa MSCRAMM surface protein originally found in the cell wall of the bacterium Staphylococcus aureus. It is encoded by the spa gene and its regulation is controlled by DNA topology, cellular osmolarity, and a two-component system called ArlS-ArlR. It has found use in biochemical research because of its ability to bind immunoglobulins. It binds proteins from many of mammalian species, most notably IgGs.
Phase variation	Phase variation is a method for dealing with rapidly varying environments without requiring random mutation employed by various types of bacteria, including Salmonella species. It involves the variation of protein expression, frequently in an on-off fashion, within different parts of a bacterial population. Although it has been most commonly studied in the context of immune evasion, it is observed in many other areas as well.

Chapter 5. Part 5: Medicine and Immunology

Site-specific recombination	Site-specific recombination, is a type of genetic recombination in which DNA strand exchange takes place between segments possessing only a limited degree of sequence homology. Site-specific recombinases perform rearrangements of DNA segments by recognizing and binding to short DNA sequences (sites), at which they cleave the DNA backbone, exchange the two DNA helices involved and rejoin the DNA strands. While in some site-specific recombination systems just a recombinase enzyme and the recombination sites is enough to perform all these reactions, in other systems a number of accessory proteins and/or accessory sites are also needed.
Quorum sensing	Quorum sensing is a system of stimulus and response correlated to population density. Many species of bacteria use quorum sensing to coordinate gene expression according to the density of their local population. In similar fashion, some social insects use quorum sensing to determine where to nest.
Antigenic shift	Antigenic shift is the process by which two or more different strains of a virus, or strains of two or more different viruses, combine to form a new subtype having a mixture of the surface antigens of the two or more original strains. The term is often applied specifically to influenza, as that is the best-known example, but the process is also known to occur with other viruses, such as visna virus in sheep. Antigenic shift is a specific case of reassortment or viral shift that confers a phenotypic change.
Coronavirus	Coronaviruses are species in the genera of animal virus belonging to the subfamily Coronavirinae in the family Coronaviridae.Coronaviruses are enveloped viruses with a positive-sense single-stranded RNA genome and a helical symmetry. The genomic size of coronaviruses ranges from approximately 16 to 31 kilobases, extraordinarily large for an RNA virus. The name 'coronavirus' is derived from the Greek κορ?να, meaning crown, as the virus envelope appears under electron microscopy (E.M).
Palmaria palmata	Palmaria palmata Kuntze, also called dulse, dillisk, dilsk, red dulse, sea lettuce flakes or creathnach, is a red alga (Rhodophyta) previously referred to as Rhodymenia palmata (Linnaeus) Greville. It grows on the northern coasts of the Atlantic and Pacific oceans. It is a well-known snack food, and in Iceland, where it is known as söl, it has been an important source of fiber throughout the centuries.
Severe acute respiratory syndrome	Severe acute respiratory syndrome is a respiratory disease in humans which is caused by the SARS coronavirus (SARS-CoV). Between November 2002 and July 2003, an outbreak of SARS in Hong Kong nearly became a pandemic, with 8,422 cases and 916 deaths worldwide (10.9% fatality) according to the World Health Organization. Within weeks, SARS spread from Hong Kong to infect individuals in 37 countries in early 2003.

Common cold	The common cold is a viral infectious disease of the upper respiratory system which affects primarily the nose. Symptoms include a cough, sore throat, runny nose, and fever which usually resolve in seven to ten days, with some symptoms lasting up to three weeks. Well over 200 viruses are implicated in the cause of the common cold; the rhinoviruses are the most common.
Molecular biology	Molecular biology is the branch of biology that deals with the molecular basis of biological activity. This field overlaps with other areas of biology and chemistry, particularly genetics and biochemistry. Molecular biology chiefly concerns itself with understanding the interactions between the various systems of a cell, including the interactions between the different types of DNA, RNA and protein biosynthesis as well as learning how these interactions are regulated.
Superinfection	In virology, superinfection is the process by which a cell that has previously been infected by one virus gets coinfected with a different strain of the virus, or another virus, at a later point in time. Viral superinfections of serious conditions can lead to resistant strains of the virus, which may prompt a change of treatment. For example, an individual superinfected with two separate strains of HIV may contract a strain that is resistant to antiretroviral treatment.
Chemokine receptor	Chemokine receptors are cytokine receptors found on the surface of certain cells that interact with a type of cytokine called a chemokine. There have been 19 distinct chemokine receptors described in mammals. Each has a 7-transmembrane (7TM) structure and couples to G-protein for signal transduction within a cell, making them members of a large protein family of G protein-coupled receptors.
Haemophilus influenzae	Haemophilus influenzae, formerly called Pfeiffer's bacillus or Bacillus influenzae, Gram-negative, rod-shaped bacterium first described in 1892 by Richard Pfeiffer during an influenza pandemic. A member of the Pasteurellaceae family, it is generally aerobic, but can grow as a facultative anaerobe. H. influenzae was mistakenly considered to be the cause of influenza until 1933, when the viral etiology of the flu became apparent; the bacterium is colloquially known as bacterial influenza.
RNA polymerase	RNA polymerase also known as DNA-dependent RNA polymerase, is an enzyme that produces RNA. In cells, RNAP is necessary for constructing RNA chains using DNA genes as templates, a process called transcription. RNA polymerase enzymes are essential to life and are found in all organisms and many viruses. In chemical terms, RNAP is a nucleotidyl transferase that polymerizes ribonucleotides at the 3' end of an RNA transcript.
Influenza	Influenza, commonly referred to as the flu, is an infectious disease caused by RNA viruses of the family Orthomyxoviridae (the influenza viruses), that affects birds and mammals. The most common symptoms of the disease are chills, fever, sore throat, muscle pains, severe headache, coughing, weakness/fatigue and general discomfort.

Chapter 5. Part 5: Medicine and Immunology

Pandemic	A pandemic is an epidemic of infectious disease that has spread through human populations across a large region; for instance multiple continents, or even worldwide. A widespread endemic disease that is stable in terms of how many people are getting sick from it is not a pandemic. Further, flu pandemics generally exclude recurrences of seasonal flu.
Symptom	A symptom is a departure from normal function or feeling which is noticed by a patient, indicating the presence of disease or abnormality. A symptom is subjective, observed by the patient, and not measured.
	A symptom may not be a malady, for example symptoms of pregnancy.
TATA-binding protein	The TATA-binding protein is a general transcription factor that binds specifically to a DNA sequence called the TATA box. This DNA sequence is found about 25 base pairs upstream of the transcription start site in some eukaryotic gene promoters. TBP, along with a variety of TBP-associated factors, make up the TFIID, a general transcription factor that in turn makes up part of the RNA polymerase II preinitiation complex.
Cancer vaccine	The term cancer vaccine refers to a vaccine that either prevents infections with cancer-causing viruses, treats existing cancer or prevents the development of cancer in certain high risk individuals. (The ones that treat existing cancer are known as therapeutic cancer vaccines).
	Some types of cancer, such as cervical cancer and some liver cancers, are caused by viruses, and traditional vaccines against those viruses, such as HPV vaccine and Hepatitis B vaccine, will prevent those cancers.
Human papillomavirus	Human papillomavirus is a virus from the papillomavirus family that is capable of infecting humans. Like all papillomaviruses, HPVs establish productive infections only in keratinocytes of the skin or mucous membranes. While the majority of the known types of HPV cause no symptoms in most people, some types can cause warts (verrucae), while others can - in a minority of cases - lead to cancers of the cervix, vulva, vagina, penis, oropharynx and anus.
Protein	Proteins are biochemical compounds consisting of one or more polypeptides typically folded into a globular or fibrous form, facilitating a biological function.
	A polypeptide is a single linear polymer chain of amino acids bonded together by peptide bonds between the carboxyl and amino groups of adjacent amino acid residues. The sequence of amino acids in a protein is defined by the sequence of a gene, which is encoded in the genetic code.
Wart	A wart is generally a small, rough growth, typically on hands and feet but often other locations, that can resemble a cauliflower or a solid blister.

	They are caused by a viral infection, specifically by human papillomavirus 2 and 7. There are as many as 10 varieties of warts, the most common considered to be mostly harmless. It is possible to get warts from others; they are contagious and usually enter the body in an area of broken skin.
Typhoid fever	Typhoid fever, is a common worldwide illness, transmitted by the ingestion of food or water contaminated with the feces of an infected person, which contain the bacterium Salmonella enterica, serovar Typhi. The bacteria then perforate through the intestinal wall and are phagocytosed by macrophages. The organism is a Gram-negative short bacillus that is motile due to its peritrichous flagella.
Toxic shock syndrome	Toxic shock syndrome is a potentially fatal illness caused by a bacterial toxin. Different bacterial toxins may cause toxic shock syndrome, depending on the situation. The causative bacteria include Staphylococcus aureus and Streptococcus pyogenes.
DNA Research	DNA Research is an international, peer reviewed journal of genomics and DNA research.
Francisella tularensis	Francisella tularensis is a pathogenic species of gram-negative bacteria and the causative agent of tularemia or rabbit fever. It is a fastidious, requires cysteine for growth, facultative intracellular bacterium. Due to its ease of spread by aerosol and its high virulence, F. tularensis is classified as a Class A agent by the U.S. government.
Livestock	Livestock refers to one or more domesticated animals raised in an agricultural setting to produce commodities such as food, fiber and labor Livestock generally are raised for subsistence or for profit.
Coagulase	Coagulase is a protein produced by several microorganisms that enables the conversion of fibrinogen to fibrin. In the laboratory, it is used to distinguish between different types of Staphylococcus isolates. Importantly, S. aureus is coagulase-positive, meaning that coagulase negativity excludes S. aureus.
Nosocomial infection	A nosocomial infection also known as a hospital-acquired infection or HAI, is an infection whose development is favoured by a hospital environment, such as one acquired by a patient during a hospital visit or one developing among hospital staff. Such infections include fungal and bacterial infections and are aggravated by the reduced resistance of individual patients. In the United States, the Centers for Disease Control and Prevention estimate that roughly 1.7 million hospital-associated infections, from all types of microorganisms, including bacteria, combined, cause or contribute to 99,000 deaths each year.

Chapter 5. Part 5: Medicine and Immunology

Soft tissue	In anatomy, the term soft tissue refers to tissues that connect, support, or surround other structures and organs of the body, not being bone. Soft tissue includes tendons, ligaments, fascia, skin, fibrous tissues, fat, and synovial membranes (which are connective tissue), and muscles, nerves and blood vessels (which are not connective tissue). It is sometimes defined by what it is not.
Gentamicin	Gentamicin is an aminoglycoside antibiotic, used to treat many types of bacterial infections, particularly those caused by Gram-negative organisms. However, gentamicin is not used for Neisseria gonorrhoeae, Neisseria meningitidis or Legionella pneumophila. Gentamicin is also ototoxic and nephrotoxic, with this toxicity remaining a major problem in clinical use.
Glomerulonephritis	Glomerulonephritis is a renal disease (usually of both kidneys) characterized by inflammation of the glomeruli, or small blood vessels in the kidneys. It may present with isolated hematuria and/or proteinuria (blood or protein in the urine); or as a nephrotic syndrome, a nephritic syndrome, acute renal failure, or chronic renal failure. They are categorised into several different pathological patterns, which are broadly grouped into non-proliferative or proliferative types.
Necrotizing fasciitis	Necrotizing fasciitis commonly known as flesh-eating disease or Flesh-eating bacteria syndrome, is a rare infection of the deeper layers of skin and subcutaneous tissues, easily spreading across the fascial plane within the subcutaneous tissue. Necrotizing fasciitis is a quickly progressing and severe disease of sudden onset and is usually treated immediately with high doses of intravenous antibiotics. Type I describes a polymicrobial infection, whereas Type II describes a monomicrobial infection.
Cryptococcus	Cryptococcus is a genus of fungus. These fungi grow in culture as yeasts. The sexual forms or teleomorphs of Cryptococcus species are filamentous fungi in the genus Filobasidiella.
Generation time	Generation time is a quantity used in population biology and demography to reflect the relative size of intervals of offspring production. Generation time usually expresses the average age of breeding females within a population. Suppose females begin breeding at age α and stop breeding at age ω, then the average age of first reproduction of a cohort of females is $$T = \frac{\sum_{x=\alpha}^{\omega} x l(x) m(x)}{\sum_{x=\alpha}^{\omega} l(x) m(x)}$$ where $l(x)$ is the hazard function and $m(x)$ is the fecundity of females aged x.

Measles	Measles, is an infection of the respiratory system caused by a virus, specifically a paramyxovirus of the genus Morbillivirus. Morbilliviruses, like other paramyxoviruses, are enveloped, single-stranded, negative-sense RNA viruses. Symptoms include fever, cough, runny nose, red eyes and a generalized, maculopapular, erythematous rash.
Rubella vaccine	Rubella vaccine is a vaccine used against rubella. One form is called 'Meruvax'.
Bordetella pertussis	Bordetella pertussis is a Gram-negative, aerobic coccobacillus capsulate of the genus Bordetella, and the causative agent of pertussis or whooping cough. Unlike B. bronchiseptica, B. pertussis is nonmotile. Its virulence factors include pertussis toxin, filamentous hæmagglutinin, pertactin, fimbria, and tracheal cytotoxin.
Pneumonia	Pneumonia is an inflammatory condition of the lung--especially affecting the microscopic air sacs (alveoli)--associated with fever, chest symptoms, and a lack of air space (consolidation) on a chest X-ray. Pneumonia is typically caused by an infection but there are a number of other causes. Infectious agents include: bacteria, viruses, fungi, and parasites.
Bacterial pneumonia	Bacterial pneumonia is a type of pneumonia caused by bacterial infection. •Fever•Rigors•Cough•Dyspnea•Chest pain•Pneumococcal pneumonia can cause HemoptysisTypes Gram-positive

Streptococcus pneumoniae (J13) is the most common bacterial cause of pneumonia in all age groups except newborn infants. Streptococcus pneumoniae is a Gram-positive bacterium that often lives in the throat of people who do not have pneumonia. |
Respiratory tract infection	Respiratory tract infection refers to any of a number of infectious diseases involving the respiratory tract. An infection of this type is normally further classified as an upper respiratory tract infection or a lower respiratory tract infection. Lower respiratory infections, such as pneumonia, tend to be far more serious conditions than upper respiratory infections, such as the common cold.
Macrolide	The macrolides are a group of drugs (typically antibiotics) whose activity stems from the presence of a macrolide ring, a large macrocyclic lactone ring to which one or more deoxy sugars, usually cladinose and desosamine, may be attached. The lactone rings are usually 14-, 15-, or 16- membered. Macrolides belong to the polyketide class of natural products.
Disseminated disease	Disseminated disease refers to a diffuse disease-process, generally either infectious or neoplastic. The term may sometimes also characterize connective tissue disease.

Chapter 5. Part 5: Medicine and Immunology

Meningitis	Meningitis is inflammation of the protective membranes covering the brain and spinal cord, known collectively as the meninges. The inflammation may be caused by infection with viruses, bacteria, or other microorganisms, and less commonly by certain drugs. Meningitis can be life-threatening because of the inflammation's proximity to the brain and spinal cord; therefore the condition is classified as a medical emergency.
Pneumococcal polysaccharide vaccine	Pneumococcal polysaccharide vaccine -- the latest version is known as Pneumovax 23 (PPV-23) -- is the first pneumococcal vaccine, the first vaccine derived from a capsular polysaccharide, and an important landmark in medical history. The polysaccharide antigens were used to induce type-specific antibodies that enhanced opsonization, phagocytosis, and killing of pneumococci by phagocytic cells. The pneumococcal polysaccharide vaccine is widely used in high-risk adults.
Amphotericin B	Amphotericin B is a polyene antifungal drug, often used intravenously for systemic fungal infections. It was originally extracted from Streptomyces nodosus, a filamentous bacterium, in 1955 at the Squibb Institute for Medical Research from cultures of an undescribed streptomycete isolated from the soil collected in the Orinoco River region of Venezuela. Its name originates from the chemical's amphoteric properties.
Blastomycosis	Blastomycosis is a fungal infection caused by the organism Blastomyces dermatitidis. Endemic to portions of North America, blastomycosis causes clinical symptoms similar to histoplasmosis. Blastomycosis can present in one of the following ways:•a flu-like illness with fever, chills, myalgia, headache, and a nonproductive cough which resolves within days.•an acute illness resembling bacterial pneumonia, with symptoms of high fever, chills, a productive cough, and pleuritic chest pain.•a chronic illness that mimics tuberculosis or lung cancer, with symptoms of low-grade fever, a productive cough, night sweats, and weight loss.•a fast, progressive, and severe disease that manifests as ARDS, with fever, shortness of breath, tachypnea, hypoxemia, and diffuse pulmonary infiltrates.•skin lesions, usually asymptomatic, appear as ulcerated lesions with small pustules at the margins•bone lytic lesions can cause bone or joint pain.•prostatitis may be asymptomatic or may cause pain on urinating.•laryngeal involvement causes hoarseness.Cause Infection occurs by inhalation of the fungus from its natural soil habitat.
Coccidioidomycosis	Coccidioidomycosis is a fungal disease caused by Coccidioides immitis or C. posadasii. It is endemic in certain parts of Arizona, California, Nevada, New Mexico, Texas, Utah and northwestern Mexico. C.

immitis resides in the soil in certain parts of the southwestern United States, northern Mexico, and parts of Central and South America. It is dormant during long dry spells, then develops as a mold with long filaments that break off into airborne spores when the rains come.

Cryptococcosis	Cryptococcosis, is a potentially fatal fungal disease. It is caused by one of two species; Cryptococcus neoformans and Cryptococcus gattii. These were all previously thought to be subspecies of C. neoformans, but have now been identified as distinct species.
Histoplasmosis	Histoplasmosis is a disease caused by the fungus Histoplasma capsulatum. Symptoms of this infection vary greatly, but the disease primarily affects the lungs. Occasionally, other organs are affected; this is called disseminated histoplasmosis, and it can be fatal if left untreated.
Conidia	Conidia, are asexual, non-motile spores of a fungus; they are also called mitospores due to the way they are generated through the cellular process of mitosis. The two new haploid cells are genetically identical to the haploid parent, and can develop into new organisms if conditions are favorable, and serve in biological dispersal. Asexual reproduction in Ascomycetes (the Phylum Ascomycota) is by the formation of conidia, which are borne on specialized stalks called conidiophores.
Ethambutol	Ethambutol is a bacteriostatic antimycobacterial drug prescribed to treat tuberculosis. It is usually given in combination with other tuberculosis drugs, such as isoniazid, rifampicin and pyrazinamide. It is sold under the trade names Myambutol and Servambutol.
Isoniazid	Isoniazid also known as isonicotinylhydrazine (INH), is an organic compound that is the first-line medication in prevention and treatment of tuberculosis. It was first discovered in 1912, and later, in 1951, was found to be effective against tuberculosis by inhibiting its mycolic acid (wax coat). Isoniazid is never used on its own to treat active tuberculosis because resistance quickly develops.
Drug resistance	Drug resistance is the reduction in effectiveness of a drug such as an antimicrobial or an antineoplastic in curing a disease or condition. When the drug is not intended to kill or inhibit a pathogen, then the term is equivalent to dosage failure or drug tolerance. More commonly, the term is used in the context of resistance acquired by pathogens.
Entamoeba histolytica	Entamoeba histolytica is an anaerobic parasitic protozoan, part of the genus Entamoeba. Predominantly infecting humans and other primates, E. histolytica is estimated to infect about 50 million people worldwide.

Chapter 5. Part 5: Medicine and Immunology

	Previously, it was thought that 10% of the world population was infected, but these figures predate the recognition that at least 90% of these infections were due to a second species, E. dispar.
Slime mold	Slime mold is a broad term describing protists that use spores to reproduce. Slime molds were formerly classified as fungi, but are no longer considered part of this kingdom. Their common name refers to part of some of these organisms' life cycles where they can appear as gelatinous 'slime'.
Dysentery	Dysentery is an inflammatory disorder of the intestine, especially of the colon, that results in severe diarrhea containing mucus and/or blood in the feces with fever, abdominal pain, and rectal tenesmus. If left untreated, dysentery can be fatal. In developed countries, dysentery is, in general, a mild illness, causing mild symptoms normally consisting of mild stomach pains and frequent passage of feces.
John Snow	John Snow was an English physician and a leader in the adoption of anaesthesia and medical hygiene. He is considered to be one of the fathers of epidemiology, because of his work in tracing the source of a cholera outbreak in Soho, England, in 1854. Early life and education Snow was born 15 March 1813 in York, England.
Epidemiology	Epidemiology is the study of the distribution and patterns of health-events, health-characteristics and their causes or influences in well-defined populations. It is the cornerstone method of public health research, and helps inform policy decisions and evidence-based medicine by identifying risk factors for disease and targets for preventive medicine. Epidemiologists are involved in the design of studies, collection and statistical analysis of data, and interpretation and dissemination of results (including peer review and occasional systematic review).
Urease	Urease is an enzyme that catalyzes the hydrolysis of urea into carbon dioxide and ammonia. The reaction occurs as follows: $(NH_2)_2CO + H_2O \rightarrow CO_2 + 2NH_3$ In 1926, James Sumner showed that urease is a protein. Urease is found in bacteria, yeast, and several higher plants.
Cryptosporidiosis	Cryptosporidiosis, is a parasitic disease caused by Cryptosporidium, a protozoan parasite in the phylum Apicomplexa. It affects the intestines of mammals and is typically an acute short-term infection.

Cryptosporidium parvum	Cryptosporidium parvum is one of several protozal species that cause cryptosporidiosis, a parasitic disease of the mammalian intestinal tract. Primary symptoms of C. parvum infection are acute, watery, and non-bloody diarrhoea. C. parvum infection is of particular concern in immunocompromised patients, where diarrhea can reach 10-15L per day.
Naegleria	Naegleria is a genus of protozoa. The genus was named after French zoologist Mathieu Naegler. Naegleria is a microscopic amoeba that can cause a very rare, but severe, infection of the brain.
Acanthamoeba	Acanthamoeba is a genus of amoebae, one of the most common protozoa in soil, and also frequently found in fresh water and other habitats. The cells are small, usually 15 to 35 μm in length and oval to triangular in shape when moving. Cysts are common.
Giardia lamblia	Giardia lamblia is a flagellated protozoan parasite that colonizes and reproduces in the small intestine, causing giardiasis. The giardia parasite attaches to the epithelium by a ventral adhesive disc, and reproduces via binary fission. Giardiasis does not spread via the bloodstream, nor does it spread to other parts of the gastro-intestinal tract, but remains confined to the lumen of the small intestine.
Metronidazole	Metronidazole is a nitroimidazole antibiotic medication used particularly for anaerobic bacteria and protozoa. Metronidazole is an antibiotic, amebicide, and antiprotozoal. It is the drug of choice for first episodes of mild-to-moderate Clostridium difficile infection.
Lumbar puncture	A lumbar puncture is a diagnostic and at times therapeutic procedure that is performed in order to collect a sample of cerebrospinal fluid (CSF) for biochemical, microbiological, and cytological analysis, or very rarely as a treatment ('therapeutic lumbar puncture') to relieve increased intracranial pressure. The most common purpose for a lumbar puncture is to collect cerebrospinal fluid in a case of suspected meningitis, since there is no other reliable tool with which meningitis, a life-threatening but highly treatable condition, can be excluded. Young infants commonly require lumbar puncture as a part of the routine workup for fever without a source, as they have a much higher risk of meningitis than older persons and do not reliably show signs of meningeal irritation (meningismus).
Sexually transmitted disease	A sexually transmitted disease also known as sexually transmitted infection (STI) or venereal disease (VD), is an illness that has a significant probability of transmission between humans by means of human sexual behavior, including vaginal intercourse, oral sex, and anal sex.

Chapter 5. Part 5: Medicine and Immunology

While in the past, these illnesses have mostly been referred to as Sexually transmitted diseases or VDs, in recent years the term sexually transmitted infections (STIs) has been preferred, as it has a broader range of meaning; a person may be infected, and may potentially infect others, without showing signs of disease. Some STIs can also be transmitted via the use of IV drug needles after its use by an infected person, as well as through childbirth or breastfeeding.

Syphilis	Syphilis is a sexually transmitted infection caused by the spirochete bacterium Treponema pallidum subspecies pallidum. The primary route of transmission is through sexual contact; however, it may also be transmitted from mother to fetus during pregnancy or at birth, resulting in congenital syphilis. Other human diseases caused by related Treponema pallidum include yaws (subspecies pertenue), pinta (subspecies carateum) and bejel (subspecies endemicum).
Substrate-level phosphorylation	Substrate-level phosphorylation is a type of metabolism that results in the formation and creation of adenosine triphosphate (ATP) or guanosine triphosphate (GTP) by the direct transfer and donation of a phosphoryl (PO_3) group to adenosine diphosphate (ADP) or guanosine diphosphate (GDP) from a phosphorylated reactive intermediate. Note that the phosphate group does not have to directly come from the substrate. By convention, the phosphoryl group that is transferred is referred to as a phosphate group.
Chancre	A chancre is a painless ulceration formed during the primary stage of syphilis. This infectious lesion forms approximately 21 days after the initial exposure to Treponema pallidum, the gram-negative spirochaete bacterium yielding syphilis. Chancres transmit the sexually transmissible disease of syphilis through direct physical contact.
Chlamydia	Chlamydia is a genus of bacteria that are obligate intracellular parasites. Chlamydia infections are the most common bacterial sexually transmitted infections in humans and are the leading cause of infectious blindness worldwide. The three Chlamydia species include Chlamydia trachomatis (a human pathogen), Chlamydia suis (affects only swine), and Chlamydia muridarum (affects only mice and hamsters).
Chlamydia trachomatis	Chlamydia trachomatis, an obligate intracellular human pathogen, is one of three bacterial species in the genus Chlamydia. C. trachomatis is a Gram-negative bacteria, therefore its cell wall components retain the counter-stain safranin and appear pink under a light microscope. Identified in 1907, C. trachomatis was the first chlamydial agent discovered in humans.
Congenital syphilis	Congenital syphilis is syphilis present in utero and at birth, and occurs when a child is born to a mother with secondary syphilis. Untreated syphilis results in a high risk of a bad outcome of pregnancy, including Mulberry molars in the fetus. Syphilis can cause miscarriages, premature births, stillbirths, or death of newborn babies.

Brugia malayi	Brugia malayi is a nematode (roundworm), one of the three causative agents of lymphatic filariasis in humans. Lymphatic filariasis, also known as elephantiasis, is a condition characterized by swelling of the lower limbs. The two other filarial causes of lymphatic filariasis are Wuchereria bancrofti and Brugia timori, which differ from B. malayi morphologically, symptomatically, and in geographical extent.
Chocolate agar	Chocolate agar - is a non-selective, enriched growth medium. It is a variant of the blood agar plate. It contains red blood cells, which have been lysed by heating very slowly to 56 °C. Chocolate agar is used for growing fastidious (fussy) respiratory bacteria, such as Haemophilus influenzae.
Prevalence	In epidemiology, the prevalence of a health-related state (typically disease, but also other things like smoking or seatbelt use) in a statistical population is defined as the total number of cases of the risk factor in the population at a given time, or the total number of cases in the population, divided by the number of individuals in the population. It is used as an estimate of how common a disease is within a population over a certain period of time. It helps physicians or other health professionals understand the probability of certain diagnoses and is routinely used by epidemiologists, health care providers, government agencies and insurers.
Trichomoniasis	Trichomoniasis, is a common cause of vaginitis. It is a sexually transmitted disease, and is caused by the single-celled protozoan parasite Trichomonas vaginalis producing mechanical stress on host cells and then ingesting cell fragments after cell death. Trichomoniasis is primarily an infection of the urogenital tract; the most common site of infection is the urethra and the vagina in women.
Opportunistic infection	An opportunistic infection is an infection caused by pathogens, particularly opportunistic pathogens--those that take advantage of certain situations--such as bacterial, viral, fungal or protozoan infections that usually do not cause disease in a healthy host, one with a healthy immune system. A compromised immune system, however, presents an 'opportunity' for the pathogen to infect.

Immunodeficiency or immunosuppression can be caused by:•Malnutrition•Recurrent infections•Immunosuppressing agents for organ transplant recipients•Advanced HIV infection•Chemotherapy for cancer•Genetic predisposition•Skin damage•Antibiotic treatment•Medical procedures•PregnancyTypes of infections |

Chapter 5. Part 5: Medicine and Immunology

Nervous system	The nervous system is an organ system containing a network of specialized cells called neurons that coordinate the actions of an animal and transmit signals between different parts of its body. In most animals the nervous system consists of two parts, central and peripheral. The central nervous system of vertebrates (such as humans) contains the brain, spinal cord, and retina.
Viral meningitis	Viral meningitis refers to meningitis caused by a viral infection. It is sometimes referred to as 'aseptic meningitis' in contrast to meningitis caused by bacteria. An example is lymphocytic choriomeningitis.
Clostridium botulinum	Clostridium botulinum is a Gram-positive, rod-shaped bacterium that produces several toxins. The best known are its neurotoxins, subdivided in types A-G, that cause the flaccid muscular paralysis seen in botulism. It is also the main paralytic agent in botox.
Tetanospasmin	Tetanus toxin is an extremely potent neurotoxin produced by the vegetative cell of Clostridium tetani in anaerobic conditions, causing tetanus. It has no known function for clostridia in the soil environment where they are normally encountered. It is also called spasmogenic toxin, tetanospasmin or abbreviated to TeTx or TeNT. The LD_{50} of this toxin has been measured to be approximately 1 ng/kg, making it second only to Botulinum toxin D as the deadliest toxin in the world.
Tetanus	Tetanus is a medical condition characterized by a prolonged contraction of skeletal muscle fibers. The primary symptoms are caused by tetanospasmin, a neurotoxin produced by the Gram-positive, rod-shaped, obligate anaerobic bacterium Clostridium tetani. Infection generally occurs through wound contamination and often involves a cut or deep puncture wound.
Transcytosis	Transcytosis is the process by which various macromolecules are transported across the interior of a cell. Macromolecules are captured in vesicles on one side of the cell, drawn across the cell, and ejected on the other side. Examples of macromolecules transported include IgA, transferrin, and insulin.
Botulinum toxin	Botulinum toxin is a protein and neurotoxin produced by the bacterium Clostridium botulinum. Botulinum toxin can cause botulism, a serious and life-threatening illness in humans and animals. When introduced intravenously in monkeys, type A (Botox Cosmetic) of the toxin exhibits an LD_{50} of 40-56 ng, type C1 around 32 ng, type D 3200 ng, and type E 88 ng; these are some of the most potent neurotoxins known.
Botulism	Botulism is metabolic waste produced under anaerobic conditions by the bacterium Clostridium botulinum, and affecting a wide range of mammals, birds and fish.

	The toxin enters the human body in one of three ways: by colonization of the digestive tract by the bacterium in children (infant botulism) or adults (adult intestinal toxemia), by ingestion of toxin from foods or by contamination of a wound by the bacterium (wound botulism). Person to person transmission of botulism does not occur.
Prion	A prion is an infectious agent composed of protein in a misfolded form. This is in contrast to all other known infectious agents (virusbacteriafungusparasite) which must contain nucleic acids (either DNA, RNA, or both). The word prion, coined in 1982 by Stanley B. Prusiner, is derived from the words protein and infection.
Encephalitozoon intestinalis	Encephalitozoon intestinalis is a parasite. It can cause microsporidiosis. It is notable as having one of the smallest genome among known eukaryotic organisms, containing only 2.25 million base pairs.
Serum albumin	Serum albumin, often referred to simply as albumin is a protein that in humans is encoded by the ALB gene. Serum albumin is the most abundant plasma protein in mammals. Albumin is essential for maintaining the osmotic pressure needed for proper distribution of body fluids between intravascular compartments and body tissues.
Atherosclerosis	Atherosclerosis is a condition in which an artery wall thickens as the result of a build-up of fatty materials such as cholesterol. It is a syndrome affecting arterial blood vessels, a chronic inflammatory response in the walls of arteries, in large part due to the accumulation of macrophage white blood cells and promoted by low-density lipoproteins (plasma proteins that carry cholesterol and triglycerides) without adequate removal of fats and cholesterol from the macrophages by functional high density lipoproteins (HDL), . It is commonly referred to as a hardening or furring of the arteries.
Endocarditis	Endocarditis is an inflammation of the inner layer of the heart, the endocardium. It usually involves the heart valves (native or prosthetic valves). Other structures which may be involved include the interventricular septum, the chordae tendineae, the mural endocardium, or even on intracardiac devices.
Lyme disease	Lyme disease, is an emerging infectious disease caused by at least three species of bacteria belonging to the genus Borrelia. Borrelia burgdorferi sensu stricto is the main cause of Lyme disease in the United States, whereas Borrelia afzelii and Borrelia garinii cause most European cases. he town of Lyme, Connecticut, USA, where a number of cases were identified in 1975.

Chapter 5. Part 5: Medicine and Immunology

	Although Allen Steere realized that Lyme disease was a tick-borne disease in 1978, the cause of the disease remained a mystery until 1981, when B. burgdorferi was identified by Willy Burgdorfer.
Pericarditis	Pericarditis is an inflammation of the pericardium (the fibrous sac surrounding the heart). A characteristic chest pain is often present. It can be caused by a variety of causes including viral infections of the pericardium, idiopathic causes, uremic pericarditis, bacterial infections of the precardium (for i.e. Mycobacterium tuberculosis), post-infarct pericarditis or Dressler's pericarditis.
Viremia	Viremia is a medical condition where viruses enter the bloodstream and hence have access to the rest of the body. It is similar to bacteremia, a condition where bacteria enter the bloodstream. Primary viremia refers to the initial spread of virus in the blood from the first site of infection.
Myocarditis	In medicine (cardiology), myocarditis is inflammation of heart muscle (myocardium). It resembles a heart attack but coronary arteries are not blocked. Myocarditis is most often due to infection by common viruses, such as parvovirus B19, less commonly non-viral pathogens such as Borrelia burgdorferi (Lyme disease) or Trypanosoma cruzi, or as a hypersensitivity response to drugs.
Plasmodium falciparum	Plasmodium falciparum is a protozoan parasite, one of the species of Plasmodium that cause malaria in humans. It is transmitted by the female Anopheles mosquito. P. falciparum is the most dangerous of these infections as P. falciparum (or malignant) malaria has the highest rates of complications and mortality.
Life cycle	A life cycle is a period involving all different generations of a species succeeding each other through means of reproduction, whether through asexual reproduction or sexual reproduction (a period from one generation of organisms to the same identical). For example, a complex life cycle of Fasciola hepatica includes three different multicellular generations: 1) 'adult' hermaphroditic; 2) sporocyst; 3) redia.
Malaria	Malaria is a mosquito-borne infectious disease of humans and other animals caused by eukaryotic protists of the genus Plasmodium. The disease results from the multiplication of Plasmodium parasites within red blood cells, causing symptoms that typically include fever and headache, in severe cases progressing to coma or death. It is widespread in tropical and subtropical regions, including much of Sub-Saharan Africa, Asia, and the Americas.

Red blood cell	Red blood cells, or erythrocytes, are the most common type of blood cell and the vertebrate organism's principal means of delivering oxygen (O_2) to the body tissues via the blood flow through the circulatory system. They take up oxygen in the lungs or gills and release it while squeezing through the body's capillaries. These cells' cytoplasm is rich in haemoglobin, an iron-containing biomolecule that can bind oxygen and is responsible for the blood's red color.
Chloroquine	Chloroquine is a 4-aminoquinoline drug used in the treatment or prevention of malaria. History Chloroquine N'-(7-chloroquinolin-4-yl)-N,N-diethyl-pentane-1,4-diamine, was discovered in 1934 by Hans Andersag and co-workers at the Bayer laboratories who named it 'Resochin'. It was ignored for a decade because it was considered too toxic for human use.
Genome	In modern molecular biology and genetics, the genome is the entirety of an organism's hereditary information. It is encoded either in DNA or, for many types of virus, in RNA. The genome includes both the genes and the non-coding sequences of the DNA/RNA. The term was adapted in 1920 by Hans Winkler, Professor of Botany at the University of Hamburg, Germany. The Oxford English Dictionary suggests the name to be a blend of the words gene and chromosome.
Pneumonic plague	Pneumonic plague, a severe type of lung infection, is one of three main forms of plague, all of which are caused by the bacterium Yersinia pestis. It is more virulent and rarer than bubonic plague. The difference between the versions of plague is simply the location of the infection in the body; the bubonic plague is an infection of the lymphatic system, the pneumonic plague is an infection of the respiratory system, and the septicemic plague is an infection in the blood stream.
Septicemic plague	Septicemic (or septicaemic) plague is a deadly blood infection, one of the three main forms of plague. It is caused by Yersinia pestis, a gram-negative bacterium. Like other forms of gram-negative sepsis, septicemic plague can cause disseminated intravascular coagulation, and is almost always fatal (the mortality rate in the medieval times was 99-100 percent).
Borrelia afzelii	Borrelia afzelii is a species of Borrelia that can infect various species of vertebrate and invertebrates.

	Among thirty Borrelia known species, it is one of 4 which is likely to infect humans causing a variant of Lyme disease. It seems fairly common that coinfection combine this Borrelia specie to another one or to other pathogens carried by the vector, which appears to be in most cases the tick.
Francisella	Francisella is a genus of pathogenic, Gram-negative bacteria. They are small coccobacillary or rod-shaped, non-motile organisms, which are also facultative intracellular parasites of macrophages. Strict aerobes, Francisella colonies bear a morphological resemblance to those of the genus Brucella.
Hepatitis	Hepatitis is an inflammation of the liver characterized by the presence of inflammatory cells in the tissue of the organ. The name is from the Greek hepar , the root being hepat- , meaning liver, and suffix -itis, meaning 'inflammation' (c. 1727). The condition can be self-limiting (healing on its own) or can progress to fibrosis (scarring) and cirrhosis.
Systemic disease	A systemic disease is one that affects a number of organs and tissues, or affects the body as a whole. Although most medical conditions will eventually involve multiple organs in advanced stage (e.g. Multiple organ dysfunction syndrome), diseases where multiple organ involvement is at presentation or in early stage are considered elsewhere.
Hepatitis A	Hepatitis A is an acute infectious disease of the liver caused by the hepatitis A virus (HAV), an RNA virus, usually spread the fecal-oral route; transmitted person-to-person by ingestion of contaminated food or water or through direct contact with an infectious person. Tens of millions of individuals worldwide are estimated to become infected with HAV each year. The time between infection and the appearance of the symptoms (the incubation period) is between two and six weeks and the average incubation period is 28 days.
Hepatitis B	Hepatitis B is an infectious inflammatory illness of the liver caused by the hepatitis B virus (HBV) that affects hominoidea, including humans. Originally known as 'serum hepatitis', the disease has caused epidemics in parts of Asia and Africa, and it is endemic in China. About a third of the world population has been infected at one point in their lives, including 350 million who are chronic carriers.
Hepatitis C	Hepatitis C is an infectious disease affecting the liver, caused by the hepatitis C virus (HCV). The infection is often asymptomatic, but once established, chronic infection can progress to scarring of the liver (fibrosis), and advanced scarring (cirrhosis) which is generally apparent after many years.

Ebola virus	Ebola virus causes severe disease in humans and in nonhuman primates in the form of viral hemorrhagic fever. EBOV is a select agent, World Health Organization Risk Group 4 Pathogen (requiring Biosafety Level 4-equivalent containment), National Institutes of HealthNational Institute of Allergy and Infectious Diseases Category A Priority Pathogen, Centers for Disease Control and Prevention Category A Bioterrorism Agent, and listed as a Biological Agent for Export Control by the Australia Group. Ebola virus was first described in 1976. Today, the virus is the single member of the species Zaire ebolavirus, which is included into the genus Ebolavirus, family Filoviridae, order Mononegavirales.
Herd immunity	Herd immunity describes a form of immunity that occurs when the vaccination of a significant portion of a population provides a measure of protection for individuals who have not developed immunity. Herd immunity theory proposes that, in contagious diseases that are transmitted from individual to individual, chains of infection are likely to be disrupted when large numbers of a population are immune or less susceptible to the disease. The greater the proportion of individuals who are resistant, the smaller the probability that a susceptible individual will come into contact with an infectious individual.
Booster dose	In medicine, a booster dose is an extra administration of a vaccine after an earlier dose. After initial immunization, a booster injection or booster dose is a re-exposure to the immunizing antigen. It is intended to increase immunity against that antigen back to protective levels after it has been shown to have decreased or after a specified period.
Drug discovery	In the fields of medicine, biotechnology and pharmacology, drug discovery is the process by which drugs are discovered or designed. In the past most drugs have been discovered either by identifying the active ingredient from traditional remedies or by serendipitous discovery. As our understanding of disease has increased to the extent that we know how disease and infection are controlled at the molecular and physiological level, scientists are now able to try to find compounds that specifically modulate those molecules, for instance via high throughput screening.
Penicillin	Penicillin is a group of antibiotics derived from Penicillium fungi. They include penicillin G, procaine penicillin, benzathine penicillin, and penicillin V. Penicillin antibiotics are historically significant because they are the first drugs that were effective against many previously serious diseases, such as syphilis, and infections caused by staphylococci and streptococci. Penicillins are still widely used today, though many types of bacteria are now resistant.
Penicillium	Penicillium is a genus of ascomycetous fungi of major importance in the natural environment as well as food and drug production.

	Members of the genus produce penicillin, a molecule that is used as an antibiotic, which kills or stops the growth of certain kinds of bacteria inside the body. According to the Dictionary of the Fungi (10th edition, 2008), the widespread genus contains over 300 species.
Folic acid	Folic acid and folate (the form naturally occurring in the body), as well as pteroyl-L-glutamic acid, pteroyl-L-glutamate, and pteroylmonoglutamic acid are forms of the water-soluble vitamin B_9. Folic acid is itself not biologically active, but its biological importance is due to tetrahydrofolate and other derivatives after its conversion from dihydrofolic acid in the liver.

Vitamin B_9 (folic acid and folate inclusive) is essential to numerous bodily functions. |
Arsenic	Arsenic is a chemical element with the symbol As, atomic number 33 and relative atomic mass 74.92. Arsenic occurs in many minerals, usually in conjunction with sulfur and metals, and also as a pure elemental crystal. It was first documented by Albertus Magnus in 1250. Arsenic is a metalloid. It can exist in various allotropes, although only the grey form has important use in industry.
Amp resistance	Amp resistance is a term for resistance to the antibiotic ampicillin. It is used as a selectable marker in bacterial transformation.
Aplastic anemia	Aplastic anemia is a condition where bone marrow does not produce sufficient new cells to replenish blood cells. The condition, per its name, involves both aplasia and anemia. Typically, anemia refers to low red blood cell counts, but aplastic anemia patients have lower counts of all three blood cell types: red blood cells, white blood cells, and platelets, termed pancytopenia.
Side effect	In medicine, a side effect is an effect, whether therapeutic or adverse, that is secondary to the one intended; although the term is predominantly employed to describe adverse effects, it can also apply to beneficial, but unintended, consequences of the use of a drug.

Occasionally, drugs are prescribed or procedures performed specifically for their side effects; in that case, said side effect ceases to be a side effect, and is now an intended effect. For instance, X-rays were historically (and are currently) used as an imaging technique; the discovery of their oncolytic capability led to their employ in radiotherapy (ablation of malignant tumours). |
| Cell wall | The cell wall is the tough, usually flexible but sometimes fairly rigid layer that surrounds some types of cells. It is located outside the cell membrane and provides these cells with structural support and protection, in addition to acting as a filtering mechanism. A major function of the cell wall is to act as a pressure vessel, preventing over-expansion when water enters the cell. |

Bacillus licheniformis	Bacillus licheniformis is a bacterium commonly found in the soil. It is found on bird feathers, especially chest and back plumage, and most often in ground dwelling birds (like sparrows) and aquatic species (like ducks). It is a gram positive, thermophilic bacterium.
Bacillus subtilis	Bacillus subtilis, known also as the hay bacillus or grass bacillus, is a Gram-positive, catalase-positive bacterium commonly found in soil. A member of the genus Bacillus, B. subtilis is rod-shaped, and has the ability to form a tough, protective endospore, allowing the organism to tolerate extreme environmental conditions. Unlike several other well-known species, B. subtilis has historically been classified as an obligate aerobe, though recent research has demonstrated that this is not strictly correct.
Cycloserine	Cycloserine is an antibiotic effective against Mycobacterium tuberculosis. For the treatment of tuberculosis, it is classified as a second line drug, i.e. its use is only considered if one or more first line drugs cannot be used. Although in principle active against other bacteria as well, cycloserine is not commonly used in the treatment of infections other than tuberculosis.
Gramicidin	Gramicidin is a heterogeneous mixture of six antibiotic compounds, gramicidins A, B and C, making up 80%, 6%, and 14% respectively, all of which are obtained from the soil bacterial species Bacillus brevis and called collectively gramicidin D. Gramicidin D are linear pentadecapeptides; that is chains made up of 15 amino acids. This is in contrast to gramicidin S which is a cyclic peptide chain. Gramicidin is active against Gram-positive bacteria, except for the Gram-positive bacilli, and against selective Gram-negative organisms, such as Neisseria bacteria.
Polymyxin	Polymyxins are antibiotics, with a general structure consisting of a cyclic peptide with a long hydrophobic tail. They disrupt the structure of the bacterial cell membrane by interacting with its phospholipids. They are produced by the Gram-positive bacterium Bacillus polymyxa and are selectively toxic for Gram-negative bacteria due to their specificity for the lipopolysaccharide molecule that exists within many Gram-negative outer membranes.
Vancomycin	Vancomycin INN ( ) is a glycopeptide antibiotic used in the prophylaxis and treatment of infections caused by Gram-positive bacteria. It has traditionally been reserved as a drug of 'last resort', used only after treatment with other antibiotics had failed, although the emergence of vancomycin-resistant organisms means that it is increasingly being displaced from this role by linezolid (Zyvox) available PO and IV and daptomycin (Cubicin) IV and quinupristin/dalfopristin (Synercid) IV.

Vancomycin was first isolated in 1953 by Edmund Kornfeld (working at Eli Lilly) from a soil sample collected from the interior jungles of Borneo by a missionary. The organism that produced it was eventually named Amycolatopsis orientalis.

Cell membrane

The cell membrane is a biological membrane that separates the interior of all cells from the outside environment. The cell membrane is selectively permeable to ions and organic molecules and controls the movement of substances in and out of cells. It basically protects the cell from outside forces.

Fusobacterium necrophorum

Fusobacterium necrophorum is the species of Fusobacterium that is responsible for Lemierre's syndrome, and appears to be responsible for 10% of all acute sore throats, 21% of all recurring sore throats, and 23% of peritonsillar abscesses with the remainder being caused by Group A streptococci or viruses.

Other complications from F. necrophorum include meningitis, complicated by thrombosis of the external jugular vein, thrombosis of the cerebral veins, and infection of the urogenital and the gastrointestinal tracts.

F. necrophorum infection usually responds to treatment with penicillin or metronidazole, but penicillin treatment for persistent pharyngitis appears anecdotally to have a higher relapse rate, although the reasons for that are unclear.

Quinolone

The quinolones are a family of synthetic broad-spectrum antibiotics. The term quinolone(s) refers to potent synthetic chemotherapeutic antibacterials.

The first generation of the quinolones begins with the introduction of nalidixic acid in 1962 for treatment of urinary tract infections in humans.

Myxopyronin

Myxopyronin is an alpha-pyrone antibiotic, the first in a new class of inhibitors of bacterial RNA polymerase (RNAP) that target switch 1 and switch 2 of the RNAP 'switch region.' Rifamycin antibacterial agents, which are first-line treatments for tuberculosis, and lipiarmycin (fidaxomicin, Dificid) also target RNAP, but target different sites in RNAP. Myxopyronin does not have cross-resistance with rifamycins and lipiarmycin. Myxopyronin may be useful to address the growing problem of drug resistance in tuberculosis. It also may be useful in treatment of methicillin-resistant Staphylococcus aureus (MRSA).

Erysipelas

Erysipelas is an acute streptococcus bacterial infection of the upper dermis and superficial lymphatics.

This disease is most common among the elderly, infants, and children.

Protein synthesis inhibitor	A protein synthesis inhibitor is a substance that stops or slows the growth or proliferation of cells by disrupting the processes that lead directly to the generation of new proteins. While a broad interpretation of this definition could be used to describe nearly any antibiotic, in practice, it usually refers to substances that act at the ribosome level, taking advantages of the major differences between prokaryotic and eukaryotic ribosome structures. Toxins such as ricin also function via protein synthesis inhibition.
Aminoglycoside	An aminoglycoside is a molecule or a portion of a molecule composed of amino-modified sugars. Several aminoglycosides function as antibiotics that are effective against certain types of bacteria. They include amikacin, arbekacin, gentamicin, kanamycin, neomycin, netilmicin, paromomycin, rhodostreptomycin, streptomycin, tobramycin, and apramycin .
Chloramphenicol	Chloramphenicol is a bacteriostatic antimicrobial. It is considered a prototypical broad-spectrum antibiotic, alongside the tetracyclines. Chloramphenicol is effective against a wide variety of Gram-positive and Gram-negative bacteria, including most anaerobic organisms.
Streptogramin	Streptogramins are a class of antibiotics. Streptogramins are effective in the treatment of vancomycin-resistant Staphylococcus aureus (VRSA) and vancomycin-resistant Enterococcus (VRE), two of the most rapidly-growing strains of multidrug-resistant bacteria. They fall into two groups: streptogramin A and streptogramin B. Members include:•Quinupristin/dalfopristin•Pristinamycin•Virginiamycin•NXL 103, an experimental streptogramin in clinical trials for the treatment of respiratory tract infections..
Secondary metabolite	Secondary metabolites are organic compounds that are not directly involved in the normal growth, development, or reproduction of an organism. Unlike primary metabolites, absence of secondary metabolites does not result in immediate death, but rather in long-term impairment of the organism's survivability, fecundity, or aesthetics, or perhaps in no significant change at all. Secondary metabolites are often restricted to a narrow set of species within a phylogenetic group.
Enterococcus faecalis	Enterococcus faecalis - formerly classified as part of the Group D Streptococcus system - is a Gram-positive commensal bacterium inhabiting the gastrointestinal tracts of humans and other mammals. It is among the main constituents of some probiotic food supplements.

	A commensal organism like other species in the genus Enterococcus, E. faecalis can cause life-threatening infections in humans, especially in the nosocomial (hospital) environment, where the naturally high levels of antibiotic resistance found in E. faecalis contribute to its pathogenicity.
Integron	An integron is a two component gene capture and dissemination system, initially discovered in relation to antibiotic resistance, and which is found in plasmids, chromosomes and transposons. The first component consists of a gene encoding a site specific recombinase along with a specific site for recombination, while the second component comprises fragments of DNA called gene cassettes which can be incorporated or shuffled. A cassette may encode genes for antibiotic resistance, although most genes in integrons are uncharacterized.
Transposable element	A transposable element is a DNA sequence that can change its relative position (self-transpose) within the genome of a single cell. The mechanism of transposition can be either 'copy and paste' or 'cut and paste'. Transposition can create phenotypically significant mutations and alter the cell's genome size.
Clavulanic acid	Clavulanic acid is a beta-lactamase inhibitor (marketed by GlaxoSmithKline, formerly Beecham) combined with penicillin group antibiotics to overcome certain types of antibiotic resistance. It is used to overcome resistance in bacteria that secrete beta-lactamase, which otherwise inactivates most penicillins. In its most common form, the potassium salt potassium clavulanate is combined with amoxicillin (co-amoxiclav, trade names Augmentin, Synulox [veterinary], and others) or ticarcillin.
Antibiotic tolerance	Multidrug tolerance or antibiotic tolerance is the ability of a disease-causing microorganism to resist killing by antibiotics or other antimicrobials. It is mechanistically distinct from multidrug resistance: It is not caused by mutant microbes, but rather by microbial cells that exist in a transient, dormant, non-dividing state. Microorganisms that display multidrug tolerance can be bacteria, fungi or parasites.
Platensimycin	Platensimycin, a metabolite of Streptomyces platensis, which is an excellent example of a unique structural class of natural antibiotics, has been demonstrated to be a breakthrough in recent antibiotic researches due to its unique functional pattern and significant antibacterial activity . This compound is a member of a class of antibiotics which act by blocking enzymes involved in the condensation steps in fatty acid biosynthesis, which Gram-positive bacteria need to biosynthesise cell membranes (β-ketoacyl-(acyl-carrier-protein (ACP)) synthase I/II (FabF/B)).

Amantadine	Amantadine is the organic compound known formally as 1-aminoadamantane. The molecule consists of adamantane backbone that is substituted at one of the four methyne positions with an amino group. This compound is sold under the name Symmetrel for use both as an antiviral and an antiparkinsonian drug.
Neuraminidase inhibitor	Neuraminidase inhibitors are a class of antiviral drugs targeted at the influenza virus, which work by blocking the function of the viral neuraminidase protein, thus preventing the virus from reproducing by budding from the host cell. Oseltamivir (Tamiflu) a prodrug, Zanamivir (Relenza), Laninamivir (Inavir), and Peramivir belong to this class. Unlike the M2 inhibitors, which work only against the influenza A, neuraminidase inhibitors act against both influenza A and influenza B. Common side effects include nausea and vomiting.
Zanamivir	Zanamivir INN ( ) is a neuraminidase inhibitor used in the treatment and prophylaxis of influenza caused by influenza A virus and influenza B virus. Zanamivir was the first neuraminidase inhibitor commercially developed. It is currently marketed by GlaxoSmithKline under the trade name Relenza as a powder for oral inhalation.
Hemagglutinin	Hemagglutinin Influenza hemagglutinin (HA) or haemagglutinin is a type of hemagglutinin found on the surface of the influenza viruses. It is an antigenic glycoprotein. It is responsible for binding the virus to the cell that is being infected.
Neuraminidase	Neuraminidase enzymes are glycoside hydrolase enzymes (EC 3.2.1.18) that cleave the glycosidic linkages of neuraminic acids. Neuraminidase enzymes are a large family, found in a range of organisms. The best-known neuraminidase is the viral neuraminidase, a drug target for the prevention of the spread of influenza infection.
Protease inhibitor	Protease inhibitors are a class of drugs used to treat or prevent infection by viruses, including HIV and Hepatitis C. Protease inhibitors prevent viral replication by inhibiting the activity of proteases, e.g.HIV-1 protease, enzymes used by the viruses to cleave nascent proteins for final assembly of new virions. Protease inhibitors have been developed or are presently undergoing testing for treating various viruses:•HIV/AIDS: antiretroviral protease inhibitors (saquinavir, ritonavir, indinavir, nelfinavir, amprenavir etc).•Hepatitis C: Boceprevir•Hepatitis C: Telaprevir

Given the specificity of the target of these drugs there is the risk, as in antibiotics, of the development of drug-resistant mutated viruses. To reduce this risk it is common to use several different drugs together that are each aimed at different targets.

Zidovudine	Zidovudine or azidothymidine (AZT) (also called ZDV) is a nucleoside analog reverse transcriptase inhibitor (NRTI), a type of antiretroviral drug used for the treatment of HIV/AIDS. It is an analog of thymidine.

AZT was the first approved treatment for HIV, sold under the names Retrovir and Retrovis. AZT use was a major breakthrough in AIDS therapy in the 1990s that significantly altered the course of the illness and helped destroy the notion that HIV/AIDS was a death sentence. |
| Clotrimazole | Clotrimazole is an antifungal medication commonly used in the treatment of fungal infections (of both humans and other animals) such as vaginal yeast infections, oral thrush, and ringworm. It is also used to treat athlete's foot and jock itch.

It is commonly available as an over-the-counter substance in various dosage forms, such as a cream, and also (especially in the case of ear infection) as a combination medicine. |
Fluconazole	Fluconazole is a triazole antifungal drug used in the treatment and prevention of superficial and systemic fungal infections. In a bulk powder form, it appears as a white crystalline powder, and it is very slightly soluble in water and soluble in alcohol. It is commonly marketed under the trade name Diflucan or Trican (Pfizer).
Griseofulvin	Griseofulvin is an antifungal drug that is administered orally. It is used both in animals and in humans, to treat fungal infections of the skin (commonly known as ringworm) and nails. It is produced by culture of some strains of the mold Penicillium griseofulvum, from which it was isolated in 1939.
Miconazole	Miconazole is an imidazole antifungal agent, developed by Janssen Pharmaceutica, commonly applied topically to the skin or to mucus membranes to cure fungal infections. It works by inhibiting the synthesis of ergosterol, a critical component of fungal cell membranes. It can also be used against certain species of Leishmania protozoa which are a type of unicellular parasite that also contain ergosterol in their cell membranes.
Nystatin	Nystatin is a polyene antifungal medication to which many molds and yeast infections are sensitive, including Candida. Due to its toxicity profile, there are currently no injectable formulations of this drug on the US market.

Index case	The index case is the initial patient in the population of an epidemiological investigation. The index case may indicate the source of the disease, the possible spread, and which reservoir holds the disease in between outbreaks. The index case is the first patient that indicates the existence of an outbreak.
Fecal-oral route	The fecal-oral route, or alternatively, the oral-fecal route or orofecal route is a route of transmission of diseases, in which they are passed when pathogens in fecal particles from one host are introduced into the oral cavity of another potential host.

There are usually intermediate steps, sometimes many of them. Among the more common causes are:•water that has come in contact with feces and is then inadequately treated before drinking;•food that has been handled with feces present;•poor sewage treatment along with disease vectors like houseflies;•poor or absent cleaning after handling feces or anything that has been in contact with it;•sexual practices that may involve feces, such as analingus.•sexual fetishes that involve feces, known collectively as coprophilia (its eating is known as coprophagia)

Some of the diseases that can be passed via the fecal-oral route include:•Poliomyelitis•Giardiasis•Hepatitis A•Hepatitis E•Rotavirus•Shigellosis (bacillary dysentery)•Typhoid fever•Vibrio parahaemolyticus infections•Enteroviruses, including poliomyelitis•Cholera•Clostridium difficile•Cryptosporidiosis•Ascariasis

Transmission of Helicobacter pylori by oral-fecal route has been demonstrated in murine models. |
| Osteomyelitis | Osteomyelitis simply means an infection of the bone or bone marrow. It can be usefully subclassified on the basis of the causative organism (pyogenic bacteria or mycobacteria), the route, duration and anatomic location of the infection.

In general, microorganisms may infect bone through one or more of three basic methods: via the bloodstream, contiguously from local areas of infection (as in cellulitis), or penetrating trauma, including iatrogenic causes such as joint replacements or internal fixation of fractures or root-canaled teeth. |
| Chain reaction | A chain reaction is a sequence of reactions where a reactive product or by-product causes additional reactions to take place. In a chain reaction, positive feedback leads to a self-amplifying chain of events.

Chain reactions are one way in which systems which are in thermodynamic non-equilibrium can release energy or increase entropy in order to reach a state of higher entropy. |

Chapter 5. Part 5: Medicine and Immunology

Oxidase test	The oxidase test is a test used in microbiology to determine if a bacterium produces certain cytochrome c oxidases. It uses disks impregnated with a reagent such as N,N,N′,N′-tetramethyl-p-phenylenediamine (TMPD) or N,N-dimethyl-p-phenylenediamine (DMPD), which is also a redox indicator. The reagent is a dark-blue to maroon color when oxidized, and colorless when reduced.
Gram-positive bacteria	Gram-positive bacteria are those that are stained dark blue or violet by Gram staining. This is in contrast to Gram-negative bacteria, which cannot retain the crystal violet stain, instead taking up the counterstain (safranin or fuchsine) and appearing red or pink. Gram-positive organisms are able to retain the crystal violet stain because of the high amount of peptidoglycan in the cell wall.
Catabolite repression	Carbon catabolite repression, is an important part of global control system of various bacteria and other micro-organisms. Catabolite repression allows bacteria to adapt quickly to a preferred (rapidly metabolisable) carbon and energy source first. This is usually achieved through inhibition of synthesis of enzymes involved in catabolism of carbon sources other than the preferred one.
Restriction fragment	A restriction fragment is a DNA fragment resulting from the cutting of a DNA strand by a restriction enzyme (restriction endonucleases), a process called restriction. Each restriction enzyme is highly specific, recognising a particular short DNA sequence, or restriction site, and cutting both DNA strands at specific points within this site. Most restriction sites are palindromic, (the sequence of nucleotides is the same on both strands when read in the 5' to 3' direction), and are four to eight nucleotides long.
Restriction fragment length polymorphism	In molecular biology, restriction fragment length polymorphism, is a technique that exploits variations in homologous DNA sequences. It refers to a difference between samples of homologous DNA molecules that come from differing locations of restriction enzyme sites, and to a related laboratory technique by which these segments can be illustrated. In RFLP analysis, the DNA sample is broken into pieces (digested) by restriction enzymes and the resulting restriction fragments are separated according to their lengths by gel electrophoresis.
Anaerobic respiration	Anaerobic respiration is a form of respiration using electron acceptors other than oxygen. Although oxygen is not used as the final electron acceptor, the process still uses a respiratory electron transport chain; it is respiration without oxygen. In order for the electron transport chain to function, an exogenous final electron acceptor must be present to allow electrons to pass through the system.
Anaerobic infection	Anaerobic infections are caused by anaerobic bacteria. Anaerobic bacteria do not grow on solid media in room air (10% carbon dioxide and 18% oxygen); facultative anaerobic bacteria can grow in the presence as well as in the absence of air.

Obligate anaerobe	Obligate anaerobes are microorganisms that live and grow in the absence of molecular oxygen; some of these are killed by oxygen. Historically, it was widely accepted that obligate (strict) anaerobes die in presence of oxygen due to the absence of the enzymes superoxide dismutase and catalase, which would convert the lethal superoxide formed in their cells due to the presence of oxygen. While this is true in some cases, these enzyme activities have been identified in some obligate anaerobes, and genes for these enzymes and related proteins have been found in their genomes, such as Clostridium butyricum and Methanosarcina barkeri, among others.
Blood culture	Blood culture is a microbiological culture of blood. It is employed to detect infections that are spreading through the bloodstream (such as bacteremia, septicemia amongst others). This is possible because the bloodstream is usually a sterile environment.
Disease surveillance	Disease surveillance is an epidemiological practice by which the spread of disease is monitored in order to establish patterns of progression. The main role of disease surveillance is to predict, observe, and minimize the harm caused by outbreak, epidemic, and pandemic situations, as well as increase our knowledge as to what factors might contribute to such circumstances. A key part of modern disease surveillance is the practice of disease case reporting.
DNA polymerase	A DNA polymerase is an enzyme (the suffix -ase is used to identify enzymes) that helps catalyze the polymerization of deoxyribonucleotides into a DNA strand. DNA polymerases are best known for their feedback role in DNA replication, in which the polymerase 'reads' an intact DNA strand as a template and uses it to synthesize the new strand. This process copies a piece of DNA. The newly polymerized molecule is complementary to the template strand and identical to the template's original partner strand.
Pantoea agglomerans	Pantoea agglomerans is a Gram-negative bacterium that belongs to the family Enterobacteriaceae. Formerly called Enterobacter agglomerans, this bacterium is known to be an opportunistic pathogen in the immunocompromised, causing wound, blood, and urinary-tract infections. It is commonly isolated from plant surfaces, seeds, fruit (e.g. mandarin oranges), and animal or human feces.
Nitrogen fixation	Nitrogen fixation is a process by which nitrogen (N_2) in the atmosphere is converted into ammonium (NH_{4+}). Atmospheric nitrogen or elemental nitrogen (N_2) is relatively inert: it does not easily react with other chemicals to form new compounds. Fixation processes free up the nitrogen atoms from their diatomic form (N_2) to be used in other ways.

Chapter 5. Part 5: Medicine and Immunology

Lactose permease	LacY proton/sugar symporter
	Lactose permease is a membrane protein which is a member of the major facilitator superfamily. Lactose permease can be classified as a symporter, which uses the gradient of H+ towards the cell to transport lactose in the same direction into the cell.
	The protein has twelve transmembrane helices and exhibits an internal two-fold symmetry, relating the N-terminal six helices onto the C-terminal helices.
Long terminal repeat	Long terminal repeats (LTRs) are sequences of DNA that repeat hundreds or thousands of times. They are found in retroviral DNA and in retrotransposons, flanking functional genes. They are used by viruses to insert their genetic sequences into the host genomes.
Energy carrier	According to ISO 13600, an energy carrier is either a substance (energy form) or a phenomenon (energy system) that can be used to produce mechanical work or heat or to operate chemical or physical processes.
	In the field of Energetics, however, an energy carrier corresponds only to an energy form (not an energy system) of energy input required by the various sectors of society to perform their functions.
	Examples of energy carriers include liquid fuel in a furnace, gasoline in a pump, electricity in a factory or a house, and hydrogen in a tank of a car.
Glutamate decarboxylase	Glutamate decarboxylase is an enzyme that catalyzes the decarboxylation of glutamate to GABA and CO_2. GAD uses PLP as a cofactor. The reaction proceeds as follows: $HOOC\text{-}CH_2\text{-}CH_2\text{-}CH(NH_2)\text{-}COOH \rightarrow CO_2 + HOOC\text{-}CH_2\text{-}CH_2\text{-}CH_2NH_2$
	In mammals, GAD exists in two isoforms encoded by two different genes - GAD1 and GAD2. These isoforms are GAD_{67} and GAD_{65} with molecular weights of 67 and 65 kDa, respectively.
Haber process	The Haber process, is the nitrogen fixation reaction of nitrogen gas and hydrogen gas, over an enriched iron or ruthenium catalyst, which is used to industrially produce ammonia.
	Despite the fact that 78.1% of the air we breathe is nitrogen, the gas is relatively unavailable because it is so unreactive: nitrogen molecules are held together by strong triple bonds. It was not until the early 20th century that the Haber process was developed to harness the atmospheric abundance of nitrogen to create ammonia, which can then be oxidized to make the nitrates and nitrites essential for the production of nitrate fertilizer and explosives.
Insertion sequence	An insertion sequence is a short DNA sequence that acts as a simple transposable element.

Insertion sequences have two major characteristics: they are small relative to other transposable elements (generally around 700 to 2500 bp in length) and only code for proteins implicated in the transposition activity (they are thus different from other transposons, which also carry accessory genes such as antibiotic resistance genes). These proteins are usually the transposase which catalyses the enzymatic reaction allowing the IS to move, and also one regulatory protein which either stimulates or inhibits the transposition activity.

Tuberculosis	Tuberculosis is a common and often deadly infectious disease caused by various strains of mycobacteria, usually Mycobacterium tuberculosis in humans. Tuberculosis usually attacks the lungs but can also affect other parts of the body. It is spread through the air when people who have the disease cough, sneeze, or spit.
EcoRI	EcoRI EcoRI is an endonuclease enzyme isolated from strains of E. coli, and is part of the restriction modification system. In molecular biology it is used as a restriction enzyme. It creates sticky ends with 5' end overhangs.
Human genome	The human genome is stored on 23 chromosome pairs and in the small mitochondrial DNA. Twenty-two of the 23 chromosomes belong to autosomal chromosome pairs, while the remaining pair is sex determinative. The haploid human genome occupies a total of just over three billion DNA base pairs. The Human Genome Project (HGP) produced a reference sequence of the euchromatic human genome and which is used worldwide in the biomedical sciences.
Retrotransposon	Retrotransposons (also called transposons via RNA intermediates) are genetic elements that can amplify themselves in a genome and are ubiquitous components of the DNA of many eukaryotic organisms. They are a subclass of transposon. They are particularly abundant in plants, where they are often a principal component of nuclear DNA. In maize, 49-78% of the genome is made up of retrotransposons.
Downstream processing	Downstream processing refers to the recovery and purification of biosynthetic products, particularly pharmaceuticals, from natural sources such as animal or plant tissue or fermentation broth, including the recycling of salvageable components and the proper treatment and disposal of waste. It is an essential step in the manufacture of pharmaceuticals such as antibiotics, hormones (e.g. insulin and human growth hormone), antibodies (e.g. infliximab and abciximab) and vaccines; antibodies and enzymes used in diagnostics; industrial enzymes; and natural fragrance and flavor compounds.

Chapter 5. Part 5: Medicine and Immunology

Phylogenetic tree	A phylogenetic tree is a branching diagram or 'tree' showing the inferred evolutionary relationships among various biological species or other entities based upon similarities and differences in their physical and/or genetic characteristics. The taxa joined together in the tree are implied to have descended from a common ancestor. In a rooted phylogenetic tree, each node with descendants represents the inferred most recent common ancestor of the descendants, and the edge lengths in some trees may be interpreted as time estimates.
Betaproteobacteria	Betaproteobacteria is a class of Proteobacteria. Betaproteobacteria are, like all Proteobacteria, gram-negative. The Betaproteobacteria consist of several groups of aerobic or facultative bacteria that are often highly versatile in their degradation capacities, but also contain chemolithotrophic genera (e.g., the ammonia-oxidising genus Nitrosomonas) and some phototrophs (members of the genera Rhodocyclus and Rubrivivax).
Amoeba proteus	Amoeba proteus, previously Chaos diffluens, is an amoeba closely related to the Giant Amoebae. These are the species that are commonly bought at science supply stores. This small protozoan uses tentacular protuberances called pseudopodia to move and phagocytose smaller unicellular organisms, which are enveloped inside the cell's cytoplasm in a food vacuole, where they are slowly broken down by enzymes.
Marine habitats	Marine habitats can be divided into coastal and open ocean habitats. Coastal habitats are found in the area that extends from as far as the tide comes in on the shoreline out to the edge of the continental shelf. Most marine life is found in coastal habitats, even though the shelf area occupies only seven percent of the total ocean area.
Algal bloom	An algal bloom is a rapid increase or accumulation in the population of algae (typically microscopic) in an aquatic system. Algal blooms may occur in freshwater as well as marine environments. Typically, only one or a small number of phytoplankton species are involved, and some blooms may be recognized by discoloration of the water resulting from the high density of pigmented cells.
Biochemical oxygen demand	Biochemical oxygen demand or B.O.D. is the amount of dissolved oxygen needed by aerobic biological organisms in a body of water to break down organic material present in a given water sample at certain temperature over a specific time period. The term also refers to a chemical procedure for determining this amount.

Brown algae	The Phaeophyceae or brown algae, is a large group of mostly marine multicellular algae, including many seaweeds of colder Northern Hemisphere waters. They play an important role in marine environments, both as food and for the habitats they form. For instance Macrocystis, a kelp of the order Laminariales, may reach 60 m in length, and forms prominent underwater forests.
Bacterial toxin	A bacterial toxin is a type of toxin that is generated by bacteria. Toxinosis is pathogenesis caused by the bacterial toxin alone, not necessarily involving bacterial infection (e.g. when the bacteria have died, but have already produced toxin, which becomes ingested) It can be caused by Staphylococcus aureus toxins. One primary classification used is to distinguish between exotoxin and endotoxin.

1. _____ is a vaccine used against rubella. One form is called 'Meruvax'.

   a. Scarlet fever serum
   b. Schistosomiasis vaccine
   c. Smallpox vaccine
   d. Rubella vaccine

2. A _____ is an animal in which only certain known strains of bacteria and other microorganisms are present. Technically, the term also includes germ-free animals, as the status of their microbial communities is also known. However, the term gnotobiotic is often incorrectly contrasted with germ-free.

   a. Gray short-tailed opossum
   b. Guinea pig
   c. Japanese Fire Belly Newt
   d. Gnotobiotic animal

3. . _____ is an optical imaging technique used to increase optical resolution and contrast of a micrograph by using point illumination and a spatial pinhole to eliminate out-of-focus light in specimens that are thicker than the focal plane. It enables the reconstruction of three-dimensional structures from the obtained images. This technique has gained popularity in the scientific and industrial communities and typical applications are in life sciences, semiconductor inspection and materials science.

   a. Cover slip
   b. Critical illumination
   c. Confocal microscopy

4. _____(SCID) is a severe immunodeficiency genetic disorder that is characterized by the complete inability of the adaptive immune system to mount, coordinate, and sustain an appropriate immune response, usually due to absent or atypical T and B lymphocytes. In humans, SCID is colloquially known as 'bubble boy' disease, as victims may require complete clinical isolation to prevent lethal infection from environmental microbes.

Several forms of SCID occur in animal species.

   a. Sinorhizobium meliloti
   b. Severe combined immunodeficiency
   c. Tetrahymena
   d. Thalassiosira pseudonana

5. _____s are small cell-signaling protein molecules that are secreted by numerous cells and are a category of signaling molecules used extensively in intercellular communication. _____s can be classified as proteins, peptides, or glycoproteins; the term '_____' encompasses a large and diverse family of regulators produced throughout the body by cells of diverse embryological origin.

The term '_____' has been used to refer to the immunomodulating agents, such as interleukins and interferons.

   a. Gc-MAF
   b. Cytokine
   c. Hepatocyte growth factor
   d. Hyperchemokinemia

1. d
2. d
3. c
4. b
5. b

You can take the complete Chapter Practice Test

for Chapter 5. Part 5: Medicine and Immunology
on all key terms, persons, places, and concepts.

Online 99 Cents

http://www.epub13.5.20451.5.cram101.com/

Use www.Cram101.com for all your study needs

including Cram101's online interactive problem solving labs in

chemistry, statistics, mathematics, and more.

Other Cram101 e-Books and Tests

Want More?
Cram101.com...

Cram101.com provides the outlines and highlights of your
textbooks, just like this e-StudyGuide, but also gives you the
PRACTICE TESTS, and other exclusive study tools for all of your
textbooks.

Learn More. *Just click*
http://www.cram101.com/

CPSIA information can be obtained at www.ICGtesting.com
Printed in the USA
LVOW03s0559270114

371097LV00002B/41/P

9 781478 422082